Reproductive
Biology of Plants

Reproductive Biology of Plants

Editors

B.M. Johri □ **P.S. Srivastava**

Springer-Verlag

Narosa Publishing House

EDITORS
Dr. B.M. Johri
Central Reference Library
University of Delhi
Delhi-110 007, India

Dr. P.S. Srivastava
Centre for Biotechnology, Faculty of Science
Jamia Hamdard, New Delhi-110 062, India

Exclusive distribution in North America (including Canada and Mexico), Europe and
Japan by Springer-Verlag Berlin Heidelberg New York.

All export rights for this book vest exclusively with Narosa Publishing House.
Unauthorised export is a violation of Copyright Law and is subject to legal action.

ISBN 3-540-67491-8   Springer-Verlag Berlin Heidelberg New York
ISBN 0-387-67491-8   Springer-Verlag New York Berlin Heidelberg
ISBN 81-7319-325-8   Narosa Publishing House, New Delhi

Printed in India

1002387738

Preface

The *Reproductive Biology of Plants* is a multiauthored edited book meant to be studied by the undergraduate and postgraduate students, and even the research scholars.

The plant viruses, bacteria, cyanobacteria, some algae and some fungi (Chapters 2 to 6) do not have reproduction comparable to other groups of plants. Therefore, specialized (and highly especialised) mode of reproduction in the above groups is discussed on the basis of their physiology, biochemistry and cell biology.

The archegoniates (bryophytes, pteridophytes and gymnosperms) have several common features, as have the seed plants (gymnosperms and angiosperms). These features are amply brought out in Chapters 8 to 11.

Reproductive biology is fully illustrated (Chapters 2 to 10), and theoretical and evolutionary tendencies are discussed.

A glossary to help the students, teachers and other readers to follow the description more meaningfully, and have a better understanding of the subject is provided.

There is plant index which should be very helpful in locating specific information.

All the contributors are experienced teachers and researchers, and have made every possible effort to present a comparative account of reproductive processes in plants.

January 2001

B.M. Johri
P.S. Srivastava

Acknowledgements

We wish to thank, and express our gratitude to:

The contributors who accepted our invitation and prepared various chapters.

Professor S.S. Bhojwani, Department of Botany, University of Delhi, for critically examining the final proofs.

Narosa Publishing House, New Delhi, especially Mr. N.K. Mehra and Mr. M.S. Sejwal, and Editors of Springer-Verlag, Heidelberg, for processing and publishing this book.

Mr. R.K. Gupta, University of Delhi, for typing.

Mr. Vishwanath (Rajpal & Sons, Delhi) for advice and encouragement.

Mr. M.L. Saini and Mr. V.N. Vashishta of Delhi University Library for facilities to complete this work.

B.M. JOHRI
P.S. SRIVASTAVA

bar

Barcode

→ Sonneveld, Tineke (Theres) Bar <u>1003 200493</u>

→ (QK 827. D4 2(2/1)) — <u>Bar 1003241095</u>

→ (QK 827.44 2(2/1))

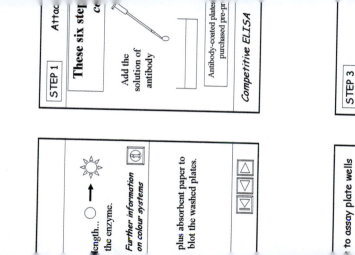

STEP 1

Attac...

These six ste...

Add the
solution of
antibody

Antibody-coated plates
purchased pre-p...

Competitive ELISA

...length...
...the enzyme.

*Further information
on colour systems*

plus absorbent paper to
blot the washed plates.

STEP 3

Add the
antigen-enzyme
conjugate

...to assay plate wells

...tigen from the sample
...mains attached to the
...ntibodies on the plate

List of Contributors

1. **Asthana, A.K.**
   Bryology Lab.
   National Botanical Research Institute
   Lucknow 226 001, India

2. **Biswas, Chhaya**
   J-1931 Chittaranjan Park
   New Delhi 110 019, India

3. **Johri, B.M.**
   Central Reference Library
   University of Delhi
   Delhi 110 007, India

4. **Krishnamurthy, K.V.**
   Department of Life Sciences
   Bharathidasan University
   Tiruchirapalli 620 024, India

5. **Krishnamurthy, V**
   15 Ramanathan Street
   T. Nagar, Chennai 600 017, India

6. **Koul, A.K.**
   Department of Biosciences
   Jammu University
   Jammu 180 001, India

7. **Kumar, Krishna**
   Department of Botany
   Gargi College
   University of Delhi South Campus
   Siri Fort Road, New Delhi 110 021, India

8. **Nath, Virendra**
   Bryology lab.
   National Botanical Research Institute
   Lucknow 226 001, India

9. **Ramachandran, Padma**
   CAS Virology
   Division of Mycology and Plant Pathology
   Indian Agricultural Research Institute
   New Delhi 110 012, India

10. **Sambali, Geeta**
    Department of Botany
    Jammu University
    Jammu 180 001, India

11. **Sen, Sumitra**
    Department of Botany
    Calcutta University
    35 Ballygunge Circular Road
    Calcutta 700 019, India

12. **Singh, Nandita**
    Department of Biotechnology
    Hamdard University
    Hamdard Nagar, New Delhi 110 062
    India

13. **Srivastava, P.S.**
    Department of Biotechnology
    Hamdard University
    Hammdard Nagar, New Delhi 110 062
    India

14. **Srivastava, Sheela**
    Department of Genetics
    University of Delhi, South Campus
    New Delhi 110 021, India

15. **Upreti, D.K.**
    Lichenology Lab
    Cryptogamic Botany
    National Botanical Research Institute
    Rana Pratap Marg, P.O. 436
    Lucknow 226 001, India

16. **Verma, Anupam**
    CAS Virology
    Division of Mycology and Plant Pathology
    Indian Agricultural Research Institute
    New Delhi 110 012, India

Contents

Contents

Reproductive Biology of Plants

B.M. Johri

The reproductive phase in the life cycle of plants is very significant and absolutely essential for diversity and the evolution of plants from simpler to more complex forms and structures.

Knowledge of the origins and diversification of land plants is based on dispersal of spores and megafossils. The megafossils are first recognized roughly 50 million years (Myr) after the appearance of land plants. Three plant-based epochs are recognized (see Crane and Kenrick 1997).

Eoembryophytic (432 Myr). Spore tetrads appear. The decay-resistant wall (sporopollenin) and tetrahedral configuration (implies meiosis) are diagnostic features of land plants.

Eotracheophytic (432–402 Myr). There is decline in diversity of tetrads, and there is dominance of individually-dispersed simple spores (hornworts, some mosses, and early vascular plants).

Eutracheophytic (398–256 Myr). Diversity of spores and megaphylls increases, and there is a substantial increase in diversity of land plants, including the early diversification of many important living groups.

Phylogenetic studies favour a single origin of land plants from charophycean green algae. A fresh water origin is likely. The absence of well-developed sporophytes, gametophytes with sexual organs of land plant-type, cuticle and sporopollenin-walled spores suggest that these innovations evolved during the transition to land. Morphological differentiation occurred both in gametophytic and sporophytic phases (Crane and Kenrich 1997). Subsequently, there was significant reduction in gametophytes and an increase in sporophytes. This will be clearly borne out by the facts mentioned in the various chapters.

Reference

Crane PR, Kenrick P (1997), Diverted development of reproductive organs: A source of morphological innovation in land plants. Pl Syst Evol **206:** 161–174

Replication of Plant Viruses

Anupam Varma and Padma Ramachandran

The year 1998 is the centenary year of the discovery of viruses. Critical experiments by Beijerinck (1898) proved for the first time that tobacco mosaic disease was not caused by a bacterium or any corpuscular organism. He called this agent 'contagium vivum fluidum'. Since then, the science of virology has come a long way and has played an important role in our understanding of modern biology. The isolation and crystallization of tobacco mosaic virus (TMV) by Stanley (1935), demonstration that TMV particles contain nucleic acid of the ribose type by Bawden and Pirie (1937), the findings of Hershey and Chase (1952) that the protein of T2 bacteriophage does not enter bacterial cells during infection, and those of Fraenkel-Conrat (1956) and Gierer and Schramm (1956) that the nucleic acid of TMV is the main infective component laid the foundations of molecular biology and biotechnology.

Viruses are nucleoproteins which contain either DNA or RNA as the genetic material. The proportion of nucleic acid and protein varies from virus to virus. Generally anisometric viruses like TMV contain 5% nucleic acid and 95% protein whereas isometric viruses like cucumber mosaic virus (CMV) have 15 to 45% nucleic acid and the remaining protein. Some viruses also contain lipids, mainly forming an envelope, as in tomato spotted wilt virus (TSWV). The other constituents are water and metallic ions. The most important component of viruses is nucleic acid, which is required for virus replication. It is, therefore, natural that the nucleic acid forms the base for defining viruses as "transmissible parasites whose nucleic acid genome is less than 3×10^8 Da in weight and that need ribosome and other components of their host cells for multiplication" (Gibbs and Harrison, 1976). This definition not only includes the large viruses belonging to Poxviridae but also the smallest known pathogens, the viroids, which are naked single stranded RNA of about 375 nucleotides capable of causing severe diseases in plants. In this chapter, we briefly discuss the process of replication of plant viruses which form the largest group of all the known viruses infecting bacteria, fungi, plants, invertebrates and vertebrates.

1. Types of Plant Viruses

The majority of plant viruses are RNA viruses. Until the late 1960s it was thought that all plant viruses were RNA viruses. It made Bawden (1964) observe: "It would be premature yet to assume this is generally true, because most viruses that infect bacteria, and some that infect animals, contain deoxy nucleic acid, and there seems to be no *a priori* reason why viruses containing deoxynucleic acid should not infect flowering plants. However, if there are any, they await discovery". He was right, the discovery came within a few years of his observations, that cauliflower mosaic virus (CaMV) is a DNA virus (Shephard et al. 1970); later, CaMV became an important tool in the hands of biotechnologists. Now we know that several DNA viruses infect plants, although the relative number of DNA viruses infecting plants compared with the RNA viruses is small (Fig. 1). On the basis of nucleic acid, plant viruses can be grouped as dsDNA, ssDNA, dsRNA and ssRNA viruses. The ssRNA viruses can be further subgrouped as ssRNA(+) and ssRNA(–). Some of the ssRNA(–) are enveloped and some naked—lacking even the coat protein (Fig. 1). The most preferred form of the genome is ssRNA(+) which must have been selected for efficient replication utilizing host machinery. The two groups of enveloped ssRNA viruses are similar to those found infecting invertebrates and vertebrates. These appear to be animal viruses which moved and established in plants during evolution. So far, no enveloped DNA virus has been found to infect plants. It is a matter of time before such viruses are also found.

Plant viruses not only vary in the nature of their genome but also in shape and size. Basically, they either have spiral symmetry, forming anisometric particles, or icosahedral symmetry, forming isometric particles; virus particles are also referred to as 'virions'. Many plant viruses have split genomes consisting of two or more distinct nucleic acid molecules encapsidated in different or similar sized particles made of the same coat-protein subunits. Among the anisometric viruses hordei-, tobra- and furoviruses have split genomes encapsidated in particles of different sizes (Fig. 1). The alfamovirus has four genomic molecules encased in particles of four different sizes 19, 36, 48 and 58 nm long and 19 nm wide. Among the isometric viruses, ilarviruses and the viruses belonging to Bromoviridae, Comoviridae, Geminiviridae and Partitiviridae have divided genomes. The total mass of nucleoprotein of different virus particles varies from 4.6 to 73 million Da. The weight of nucleic acid alone is between 1–3 million Da per virus particle for most of the viruses; sometimes it may be as high as 16×10^6 Da, as in the case of wound tumor virus, of the reovirus group. The protein shell which constitutes the rest of the mass of the virus particle, is made up of one kind of coat-protein subunit. The subunits of coat-protein are packed in regular arrays either to give icosahedral or spiral symmetry. The amino acid sequence is identical in the coat-protein subunits of a given virus, but it varies from virus to virus and even strains of the same virus. The interaction between the protein and viral nucleic acid determines the stability of the virus particle.

2. Strategies for Replication

Plant viruses, like all other viruses, differ greatly from other microorganisms, as the viruses completely depend on the host machinery for replication. The replication of a virus starts the moment an infective particle enters a susceptible cell. A general course of events in the life cycle of a virus is: (1) entry of the virus particle into living susceptible cell, (2) disassembly or 'unwinding of the virus particle', (3) transcription or synthesis of "mRNA" of viruses other than those with positive sense ssRNA, (4) translation or synthesis of proteins required for virus replication, and (5) maturation of virus particles and their movement to the neighbouring cell (Fig. 2).

The main strategies which the viruses use for replication are:

1. *Subgenomic (sg) nucleic acid.* The synthesis of one or more subgenomic nucleic acid molecules enables the production of proteins in desired amounts.
2. *Polyprotein.* The viral genome may code for one polyprotein from the whole genome using single open reading frame (ORF). The polyprotein is then cleaved at specific sites by a viral-coded protease/s to give a final gene product.
3. *Multipartite genome.* In a large number of viruses the genome is in the form of two or more molecules which may be encapsidated in one particle or different particles.
4. *Readthrough proteins.* The termination codon of the 5' gene may be 'leaky' and allow a proportion of ribosomes to carry on translation to another stop codon downstream from the first, giving rise to a second larger functional polypeptide.
5. *Transframe proteins.* Two proteins may commence at the same 5' AUG codon by a switch of reading frame to give a second longer transframe protein.

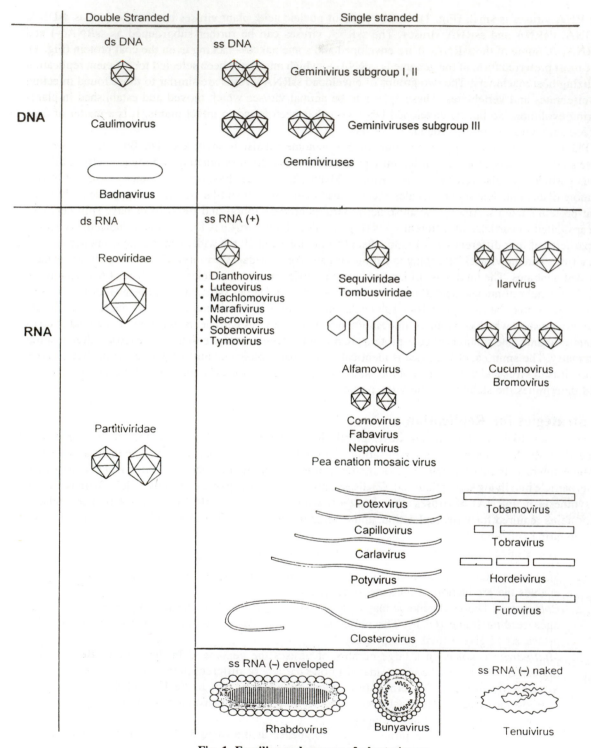

Fig. 1. Families and groups of plant viruses

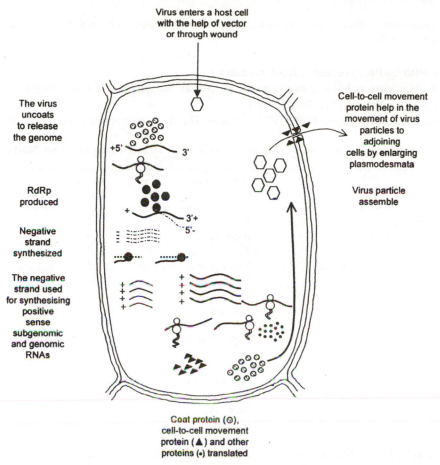

Fig. 2. Replication of a ss-positive sense RNA virus in a plant cell and its movement to the adjoining cell

These strategies have evolved for utilising eukaryotic protein synthesizing systems and achieving genetic flexibility, as in the case of the multipartite genome, which provides the opportunity for separation of gene function, packaging of the genome into different particles, and above all production of natural recombinants.

3. Replication

Like other viruses, genome organisation of plant viruses is based on economic use of genomic nucleic acids by minimising non-coding regions, overlapping ORFs, readthrough polypeptides that are coterminal with a smaller protein, and 5′ and 3′ non-coding sequences.

The viral genome products are of two types—structural and non-structural proteins. Structural proteins are mainly the coat-protein and those associated with lipoprotein membranes of the enveloped viruses. The non-structural proteins coded by viral genomes are proteases required for cleaving virus-coded polyproteins, polymerases like RNA-dependent RNA polymerase (RdRp) or RNA-dependent DNA-polymerase (RdDp), also known as reverse transcriptase as in cauliflower mosaic virus, cell-to-cell movement proteins, transmission protein and 5′ VPg (virus protein linked to genome) which can also be considered as a structural protein. In viruses with multipartite genomes, 5′ VPgs are present in all the genomic RNAs. Plant viruses are not known to produce cell recognition proteins as are found in viruses infecting bacteria and animals. The plant viruses which circulate in their insect vectors are, however, expected to have surface recognition proteins, but these await identification.

3.1   Viruses with ss-Positive Sense RNA Genomes

A majority of plant viruses have ss-positive sense RNA as the genome. These viruses use one or more genome strategies for their replication (Table 1).

RdRp is essential for the replication of RNA viruses. The RNA polymerase is required for (1) synthesis

Table 1.   Genome strategies used by ss-positive sense RNA viruses (Adapted from Matthews 1992)

Strategy	Virus group	No. of ORFs	No. of proteins coded
One strategy; polyprotein	Potyvirus	1	8
One strategy; subgenomic RNA	Potyvirus	5	4–5
	Tombusvirus	≥ 3	≥ 3
Two strategies; subgenomic RNA plus	Tobamovirus	5	4–5
readthrough or frameshift protein	Luteovirus	6	6
	Carmovirus	5	7
Two strategies; subgenomic RNAs	Tymovirus	3	3–4
and polyprotein	Sobemovirus	4	4–5
Two strategies; multipartite genome	Comovirus	2	9
and polyprotein		2	5
Two strategies;	Bromovirus	4	4
subgenomic RNAs and	Cucumovirus	4	4,
multipartite genome	Alfamovirus		
	Ilarvirus	4	4
	Hordeivirus	4	4
		7	7
Three strategies; subgenomic RNAs,	Tobravirus	5	5–6
multipartite genome and readthrough	Furovirus	9	6–9
protein (or frameshift)	Dianthovirus		4

of complementary RNA (cRNA) of negative polarity and (2) cRNA directed synthesis of viral RNA (vRNA). Previously, it was believed that RdRp present in uninfected plant cells is activated by viral infection and is used in viral multiplication but now it has been shown that viral genomes contain RNA polymerase genes.

The viral RdRps are difficult to obtain in an active form and in large quantities. Both *in vivo* and *in vitro* studies have shown the synthesis of sgRNA by a viral RdRp, starting at an internal site, the sub genomic promoter (SGP). In general, the sgRNAs are the mRNAs for the 3′ proximal genes of polycistronic viral RNAs and are identical in sequence to the 3′ end of the genomic (g) RNA. The 5′ end of the sgRNA has the same starting nucleotide as in the genomic RNA.

Synthesis of sgRNA often exceeds the amount necessary for the production of the corresponding gene. When an appropriate SGP region is inserted into a new location on the viral RNA, a new sgRNA is produced, thus revealing the functionality of the SGP. This has been demonstrated for brome mosaic virus (BMV), TMV, CMV and alfalfa mosaic virus (AlMV). These experiments also showed that the expression of sgRNA was dependent on the position of the corresponding SGP on the gRNA. Higher levels of sgRNA lying closer to the 3′ end of the RNA are observed for TMV, BMV and CMV, whereas the opposite occurs for AlMV. Expression of sgRNA follows a time-dependent regulation as in the case of

TMV, in which expression of the 30 kDa movement protein from sgRNA is low and time-dependent, whereas the synthesis of coat protein from sgRNA2 is delayed but then expressed at a constant and very high rate. However, the mechanism of sgRNA synthesis is still poorly understood. It awaits characterisation of the viral RdRp and identification of other factors that may be involved in this process, either from the host or of viral origin.

Since sgRNAs share the same 3′ sequence as, and are more abundant than, the gRNAs, they may interfere with virus replication. Therefore, double-stranded (ds) RNAs, which represent replicative forms corresponding to sgRNAs have not been observed. A detailed study of the dsRNA forms of potato virus X (PVX) and barley stripe mosaic virus (BSMV) sgRNAs revealed that the 3′ end of the (−) strand of the gRNA has an extra G unpaired residue compared with the (+) strand, this extra nucleotide being necessary for the synthesis of (+) strand of RNA. On the contrary, the 3′ end of the (−) strand of the sgRNA in the ds form lacks this extra nucleotide and both strands are perfectly matched, which means the sgRNA cannot be synthesised from a sgRNA template. The 5′ end of the gRNA can be constituted by a cap, a viral encoded protein designated VPg or a poly protein. The first nucleotide of sgRNA is often the same as for the gRNA and this may reflect the necessity for the sgRNA to have the same 5′-end structure as the gRNA.

In the case of TMV where two sgRNAs are produced, the highly expressed sgRNA corresponding to the CP is capped, whereas the less abundant sgRNA of the 30 kDa protein is probably uncapped. sgRNAs are often found in the virion particles, albeit in varying amounts. They are only detected by hybridization with radioactive probes or by their capacity to support reasonable levels of protein synthesis. The sgRNAs are very efficient mRNAs; their translational strategies are the same as those reported for gRNAs. sgRNAs are synthesised by a large number of viruses and serve many different purposes: from high level gene expression of CP genes to time- and amount—limited protein synthesis. These make sgRNAs very interesting tools for studying the genes under the control of viral polymerase. They could also be used to encode resistance genes that would be expressed only upon infection.

Translation and Synthesis of Proteins

The process of translation is composed of three phases: (1) initiation, in which an 80S ribosome assembles at the start codon of mRNA and initiates the formation of the first peptide bond of the nacent protein; (2) elongation, which involves the translocation of the ribosome down the open reading frame, resulting in the production of the protein; and (3) termination, which results in the release of both the completed protein and the 80S ribosome.

During initiation, the initiation factors associate with the 5′ terminal cap structure and the poly (A) binding protein at the poly (A) tail at the 3′ terminus. After binding of the initiation factor, the 40S ribosomal subunit close to the 5′ terminus scans down the 5′ leader in search for the initiation codon, AUG. Once the 40S subunit has located the initiation codon, the 60S subunit joins to create the translationally competent 80S ribosome. Thus it has been observed that translation occurs through a large complex of 5′ terminal cap, the cap factors, PAB protein and the poly (A) tail.

Plant viral mRNAs though not cellular RNAs, are translated by the translational machinery of the cell and therefore must compete with cellular RNAs. Although many plant viral mRNAs are polyadenylated, there are several that terminate in an alternative structure. The best studied example is of TMV. The TMV gRNA serves as a capped, non-polyadenylated mRNA that is very efficiently translated. Instead of a poly(A) tail, TMV RNA terminates in a highly structured 3′ untranslated region (UTR). The 3′ UTR is composed of 5 RNA pseudo-knots and has 177 bases. It consists of two domains, a tRNA-like domain and an upstream pseudo domain (UPD), the two showing dependence. The UPD sequences are known to be conserved in eight viruses. The viral mRNAs have thus evolved functional alternatives to a poly (A) tail that nevertheless requires interaction with the 5′ cap. Besides the 5′ cap and poly (A), two other strategies that allow plant viruses in general to maximise their translational capacity and to translationally regulate expression of genes are overlapping ORFs and in frame ORF.

Various types of regulation can occur at the level of elongation of protein synthesis. One of them, frame shifting, occurs from movement of ribosomes either in the 5' direction (-1 frame shift) or in the 3' direction (+1 frame shift) on the mRNA. Two proteins (the frame and the trans frame proteins) are produced that are identical from the N terminus to the frame shift point, but differ beyond that point. The frame protein is always more than the transframe protein. Frame shift usually permits the expression of the viral replicase; in most cases the RdRp. This mechanism has been demonstrated in at least 14 plant viruses belonging to different groups.

Finally, the termination of protein synthesis occurs due to the presence of an inframe termination codon in a mRNA. Interestingly, the sequence UGA appears to be a universal termination codon, as it has been found for many groups of plant viruses. Presence of 'readthrough' termination signals has also been suggested for termination of proteins.

The first ORF from the 5' terminus of the genome, or genome segment, may terminate in a stop codon that is sometimes suppressed. This suppression allows the ribosome to more to the next stop codon, giving rise to a readthrough protein of greater length. The synthesis of readthrough proteins depends primarily on the presence of appropriate suppressor tRNAs. The proportion of readthrough proteins produced may be modulated by nucleotide sequences around the suppressed or termination codon. A sequence containing two or three A residues immediately 5' to the UAG stop codon has been found in several viruses; 5'-CAAUGA-5'.

Assembly

The coat protein subunits get packaged with the viral nucleic acid to make the virus particles (Fig. 3). This occurs by a specific recognition process between coat protein and viral nucleic acid. The classical experiments of Fraenkel-Conrat and Williams (1955) gave the first insight into the assembly process. Their work showed that TMV coat protein and TMV RNA can be reassembled into intact virus particles in vitro. Reconstitution of virus rods gave increased specific infectivity which was resistant to RNase attack. Most of our knowledge on assembly of virus comes from studies made using TMV where the origin of assembly is at a specific internal sequence of nucleotides. In TMV, this sequence lies outside the coat protein gene in a 5' direction. The initiating sequence consists of three triplets (AAG AAG UCG) which form a loop at the end of a base paired stem containing about 14 base pairs. The nine-base loop combines first with a 20S aggregate of coat, also called 20S double disk, from the lower side of the disk, but how the correct side for entry is chosen is not yet known. The loop opens up as it intercalates between the two layers of subunits. This protein RNA interaction causes the disk to switch to the spiral lock washer form (often referred to as a protohelix). Both RNA tails protrude from the same end. The lock washer-RNA complex is the beginning of the helical rod. A second double disk can add to the first on the side away from the RNA tails. As it does so, it switches to the spiral form and two more twins of the RNA become entrapped. Growth of the spiral rod continues in the 5' direction as the loop of RNA receives successive disks, and the 5' tail of the RNA is drawn through the axial hole. It has been generally accepted that rod assembly and rod extension is faster in the 5' than in the 3' direction. The 20S aggregate hypothesis has, however, not yet been documented in *in vivo* studies. The process of assembly of virus differs from group to group as is the case with transcription and translation.

Potyviruses

Potyviruses represent the largest group of plant viruses causing economically important diseases (Varma 1988). This is the only group which uses polyprotein as the only strategy for replication (Fig. 4). Viruses

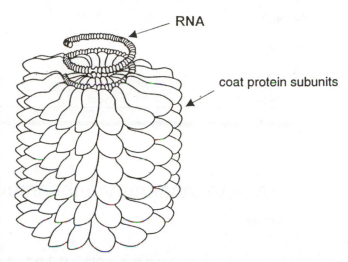

RNA

coat protein subunits

Fig. 3 **Diagramatic representation of part of the TMV particle showing spiral arrangement of the coat protein subunits in relation to the RNA. Each protein subunit is shaped like a shoe. In successive turns of the helix, the subunits form a groove in which the RNA is located. The spiral has a pitch of 2.3 nm. Each turn of the spiral has $16^1/_3$ subunits accommodating 49 nucleotides (3 nucleotides per protein subunit). The particle structure forms an axial canal of 4 nm diameter. (Adapted from Klug and Caspar 1960)**

belonging to this group have monopartite genomes of about 9500 nucleotides coding for a single polyprotein which is cleaved to yield three proteinases, apart from a protein each for cell-to-cell movement, insect transmission, replication, polymerase and coat protein. Initial proteolytic cleavage occurs while the enzyme is still a part of the polyprotein (cis cleavage), and later cleavage (trans cleavage) is made by the enzymes cleaved initially. A single polyprotein strategy for the replication of the virus means that only one coat protein subunit is formed from one polyprotein. To form one full particle of a potyvirus about 2000 coat-protein subunits are required. Thus the virus produces equal numbers of non-structural proteins coded by its genome which accumulate in the cells, forming the inclusions characteristic of potyviruses.

Potexviruses
To avoid overproduction of non-structural proteins, potexviruses use the strategy of sgRNA for producing coat-protein subunits which are required in large numbers. These viruses have a genome of about 5800 nucleotides with five ORFs coding for a polymerase, a cell-to-cell movement protein, fungal transmission protein, a coat protein and another protein of unknown function. In white clover mosaic virus, a subgenomic mRNA is formed to produce coat-protein only, whereas in potato virus X a sgRNA containing four ORFs, including the one for coat protein, is found.

Tobamoviruses
Type member TMV is the most studied plant virus. It has a genome of about 6400 nucleotides containing five ORFs, a cap at the 5′ end and a tRNA-like structure at the 3′ end. It uses the strategy of production of RNAs and read through proteins, due to a leaky termination codon (UAG) of the ORF1, which uses both ORF1 and ORF2 to form a larger protein (Fig. 5).

Comoviruses
The genome of the type member cowpea mosaic virus is bipartite, consisting of two strands with M_r of 1.22×10^6 (component M) and 2.04×10^6 (component B). Both the strands have one ORF each. Their products are then processed to make functional proteins. The smaller genomic component initially produces polyproteins of two sizes, 105 KDa and 95 KDa, as one is initiated at nucleotide 161 and the other at nucleotide 512. Both the genomic components are essential for infection. The larger B component codes for proteins required for replication, and the smaller M component codes for the structural proteins.

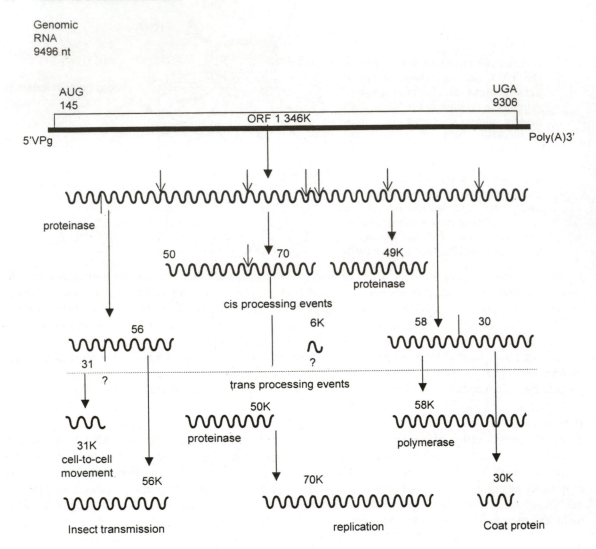

Fig. 4. Organization and expression of a Potyvirus genome (TEV)

3.2 Viruses with ss-Negative Sense RNA Genomes

The plant rhabdoviruses consist of large enveloped particles consisting of negative sense ssRNA. The genome of sonchus yellow net virus (SYNV), which is an important rhabdovirus, codes for five proteins for which individual discrete mRNAs are transcribed from the negative sense genome. The first from the 3′ end of the genomic RNA codes for the nucleocapsid protein (N) while the second gene encodes another structural protein M2. Transcriptase activity using viral RNA as a template has also been demonstrated in rhabdoviruses. The mechanism by which the negative sense strand is converted to a positive sense strand is not exactly known. Cytological studies show that the newly made virus particles accumulate in the perinuclear space, with some particles scattered in the cytoplasm of infected cells.

3.3   Viruses with Ambisense RNA Genomes

Some viruses, have RNAs of both (+) and (−) polarity; some proteins are translated from the genome-sense

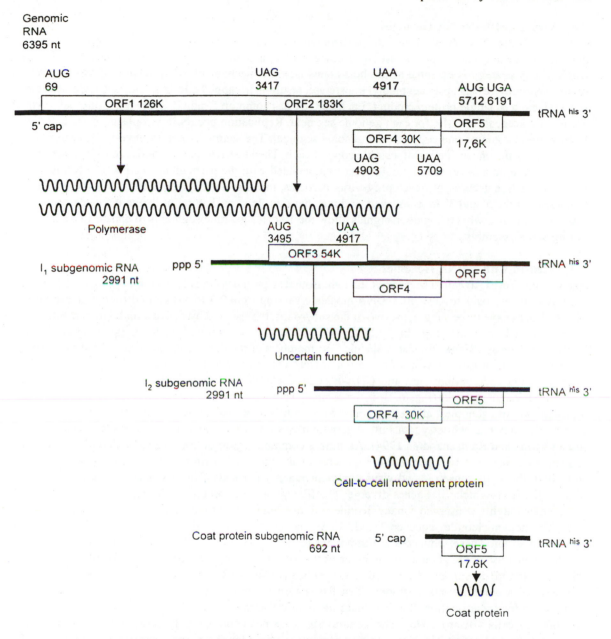

Fig. 5. Organization and expression of a Tobamovirus genome

RNA and some from the opposite strand. The tospoviruses like TSWV, use this ambisense strategy. They have three genomic segments designated, L (a. 8200 nucleotides), M (a. 5400) and S (a. 3000). Each of these is found in a circular form in the particles as a result of terminal inverted repeat sequences which are probably also involved in replication. The M RNA is (–)ve sense and has a nine-base long complementary sequence at the 5′ and 3′ termini. The S RNA of TSWV has an ambisense strategy, one protein is translated in the genomic sense, and the other from the virion-complementary strand, both are translated from sgRNAs. Replication occurs in the cytoplasm (de Haan et al. 1990).

3.4   Viruses with dsRNA Genomes

Viruses with dsRNA genomes include phytoreoviruses (12 genome segments) and cryptoviruses (two genome segments). The isometric particles include RNA transcriptase to produce mRNA from the genomic dsRNA; this strategy is essential as the host plants lack the mechanism of replication of dsRNA. Wound tumor phytoreovirus encapsidates methyl transferase, resulting in capped mRNAs, and each genome segment is translated to a single protein product. Each dsRNA has common 5' and 3' sequences, internal to which are inverted repeats specific for each genome segment. Replication occurs in viroplasm in the cytoplasm. Each particle contains one copy of each genome segment. The segment-specific inverted repeats may act as signals for the encapsidation of a single copy of each. The negative sense strands are synthesized by the viral replicase on a (+ve) sense template that is associated with the particulate fraction. dsRNA is formed within the nascent cores of developing virus particles and that dsRNA remains within these particles. Sequences at the 5' and 3' are conserved, and may have a role in packaging of the genome segments. Still, it is not clear as to what recognition signals allow the 10 or 12 genome segments to appear in each particle during virus assembly.

3.5   Viruses with ssDNA Genomes

The geminiviruses, so called because of the twin isometric particles (Fig. 1), contain circular ssDNA. Each twin particle encapsidates a single ssDNA molecule varying from 2.5 to 3.0 kb in different geminiviruses. These viruses have three subgroups; two of the subgroups, represented by maize streak virus infecting only monocots, and beet curly top virus infecting only dicots are transmitted by leafhoppers and contain an ssDNA genome as a single circular molecule. The members of the third group are transmitted by whiteflies and their genome mostly consists of two circular DNAs referred to as DNA-A and DNA-B; DNA-A codes for replication and coat proteins, and DNA-B codes for proteins required for cell-to-cell movement and systemic spread of the virus (Fig. 6). Both the components are required for infection (Fig. 7). In geminiviruses with monopartite genomes, all the proteins are coded by one molecule. Representatives of this subgroup, commonly known as whitefly transmitted geminiviruses (WTGs), cause severe diseases of crop plants in India (Varma and Ramachandran 1994). All have a common region of 200 to 250 nucleotides of a similar sequence between the two genomic components of the virus. All geminiviruses have genes transcribed from both the virus sense (+) and virus complementary (−) strands. These viruses contain an intergenic region (IR) from which viral genes diverge. The IR region is important for the replication of viral DNA, and contains highly conserved nonanucleotide and inverted repeat sequences capable of forming a hairpin loop. The nonanucleotide sequence TAATATTAC has been found in all geminiviruses which have been sequenced so far. These sequences are important for the replication of the geminiviruses.

The replication of the geminivirus genome occurs by a rolling-circle mechanism. On infection, the viral ssDNA (+strand) is converted into a dsDNA replicative form (RF) through host-directed RNA-primed synthesis of a complementary strand. Then RF is further amplified and later only the viral ssDNA is synthesized for encapsidation. Replication of the viral DNA and its encapsidation occurs in the nucleus of the infected cells (Bisaro 1996). The geminiviruses do not code for polymerase and rely on the host machinery for replication. However, the host proteins required for this purpose remain to be identified. The virus codes for proteins required for replication initiation, enhancement and quantitative regulation of RF and ss forms.

3.6 Viruses with ds-DNA Genomes

Two virus groups viz. caulimovirus and badnavirus, also called pararetroviruses, comprise double stranded DNA genomes. They are of great molecular biology interest because they could be effective gene vectors for plants and also for their replication strategy using an RNA intermediate through reverse transcription.

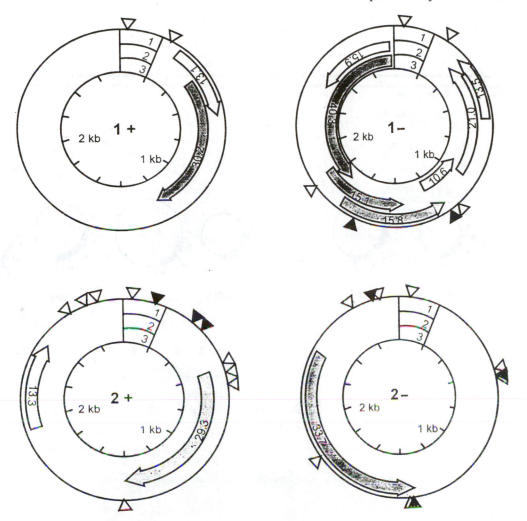

Fig. 6. **Genome organization of African cassava mosaic virus, one of the most studied whitefly transmitted geminiviruses (WTGs).** *Open arrows* **indicate ORFs and direction of transcription and coding capacity. The** *shaded ORFs* **are found in other WTGS. The** *boxed region* **to the** *right* **of the 0–kb position, is the 200-base common intergenic region.** *Open triangles* **indicate promoter sequences (TATA boxes) and the** *closed triangles* **polyadenylation signals (AATAAA sequences). (Adapted from Townsend et al. 1985)**

Cauliflower mosaic virus (CaMV) is the best understood virus (Mason et al. 1987). Virus replication takes place in the viroplasm. The 8 kb viral DNA has three discontinuities—one in the inner strand which is the coding strand and two in the outer which is the non-coding strand. The genome has eight genes with very little overlap except for gene VIII (Fig. 8). All of the gene products are transcribed from the (+) strand to form two RNAs; 35S and 19S. The 35S RNA extends around the whole genome and contains a direct terminal repeat of 180 nucleotides, while the 19S RNA shares the same 3′ terminus with the 35S RNA.

ORF 1 has been implicated in cell-to-cell movement. Its similarity to the 30 kDa TMV movement protein has been shown. It has been proposed that CaMV might move through the plasmodesmata as a 35S RNA-ORF I protein complex (Citovski et al. 1991). The product of gene IV functions in aphid transmissibility,

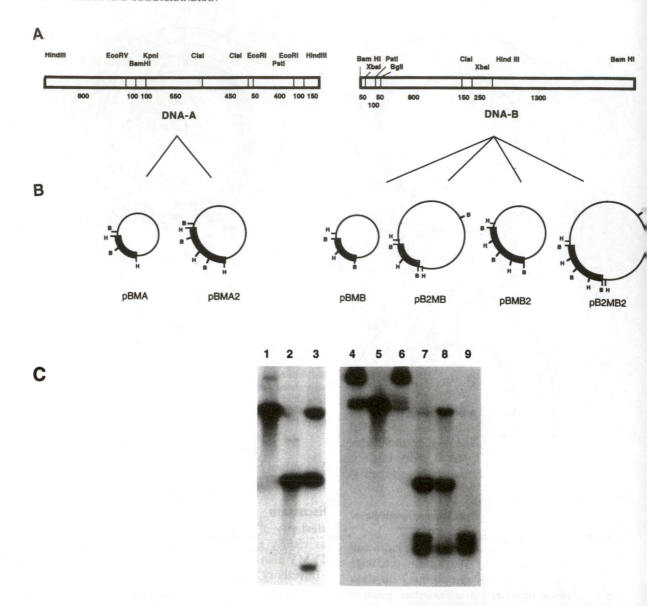

Fig. 7.   **(A) Physical map of MYMV DNA-A and DNA-B. (B) pBin 19 and MYMV DNA-A or B recombinants. Various monomeric and dimeric clones of MYMV DNA-A and DNA-B. (C) Southern analysis of MYMV DNA-A and DNA-B Clones. Lanes 1 to 3: pBMA2 restricted with *Xba* 1 (lane 1; fragment size 15.4 kb), *Hind* III (lane 2; fragment size 2.7 kb), and *Bam* HI (lane 3; fragment sizes 11.8, 2.7 and 0.9 kb). Lanes 4, 7: pB2MB2 restricts with *Eco* RI (lane 4; fragment size 25.4 kb) and *Hind* III (lane 7; fragment sizes 2.7, 1.4 and 1.3 kb). Lanes 5, 8: pBMB2 restricted with *Eco* RI (lane 5; fragment size 15.4 kb) and *Hind* III (lane 8; fragment sizes 11.4, 2.7 and 1.3 kb). Lanes 6, 9: pB2MB restricted with *Eco* RI (lane 6; fragment size 22.7 kb) and *Hind* III (lane 9; fragment sizes 1.4 and 1.3). Lanes 1 to 3 were probed with MYMV DNA-A and lanes 4 to 9 with MYMV DNA-B. Infection is achieved only when pBMA2 and pB2MB as pB2MB2 are used for agronifection. (Mandal et al. 1997).**

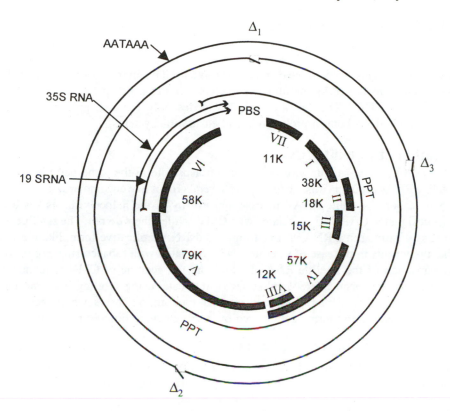

Fig. 8. Genetic organization of cauliflower mosaic virus. The 8-kbp viral DNA has three single stranded interruptions, one (D1) in the coding strand, and two (D2 and D3) in the noncoding strand. The genome has 8 potential ORFs (I to VIII) coding for proteins varying from 11 to 79 kDa. The 19 S and 35 S RNAs have different promoters, but share the same 3' termini. The two mRNAs are translated from a fully ds form of the DNA. (Adapted from Pfeiffer et al. 1987).

while ORF III codes more for structural proteins. The ORF II product is a 75 kDa protein precursor of the 42 kDa coat protein of the virus. ORF V is the largest ORF in the genome. It has been shown to give rise to a protein similar in molecular weight to reverse transcriptase. In *in vitro* studies it has been shown to accumulate significant levels of reverse transcriptase. The ORF VI gene product, a 58 kDa protein, plays a major role in disease induction, symptom expression and controlling host range; this protein is also a major component of the viroplasms. For ORF VII and VIII, specific functions have not yet been assigned.

The infecting dsDNA virion reaches the nucleus. The gaps of both (+) and (−) strands are removed and the supercoiled DNA forms a minichromosome in association with histones. These serve as a template and, in the presence of RNA polymesase II, transcribe two RNAs; 35S and 19S, which are poly-adenylated before migrating to the cytoplasm. The 19S RNA translates only one protein of 58 kDa whereas the 35S RNA is unusual in that at least 5 to 7 proteins are translated from it.

The viral DNA synthesis takes place on a 35S RNA template using viral reverse transcriptase. The synthesis of the (−) DNA strand continues until it reaches the 5' end of the 35S RNA. A switch of the enzyme to the 3' end to complete the copying is done by the 180 nucleotide direct repeat sequence at each end to the 35S RNA. The RNA template gets degraded by RNase H activity leaving two purine-rich segments which act as primer for the synthesis of the second (+) strand. Both progeny DNA synthesis and particle maturation take place in the viroplasm.

4.   Replication of a Satellite Virus

Satellite tobacco necrosis virus (STNV) is the smallest known plant virus with particles of 18 nm diameter. For replication, STNV is completely dependent on tobacco necrosis virus (TNV)-a ss-positive sense RNA virus. STNV RNA is highly stable *in vivo* and can survive for days in the absence of the helper TNV. Like TNV, STNV is also transmitted by the fungus *Olpidium brassicae*. RdRp coded by TNV appears to help in the replication of STNV. The genome of STNV codes only for coat protein, and is preferred for translation over TNV RNA both in prokaryotic and eukaryotic systems.

5.   Replication of Satellite RNAs

Satellite RNAs differ from satellite viruses as they encaspsidate in the same particle as the helper virus genome. Satellite RNAs occur in viruses belonging to several groups like cucumovirus, nepovirus, tombusvirus and sobemovirus. Some satellite RNAs increase the severity of the helper virus as has been found with satellite RNA designated CARNA5 associated with CMV, while others do not. The satellite RNAs replicate either like the ss-positive sense RNA viruses forming a dsRNA replicative form, like the satellite RNA of CMV, or like the viroids in the case of satellite RNA associated with tobacco ringspot virus (TRSV). The 5′ and 3′ organisation of the satellite RNA of CMV and the genome of CMV is identical, indicating a common replicaton mechanism. STRSV, unlike TRSV, replicates using a rolling circle mechanism, forming intermediary multimeric tandem repeats and circular monomeric forms which get linearised before encapsidation. The multimeric forms get cleaved by a self cleaving domain-ribozyme.

6.   Defective Interfering Nucleic Acids

Defective interfering (DI) nucleic acids are found in both RNA and DNA viruses. DIs are derived from the viral genome from large deletions while retaining the 5′ and 3′ sequences. The DI RNA of tomato bushy stunt virus (TBSV) is a good example. Six deletions varying from 3 nucleotides to > 3000 nucleotides in a 4.8 kb viral genome results in DI RNA of 396 nucleotides which replicates using the same strategy as TBSV, and could represent up to 60% of virus-specific RNA in infected leaves, although a very small proportion (about 40%) of the DI RNA gets encapsidated. It reduces virus replication and results in attenuation of disease symptoms. This effect could be used to advantage for developing resistance in plants to virus infections, although so far this approach has not been put to commercial use for biosafety considerations of risks of undesirable mutations (Wilson et al. 1998).

7.   Viroids

Viroids are circular ssRNA molecules of up to 375 nucleotides, with base-paired regions alternating with unpaired loop-like regions giving a rod-like appearance to the molecules. They do not code for any protein and replicate using host machinery for nucleic acid synthesis. Some viroids cause economically important diseases in plants whereas others appear to have been selected to provide the desired plant type, particularly where dwarfing is required.

The viroids replicate using a rolling-circle mechanism, first forming multimeric (–) linear strands which are then copied into multimeric (+) strands. The multimeric strands are self-cleaved ribozymically, forming monomers which ligate to form circular molecules. Host enzymes are used for nucleic acid synthesis and ligation.

8.   Cell-to-Cell Movement of Plant Viruses

Virus particles are initially introduced into host cells by mechanical damage or through the vectors. Upon entry into a susceptible cell, specific virus-host interaction takes place resulting in the replication of the virus and formation of large numbers of virus particles which must move to adjacent cells and cause systemic infection (Fig. 2). Systemic virus spread in the host is an active process involving two phases; (1) cell-to-cell movement, and (2) long-distance spread in plants.

For a long time it was believed that plant viruses spread passively by diffusion though plasmodesmata into surrounding healthy cells. Plasmodesmata are trans-cell wall channels which link plant cells, forming an intercellular continuum (Robards and Lucas 1990). The basic structure of a plasmodesma is a 28 nm canal lined with plasma membrane. A thin strand of modified endoplasmic reticulum, the desmotubule, runs from cell-to-cell through the center of this canal. There is a neck region formed by a constriction of the plasmodesmatal canal at its ends. At this junction, a ring of large particles are found just below the plasma membrane. This outer ring structure was suggested as functioning as a sphincter, involved in regulation of plasmodesmatal permeability. In a cross-section of the plasmodesmatal canal, the desmotubule is seen surrounded by a nine particle subunit of 5 nm diameter. While the desmotubule and the 5 nm subunits are impermeable, intercellular transport occurs through the spaces; 1.5–3.0 nm gaps between the subunits. This means that the size limit of the plasmodesmatal channel is far below the size of virus particles (17–80 nm) or even free folded nucleic acids (10 nm; Gibbs 1976; Citovski and Zambryski 1991).

8.1 Virus Movement Proteins

It is now generally accepted that many viruses encode proteins, called transport or movement proteins, which are involved in the modification of the structure of plasmodesmata in such a way as to enable virus infection products to pass from infected to uninfected adjacent cells. Evidence has come from various kinds of experimentation; mutational analysis, behaviour in transgenic plants, complementation, sequence analysis of gene products, cytological observations and others.

Viral movement proteins specifically localize to and interact with host cell plasmodesmata. Two types of interactions are proposed; (1) suppression of plant cell defense system(s), and (2) enlargement (gating) of the plamodesmatal channel (Atabekov and Dorokhov 1984). Evidence has come for the second type of interaction from studies conducted with plants transgenic for the P30 movement protein of TMV. After microinjection into mesophyll cells, movement of fluorescent dye between cells was monitored by fluorescence microscopy. It was found that the size exclusion limit (SEL) of plasmodesmata in plants that express P30 was 3-4 times higher (2.4-3.1 nm) than in the control plants (0.73 nm). This increase in plasmodesmatal permeability still could not account for transfer of free folded TMV RNA molecules which are about 10 nm wide. Ultrastructural studies of TMV-infected plant tissue did not reveal significant changes in the size of plasmodesmata, nor could virus particles be detected within them. On the other hand, infection by CaMV results in modification of plasmodesmata with virions observed within them. These results indicate adoption of different strategies for cell-to-cell movement by different groups of viruses.

Three strategies for cell-to-cell movement of plant viruses are possible (Fig. 9). In the first strategy the movement protein (MP) binds to the viral nucleic acid and ushers through the plasmodesmata to the neighbouring cells. The MP in this process also protects the viral nucleic acid from host cell nucleolytic activity (Fig. 9B). In the second strategy, MP interacts with the cell wall, increasing the size of plasmodesmata to allow free movement of viral nucleic acid (Fig. 9C), and in the third strategy the whole virus particle moves through a conduit formed by interconnecting endoplasmic reticulum (Fig. 9D).

The mechanisms by which the movement protein reaches the plasmodesmata and increases the molecular size exclusion limit (SEL) of plasmodesmata needs to be determined. This would help in the development of virus-resistant plants and contribute to the understanding of viral pathogenesis. In the viruses with divided genomes, one nucleic acid species is necessary for the movement of the other nucleic acid species. In bipartite geminiviruses, the viruses are encoded for information for virus movement on their small piece of DNA which also encodes viral proteins. There are certain viruses whose genomes do not contain code for a movement protein and these are mediated by another (helper) virus. For example, potato leaf roll virus

(PLRV) which is restricted to phloem cells, can move to mesophyll cells mediated by helper virus PVV. Another classic example is cryptic virus which cannot be transmitted mechanically nor by a vector. They have bipartite dsRNA genomes and each segment codes for only one protein, i.e. one segment encodes an Rd Rp for replication. These viruses have no movement proteins for cell to cell movement. They appear to systemically 'infect' plants by spreading through host cell divisions. The viroids, which probably do not encode any protein, also seem to spread from cell-to-cell along with host cell division.

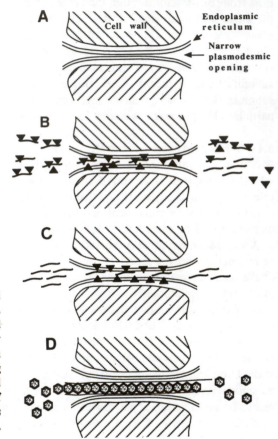

Fig. 9. **Cell-to-cell movement of plant viruses. A. Normal plasmodesma with narrow plasmodesmic opening. B. Virus moving from infected cell to un-infected cell as nucleic acid (wavy lines) ushered by movement protein (black triangles) through enlarged plasmodesmic opening. C. Virus moving from cell-to-cell as nucleic acid through enlarged plasmodesmic opening as a result of action of movement protein. D. Virus moving as particles through tubules formed by endoplasmic reticulum.**

Long-Distance Movement. The process of systemic invasion of plants by viruses is often referred to as long-distance movement of the viruses. This includes both systemic infection resulting from slow cell-to-cell movement and movement through vascular tissue. There are, however, certain differences from cell-to-cell movement. Plasmodesmata are present at every interface from mesophyll cells to vascular cells, although their number and structure vary greatly. The MP increases the SEL of plasmodesmata between mesophyll cells. But experimental evidence is not available on the mechanism of movement through the plasmodesmata linking bundle sheath and phloem-parenchyma cells, circulation in the sieve elements and movement back to the mesophyll cells.

Involvement of the coat protein (CP) in vascular movement is obvious: it does influence vasular movement. For some viruses, the CP is not required for cell-to-cell movement but is an essential co-factor for vascular movement. Site-directed mutagenesis performed in the CP open reading frame of a number of viruses has shown that the resulting mutants are unable to move throughout the plant. In TMV, CP is required for efficient systemic spread in the plant, but not for cell-to-cell movement. Similar observations have also been made for turnip yellow mosaic tymovirus in which absence of CP expression abolishes vascular transport (Bransom et al. 1995). In tobacco etch potyvirus (TEV), vascular movement of TEV mutants defective in CP could be restored by a transgene expressing the CP. Removal of either region results in the inability for vascular transport, while slow cell-to-cell movement continues. Further, it has been demonstrated that the terminal regions of CP interact with host factors for movement through the vascular system (Dolja et al. 1994). In contrast, whitefly transmitted geminiviruses (WTGs) do not require CP for systemic spread of the virus. In African cassava mosaic geminivirus (ACMV) and tomato golden mosaic geminivirus (TGMV), large deletions in the CP gene do not affect systemic spread if the mutated DNA size is maintained.

Besides CP, other factors such as replicate protein in TMV, non-structural proteins such as helper component proteinase (HC-Pro) in TEV, 5′ leader sequence in barley stripe mosaic virus (BSMV) and host specific factors have also been shown to play a role in vascular transport of viruses. The need for CP in vascular movement reveals that, in a majority of viruses, movement in the vascular system is as intact particles. This would also mean that plasmodesmata of vascular and mesophyll tissues differ (Ding et al. 1995).

When a virus infects a plant, it moves from one cell to another and multiplies in most, if not all such cells. The first intact particle appears in plant cells approximately 10 h after infection. In leaf parenchyma cells, the virus moves approximately 1 mm or 8 to 10 cells/day. Viruses reach the phloem through which they are rapidly transported over long distances within the plant. Most viruses require 2 to 5 days or more to move out of an inoculated leaf. From the phloem, the spread is very fast towards the growing region and often food storage organs such as tubers, rhizomes and roots.

The two ways viruses manifest themselves in the plant are through systemic infections and local lesion symptoms. Mosaic type symptoms are generally tissue restricted. It has been estimated that mosaic affected plant cells have 100000 to 1000000 virus particles per cell. Often segments of plant tissue in an infected plant may remain virus-free e.g. in the case of fruit trees affected by viruses. The meristematic tips in a majority of plants are not invaded by viruses and hence have led to raising virus-free plants through a technique of in vitro culture of meristems (Kassanis and Varma 1967). In local lesions, generally produced on leaves, there is a concentric zone where the virus is present. The reasons for such delimiting are not yet known, although it may be due to lack of virus movement.

9. Future Prospects

Tremendous progress has been made in our understanding of the replication of viruses since the discovery of viruses a hundred years ago. We now know the diverse strategies which are used for virus replication. Most of the plant viruses are vectored by insects, some of them also replicate in their vectors, indicating adaptation of the replication strategy which could function both in plant and insect systems. Structure and replication mechanism of viruses like reoviruses and rhabdoviruses suggest insect origin of these viruses, whereas most of the others have apparently evolved from plant systems (Nuss and Dall 1989). Some plant viruses also resemble animal viruses in amino acid sequence and gene arrangement, but fortunately plant viruses do not replicate in animal systems, which would have been disastrous.

Some good evidence in favour of plant virus origin has come from the acquisition of satellite transgene by CMV. When the transgenic plants expressing CMV satellite were inoculated by satellite-free CMV, the progeny virus was found to acquire the satellite RNA as well (Harrison et al. 1987). In any case, like the other pathogens, the palnt viruses must have also coevolved with their hosts.

Although good advances have been made in our understanding of the replication of viruses, a great deal is yet to be understood. We know very little about the mechanism of uncoating as the virus enters a susceptible cell and how the replication of viruses is initiated. Precise functions of the virus-coded proteins are also not fully understood, and even less is known about the interaction of these proteins with host proteins and how the host system is directed to undertake replication of an alien nucleic acid. We also want to know why infections with some viruses result in proliferation of cells—like the formation of enations— but not by other viruses. What is the mechanism which restricts some viruses to certain tissues and how is the overall amount of virus produced in a cell regulated? What happens to the viruses like WTGs which are not found in old symptomatic leaves? The timing and site of replication and the role of host cell membranes in the entire process needs to be determined for various virus-host combinations.

The times ahead are challenging and exciting since the latest developments in molecular biology will help in elucidating the questions raised above, leading to opening up of new approaches for the application of biotechnology in developing plants resistant to viruses.

10. Acknowledgements

The authors would like to thank Dr. Narayan Rishi, Professor, HAU, Hisar, for his valuable suggestions, and Mr. Rajesh Bhatia and Ms. Minu Tiwari for their help in typing the manuscript.

References

Atabekov JG, Dorokhov Yu L (1984) Plant virus specific transport function and resistance of plants to viruses. Adv Virus Res 29: 313–364

Bawden FC (1964) Plant viruses and virus diseases, 4th edn. Ronald Press, New York, 361 pp.

Bawden FC, Pirie NW (1937) The isolation and some properties of liquid crystalline substances from solanaceous plants infected with three strains of tobacco mosaic virus. Proc R Soc Lond Ser B Biol Sci 123: 274–320

Beijerinck MW (1898) Over een contagium vivum fluidum also oorzaak van de vlekziekte der tabaksbladen. Versi Gewone Vergad Wis Natuurk. Afd K Akad Wet Amsterdam 7: 229–235

Bisaro DM (1996) Geminivirus DNA replication. In: DNA replication in eukaryotic cells. Cold Spring Harbor Laboratory Press, Cold Spring Harbor, New York, pp 833–854

Bransom KL, Weiland JJ, Tsai CH, Dreher TW (1995) Coding density of the turnip yellow mosaic virus genome. Roles of the overlapping coat protein and readthrough coding regions. Virology 206: 403–412

Citovski V, Zambryski P (1991) How do plant virus nucleic acids move through intercellular connections? Bioassays 13: 373–377

Citovski V, Knorr D, Zambryski P (1991) Gene I. Potential movement locus of CaMV encodes an RNA binding protein. Proc Natl Acad Sci USA 88: 2476–2480

De Haan P, Wagemakers L, Peters D, Goldbach R (1990) The SRNA segment of tomato spotted wilt virus has an ambisense character. J Gen Virol 71: 1001–1007

Ding XS, Shintaku MH, Arnold SA, Nelson RS (1995) Accumulation of mild and severe strains of tobacco mosaic virus in minor veins of tobacco. Mol Plant Microbe Interact 8: 32–40

Dolja VV, Haldeman R, Robertson NL, Dougherty WG, Carrington JC (1994) Distinct functions of capsid protein in assembly and movement of tobacco etch poty virus in plants. EMBO J 13: 1482–1491

Fraenkel-Conrat H (1956) The role of nucleic acid in the reconstitution of active tobacco mosaic virus. J Am Chem Soc 78: 882–883

Fraenkel-Conrat H, Williams RC (1955) Reconstitution of active tobacco mosaic virus from its inactive protein and nucleic acid components. Proc Natl Acad Sci USA 41: 690–698

Gibbs AJ (1976) Viruses and plasmodesmata. In: Gunning BES, Robands AW (eds) Intercellular communication in plants: studies on plasmodesmata. Springer Berlin Heidelberg New York, pp 149–169

Gibbs AJ, Harrison BD (1976) Plant virology, the principles, Edward Arnold London

Gierer A, Schramm G (1956) Infectivity of ribonucleic acid from tobacco mosaic virus. Nature 177: 202

Harrison BD, Mayo MA, Baulcombe DC (1987) Virus resistance in transgenic plants that express cucumber mosaic virus satellite RNA. Nature 328: 799–802

Hershey AD, Chase M (1952) Independent functions of viral proteins and nucleic acid in growth of bacteriophage. J Gen Physiol 36: 39–56

Kassanis B, Varma A (1967) The production of virus-free clones of some British potato varieties. Ann appl Biol 59: 447–450

Klug A, Caspar DLD (1960) The structure of small viruses. Adv Virus Res 7: 225–325

Leisner SM, Turgeon R (1993) Movement of virus and photoassimilate in the phloem: a comparative analysis. Bio Essays 15: 741–748

Mandal B, Varma A, Malathi VG (1997) Systemic infection of *Vigna mungo* using the cloned DNAs of the blackgram isolate of mungbean yellow mosaic geminivirus through agroinoculation and transmission of the progeny virus by whiteflies. J Phytopathol 145: 505–510

Mason WS, Taylor JM, Hull R (1987) Retroid virus genome replication. Adv Virus Res 32: 35–96

Matthews REF (1992) Fundamentals of plant virology. Academic Press, New York, 403 pp

Nuss DL, Dall DJ (1989) Structural and functional properties of plant reovirus genomes. Adv Virus Res 38: 249–306

Pfeiffer P, Gordon K, Fütterer J, Hohn T (1987) The life cycle of cauliflower mosaic virus. *In*: Von Wettstein D, Chua NH (eds) Plant molecular biology. Plenum, New York, pp 443–458

Robards AW, Lucas WJ (1990) Plasmodesmata. Annu Rev Plant Physiol 41: 369–419

Shepherd RJ, Brueing GE, Wakeman RJ (1970) Double stranded DNA from cauliflower mosaic virus. Virology 41: 339–347

Stanley WM (1935) Isolation of a crystalline protein possessing the properties of tobacco mosaic virus. Science 81: 644–645

Townsend R, Stanley J, Curson SJ, Short MN (1985) Major polyadenylated transcripts of cassava latent virus and location of the gene encoding the coat protein. EMBO J 4: 33–37

Varma A (1988) Important filamentous viruses in the Indian sub-continent. *In*: Milne RG (ed) Filamentous viruses. Plenum Publishing, New York, pp 371–388

Varma A, Ramachandran P (1994) Plant viruses. *In*: Johri BM (ed) Botany in India. History and progress. Oxford and IBH Publ, New Delhi, pp 81–107

Wilson TMA, Cruz SS, Chapman S (1998) Viruses of plants in the service of man: from crop protection to biotechnology. *In*: Chopra VL, Singh RB, Varma Anupam (eds) Crop productivity and sustainability-shaping the future. Oxford, IBH Publishing, New Delhi, pp 244–257

3

Reproductive Biology of Bacteria

Sheela Srivastava

1. Introduction

One of the striking features of biological evolution has been differentiation of sex. It is generally assumed that the fundamental function performed by sex is reproduction. A brief survey of the biological systems, however, clearly reveals that most plants, many of the simpler animals, and essentially all microorganisms have effective means of asexual reproduction. This process is highly efficient in the propagation of the species. Sex has a second major biological function not performed by asexual reproduction. This consists of providing within the families, populations, and species the genetic variability without which no evolutionary success would have been possible, and is accomplished by genetic exchange phenomenon brought about during sexual reproduction. This evolutionary significance has led to its establishment amongst the majority of the life forms, including bacteria.

Sexual reproduction, as the term is used by biologists, does not necessarily imply a clear-cut distinction of male and female sexes. In fact, it was this lack of morphological identity that led to the idea that microorganisms and many lower life forms do not reproduce sexually. Several of these organisms do not use sexual reproduction as a means of propagation but employ it to create the essential genetic variability. However, even if we cannot distinguish them morphologically, experimental evidence has clearly proved that the two mating partners are physiologically different (mating types). Sexual reproduction can, thus, be defined as a situation in which there is mingling of the nuclear material from two different cells and a probability exists for the exchange of their genetic material (recombination). The generation of recombinants, therefore, provides one of the most convincing pieces of evidence for the existence of sexual reproduction (Sinha and Srivastava 1983).

Although the world of microorganisms was brought to light more than three centuries ago, following the invention of the microscope, Louis Pasteur is credited with bringing them into the realm of experimental science. The role played by microbes in human affairs for a long time overshadowed their importance in the study of some fundamental biological problems. In fact, it was with genetics that microbiology had the greatest difficulty in finding common ground. For many years, both microbiologists and geneticists tacitly agreed that the principles and laws of genetics cannot be applied to the organisms that do not reproduce sexually. It was only after the classical work of Beadle and Tatum (1941) on the biochemical genetics of *Neurospora* that bacterial genetics got really started. Within a few years, a number of mutants of bacteria and bacterial viruses (bacteriophages) were collected by Delbrück, Luria, and Tatum.

The entire edifice of classical genetics is based on transmission of characters from generation to generation. Such ideas originated from controlled hybridization between races or varieties of the same species which differ in certain hereditary characters. For a long time, this approach seemed inapplicable to microorganisms in general, and to bacteria in particular. Firstly, they appeared to reproduce only vegetatively and because of their low degree of organization, the distinction between somatic and reproductive parts or between genotype and phenotype was almost impossible to make.

So, although all ingredients were available, hybridization experiments could not be conducted in bacteria until 1946. Prior to this time, sexual phenomena were looked for, described, and even imagined on a purely morphological basis by microscopic observations of stained preparations. Even when many of these observations strongly suggested a fusion or conjugation, they could not be properly evaluated in the face of the lack of adequate genetic support. Attempts were made to cross bacteria by Sherman and Wing (1937), followed by Gowen and Lincoln (1942). These experiments clearly stated the conditions and were designed to demonstrate the existence of a sexual process in bacteria. The results obtained, however, were negative as the characters chosen were not easily selectable for recombination, and at best could demonstrate only the frequent classes of genetic recombinants.

2.   The Historical Cross

It should be stressed that in 1946, when Lederberg and Tatum discovered genetic recombination and, therefore, sexual reproduction in bacteria, it was not a casual observation but the outcome of a carefully planned investigation.

Lederberg and Tatum laid emphasis on the following parameters:

1. The phenomenon of recombination in bacteria is going to be rare and, therefore, it is essential to select. Also, one must be able to distinguish such recombinations from spontaneous mutations.
2. The bacterium chosen by them was *Escherichia coli* strain K-12, which had been used for many years as a standard laboratory strain in bacteriology courses at Stanford University. The strain could grow on a simple synthetic medium and had already been used by Tatum earlier for inducing biochemical mutations.

The use of strains which differ in their growth factor requirements appeared particularly favourable for the selection of possible genetic recombinants, more so if they were double or triple requirers (carrying mutational lesions that led to these requirements). These characters are highly stable, although each one of them can revert to wild type condition independently (Lederberg 1950, 1951). A simultaneous reversion of two or more such characters is highly improbable and will occur at a frequency of $\sim 10^{-12}$.

They then set about doing the cross between the strains 58–161 requiring methionine and biotin; *met, bio* with W677 (requiring threonine, leucine, and thiamine; *thr, leu, thi*). When cells of either strain were spread in large numbers on a synthetic agar medium devoid of any of these growth factors (minimal medium), no colonies showed up. On the other hand, when the same number of cells were mixed and then spread together on the minimal medium about 100 colonies appeared. These colonies could be further subcultured on minimal agar confirming that some sort of reassortment of genetic material between the two genotypes had occurred. So if the two polyauxotrophic mutants were mixed, *met, bio, thr$^+$, leu$^+$, thi$^+$* and *met$^+$, bio$^+$, thr, leu, thi*, for example, and exchanged their genetic material, one class of the possible selectable recombinant will be *met$^+$, bio$^+$, thr$^+$, leu$^+$, thi$^+$*, that should be able to grow on minimal medium. This was indeed observed (Fig. 1). The frequency at which the prototrophs were formed was low (10^{-6}–10^{-7}) relative to the original cell number), but detectable. Since the process involved the coming together of the two single-celled organisms, it came to be called conjugation (Lederberg and Tatum 1946a, b, Tatum and Lederberg 1947).

2.1   Nature of the Bacterial Cross

In order to ascertain the underlying process of the reassortment of genetic characters, a simple but ingenious

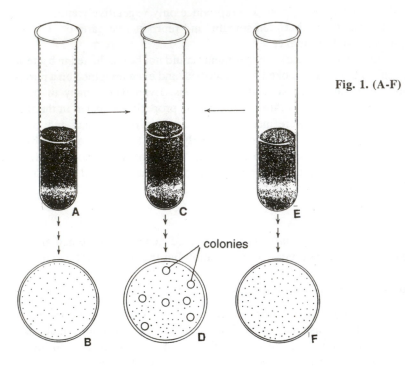

Fig. 1. (A-F)   Genetic recombination. Demonstration of sexual reproduction in *Escherichia coli*. Both cell types A and E fail to grow on an unsupplemented nutrient medium (MM) since each carries mutations in the genes required to synthesize constituents needed for cell growth (B, F). However, when they are mixed for a few hours (C) and then plated, a few colonies appear on the MM agar plate (D). These colonies constitute the recombinants in which the exchange of genetic material occurs between the two cell types enabling them to synthesize all the required nutrients (After Lederberg and Tatum 1946a, b).

experimental set-up was designed by Davis (1950). He placed the two multiple auxotrophic strains of *E. coli* in the two arms of a U-tube where a sintered glass filter was placed, separating the two strains. While the liquid could pass through the filter, the cells could not (Fig. 2). When the material from two arms was periodically plated to check for recombinants, none appeared. If the filter was removed so that the cells were free to move around, recombinants reappeared on plating. So it appeared that cell to cell contact is necessary for recombinants to form. This proved conclusively that bacterial sexual reproduction can be classified as conjugation.

It was earlier demonstrated by Lederberg (1947) that if the two parental strains differ in unselected markers, their distribution amongst the recombinants was non-random. This tendency of an unselected allele to remain associated with a selected one is a classic measure of their linkage and helps to define a linkage group. By this criterion, all the loci (genes) used in these early crosses fell onto one linkage group (Clowes and Rowley 1954).

3.   Sexual Compatibilites

Lederberg et al. (1952) observed that certain derivatives of the strain 58–161 (*bio, met*) could not produce recombinants with the standard W677 (*thr, leu, thi*), and thus turned infertile. They inferred that in the absence of a clear morphological difference, bacteria have still evolved the requirements for mating types. On this basis, these strains can be identified into F^+ (the fertile ones) and F^- (the infertile strain). For any cross to be successful it must involve one F^+ and one F^- strain (although $F^+ \times F^+$ can also produce low frequency recombination).

Hayes (1952a, b) established these distinctions on the basis of additional functional differences too. He demonstrated that the treatment of the F^+ strain with UV-irradiation or antibiotic streptomycin stimulates recombination but that treating the F^- strain in such a way strongly inhibits the process. He suggested that

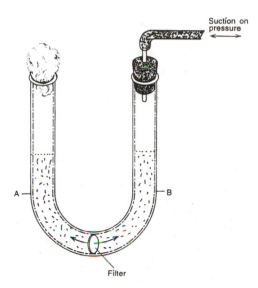

Fig. 2. Davis' U-tube experiment to demonstrate the essentiality of physical contact between the cells for recombination (conjugation). Two strains A and B genetically unable to synthesize the required metabolites (auxotrophs) were placed separately in two arms of the U-tube, separated by a sintered glass filter. While liquid can pass through the filter, by applying suction or pressure, bacterial cells cannot. The cells plated after several hours of incubation, from either arm, did not produce any colonies on MM agar (after Davis 1950).

F^+ acts as the donor of the genetic material whereas F^- receives it. The F^+, therefore, is alternatively called the donor or male strain and F^- as the recipient or female, with the transfer taking place unidirectionally. It also became clear that the survival or viability of the recipient strain is absolutely essential, as this strain is responsible for producing the progenies.

Lederberg et al. (1952) tested the compatibility of the progenies of the $F^+ \times F^-$ cross and found all of them compatible, that is F^+. It is as if the F^+ character is infectiously transferred, independently of any other genetic marker. Sexual conversion is thus the most striking feature of the $F^+ \times F^-$ cross with the frequency of conversion 10^6-fold higher than recombination (Fig. 3). Moreover, if F^+ cells are grown to a high cell density, some cells start behaving as though they are F^-, but the genetic markers remain unchanged. Such a situation is defined as the 'formation of phenocopies'.

4.   Analysis of Recombinants and Nature of Chromosome Transfer

When prototrophic recombinants were analysed by Hayes (1953), they showed a bias towards the unselected markers of the F^- strain. Two interpretations were provided to explain these results. Firstly, it may be assumed that the donor chromosome is completely transferred but a large part of it is eliminated and only a small portion is utilized (Lederberg 1949, 1955a, b, 1956). The very wasteful nature of this mechanism led to its low acceptance amongst the scientific community. The second, which had experimental support, suggested that only a part of the donor chromosome is indeed transferred to the recipient. If it is utilized recombinants are produced but it is lost if this does not happen (Fig. 4). The structure so formed after conjugation is not a real zygote but a partial diploid, merozygote or a heterogenote (Zelle and Lederberg 1951).

Hayes (1953a) explained that the ability of F^+ cells to transmit a mating type very efficiently, and to act as genetic donors less efficiently is based on the possession of a sex factor or F factor. This factor is rapidly (infectiously) transferred, and occasionally confers the ability to promote the transfer of genetic markers, but prevents the cell carrying it to act as a recipient.

5.   High Frequency Transfer Donors and Nature of Chromosome Transfer

In 1952, Lederberg et al. and in 1953b Hayes discovered, a new class of donor from 58–161 F^+, on the

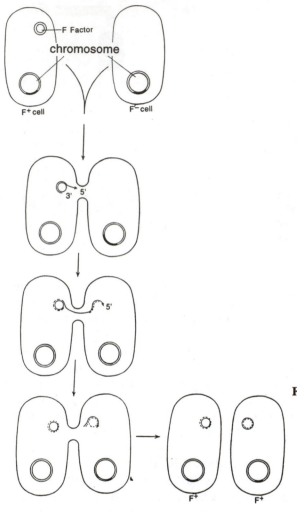

Fig. 3. The outcome of an F$^+$ (donor/male) × F$^-$ (recipient/female) mating in bacteria. The circular F factor (plasmid) is nicked and transferred in a single-stranded state to the F$^-$cell, where the complementary DNA strand is synthesized. The resulting cells (exconjugants) are both F$^+$ or functionally male; F$^-$ is converted to the F$^+$ state (infectious transfer of F factor) (After Lederberg et al. 1952; Hayes 1953a)

basis of its ability to promote high frequency recombination (Hfr). When this Hfr was crossed with W677 F$^-$, it gave 1000-times more prototrophs than the usual F$^+$ × F$^-$ cross. In compatibility tests, it behaved like F$^+$, but sometimes lost its ability to promote efficient genetic transfer, thus reverting to F$^+$ condition. This process suggests that Hfrs are derived from F$^+$ by some change of their state and may revert by reversing the change. The behaviour of Hfr depends upon the marker selected, as only some markers show high frequency of recombination, others are transferred at varying efficiency of high to low shown in a typical F$^+$ × F$^-$ cross.

The actual nature of the chromosome transfer was elucidated by Wollman et al. (1956) and Jacob and Wollman (1961) by the technique called an 'interrupted mating experiment'. At various times after mixing the parental populations, the culture is agitated vigorously in a blender before it is diluted and plated on selective media.

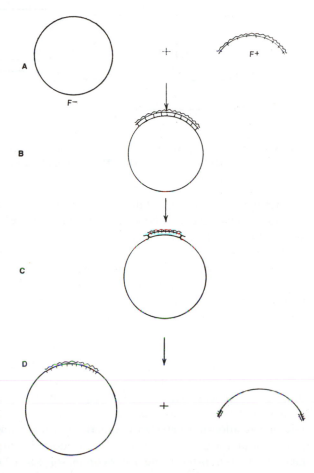

Fig. 4 A-D Possible mechanism by which $F^+ \times F^-$ mate to produce recombinants. A donor fragment pairs with a homologous region of the F^- chromosome, and replaces it by crossing over (After Hayes 1955, 1957).

When such colonies were analyzed, different donor markers first appeared in the progenies only after different periods had been allowed before mating was brought to halt. In a cross between Hfr $thr^+ leu^+ gal^+ \lambda^+ str^s \times F^- thr^- leu^- gal^- \lambda^- str^r$, $thr^+ leu^+$ appeared at 8 and 9 min but gal^+ and λ infectious centres only at 25 and 26 minutes, respectively (Hayes 1955, 1957, 1968).

A theoretical representation of such a sequential transfer is shown in Fig. 5. With each Hfr, the transfer was initiated at a specific point (origin of transfer or *oriT*) and then followed linearly for the time two cells remained together (Fig. 6). The observed kinetics suggest that there is a distinct relationship between the time at which a given marker is transferred from Hfr to the F^- cell and its map location on the donor chromosome. Alternatively, the closely linked markers appear at the same time and distantly located ones at different times.

Later, several other markers were used in appropriate crosses and their time of entry was ascertained by interrupted matings. This procedure allowed the construction of what is popularly known as the temporal map of *E. coli*, in which the genes are placed on the time of entry scale (in min; Fig. 7A, B). The correlation between the time of first appearance and the ultimate frequency of transfer suggests that matings suffer spontaneous separation of the donor-recipient pairs at a certain period after the conjugation has begun. Breakage of the donor chromosome then has the same effect as that achieved in an interrupted mating

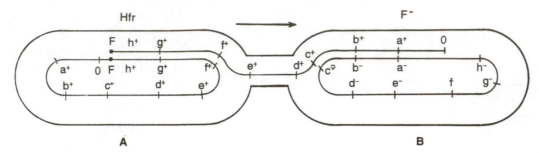

Fig. 5. (A, B) Diagrammatic representation of sequential gene transfer during bacterial conjugation between Hfr (A) and F⁻ cells (B). *O* origin of transfer or *ori*T, F fertility factor (After Jacob and Wollman 1961).

experiment. Since the probability of a spontaneous break depends upon the period for which the two cells maintain contact, the likelihood of a break must be greater the farther a marker lies from *oriT* and the same should be reflected in the frequency with which it enters the recipient.

Another interesting feature of the bacterial chromosome was illustrated when different Hfrs were discovered, described and used in interrupted matings. Jacob and Wollman (1961) showed that each Hfr has a characteristic order of transferring the markers which is sequential and linear; but different markers are transferred at high frequency by different Hfrs (Fig. 8). If these sequences are compared, all markers of *E. coli* were linked in a single group, but the sequences of all Hfr donor chromosomes cannot be arranged in a simple linear map. Two features came out clearly from this exercise: (1) genetic transfer may take place in either direction, and (2) except for the loci separated by the point of origin, all markers show the same relationship in location in all Hfrs.

The sequence in which markers are transferred by different Hfrs can be explained by the construction of a circular genetic map (Fig. 9), as suggested by Jacob and Wollman (1961) and Hayes (1968). It is thus clear why the markers at opposite extremities of the map of one Hfr are closely linked in the other, and how the relationship between the markers is maintained. Consistent with the later experiments and to provide a uniform basis of marker localization, Bachmann et al. (1976) revised the estimate of the time taken for total transfer of the donor chromosome to 100 minutes. *E. coli* genetic map currently in use has been standardized for 100 minutes and the markers have been displayed on the basis of temporal units from a specific *oriT*. However, the number and location of so many markers has become available now that a linear map has replaced the conventional and essentially circular map.

Genes can also be mapped on the basis of more familiar recombination units (% recombination). This method relies on selecting for the Hfr marker which enters late (distal marker) and then testing these individuals for the markers that must have entered ahead (proximal markers). If these recombinants showed the proximal recipient marker, it must have been due to a crossover separation between the two and should be reflective of their relative distances. The recombination frequency can be calculated:

$$RF = \frac{\text{Total number of recombinants}}{\text{Total number of population}}$$

In *E. coli*, 1 min on the map is equivalent to 20% recombination calculated in this manner.

Sometimes, the loci are so close that it becomes difficult to order them in relation to the third locus. So if the recombination frequency (RF) between a-b can not be experimentally distinguished from that of a-c, for example, it is difficult to say whether the gene order is abc or acb. One technique often used is to make a pair of reciprocal crosses using the same marker genotypes both as donor and recipient, e.g. a⁺b⁺c × abc⁺ If prototrophs are scored with the reciprocal cross giving dramatically different frequencies,

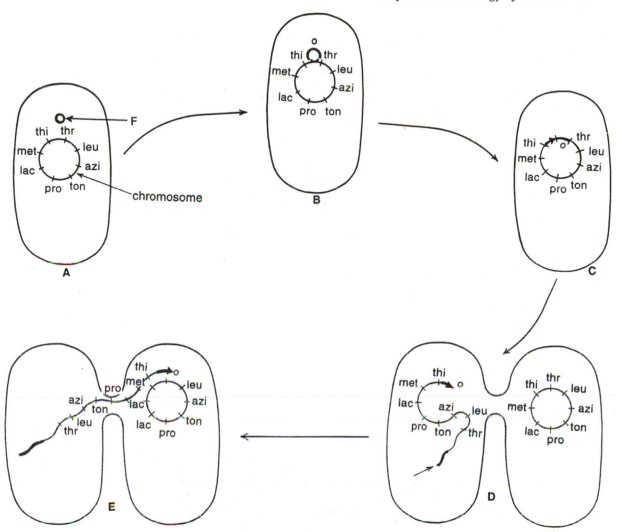

Fig. 6   A-E Hfr-mediated sequential linear gene transfer. F factor integrates into the chromosome (A-B) to generate *Hfr* which conjugates with an F⁻ cell (C) . The Hfr chromosome is nicked at *oriT*, located on the F factor (D). The *oriT* is oriented for transfer to F⁻ cell first (dragging along the chromosome) (E). As the cells remain together, more and more of the donor chromosome is transferred linearly and sequentially. The main body of the F factor is at the tail end which normally does not enter the F⁻ cell, and therefore no conversion occurs (after Hayes 1953b, 1968; Jacob et al. 1960).

the order is acb. But if no significant difference is observed in frequencies between the two reciprocal crosses, then the order must be abc. This is based on the principle that the first order will require double and quadruple crossovers whereas the second one will be generated by double crossovers in both situations.

It should be noted that the mere transfer of male genetic material does not ensure the production of recombinants. Experimental evidence suggests that 100% Hfr cells can sponsor zygote formation if all of them are able to find mates in the population. Conjugation promoted by Hfr donors is therefore a very common occurrence, contrasted with the rare events sponsored by F⁺ cells.

Although the most proximal markers will get transferred to all the zygotes, their appearance amongst

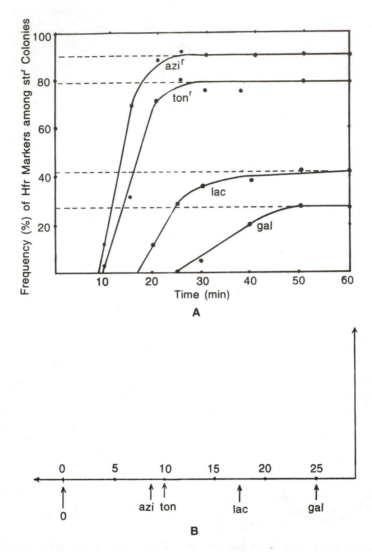

Fig. 7 (A-B) Genes mapped on *E. coli* on the basis of time (in minutes) at which they enter a recipient cell. This was demonstrated by Wollman, Jacob and Hayes using the 'interrupted mating experiment'. A multiple mutant $F^-(azi^r\ ton^r, lac^-, gal^-, str^r)$ was mated with a wild-type *Hfr* (azi^s, ton^s, lac^+, $gal^+\ str^s$). At different times after the cells were mixed, the samples were withdrawn and whirled on a Waring blender (to distrupt the mating). The cells were then plated on MM containing streptomycin. The resulting colonies were checked for the other four markers (stated above). The figure (A) shows the time at which each of the donor markers begin to appear amongst the recombinants. (B) This forms the basis for the construction of a linkage map of the *E.coli* chromosome. The unit of distance in given in min. (After Wollman et al. 1956).

the progenies will be guided by the efficiency of integration. For example, HfrH is able to promote 100% zygote formation but only ~20% of prototrophs exhibit the early markers, $thr^+\ leu^+$; the efficiency of integration of these markers is thus 20%. In other words, in the majority of cases, zygote formation is not succeeded by genetic recombination.

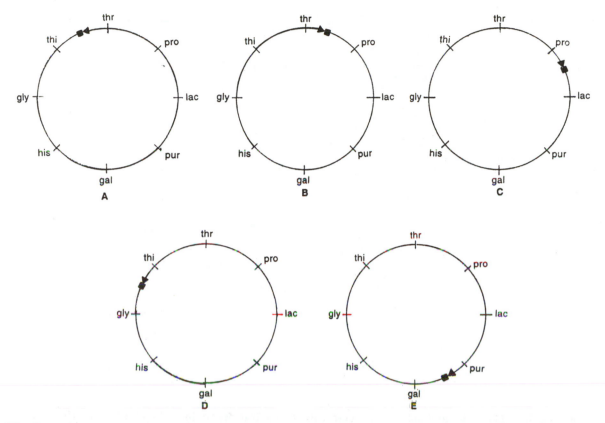

Fig. 8   A-E The generation of different Hfr strains by the insertion of F plasmid (denoted by ■◄) at different points and orientation. This leads to a characteristic orientation of gene transfer (after Jacob et al. 1960; Broda 1967).

The distance between two loci on *E. coli* chromosome, therefore, can be expressed as minutes of transfer, units of recombination, or length of DNA. Although temporal mapping is still the most popular, physical distances can be measured as the entire *E. coli* chromosome of 2.8×10^9 Da or 4.6×10^6 base pairs can be taken as equivalent to 100 min of transfer. The rate of transfer calculated will be 30×10^6 Da or 50,000 base pairs per min. Alternatively, it will take 2 seconds to transfer a gene of 10 Da. However, the experimental data suggest that the recombination frequency per unit distance may not be homogeneously distributed.

Based on strong genetic evidence derived from detailed linkage analyses, Taylor and Adelberg (1961) were able to suggest that bacterial chromosome must be circular. This conclusion won final support from the autoradiographic data of Cairns (1963) in which the replicating *E. coli* chromosome was directly visualized as a tangled circle.

The non-infectious nature of Hfr in contrast to the freely transferable nature of F+ indicated that F factor in these two conditions may exist in different states. It also suggests that while the F factor in its free, autonomous F+ state is the first one to get transferred, in Hfr it behaves as a part of the bacterial chromosome and rarely gets transferred to the recipient (Fig. 10a). The progenies which do become Hfr inevitably possess the markers that are transferred at the lowest frequency. On the basis of these data, it can be established that in Hfr, the F factor is integrated in the bacterial chromosome. The transfer also begins at

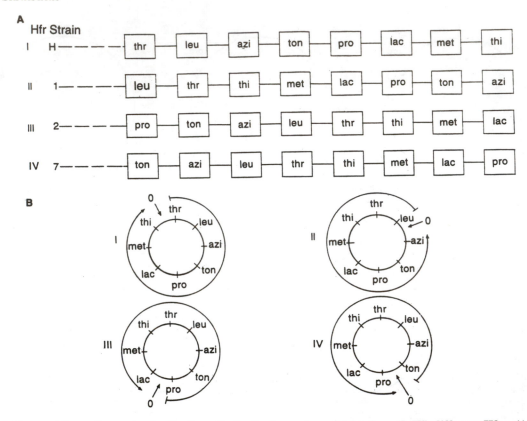

Fig. 9 (A, B) The order and polarity (hypothetical) of gene transfer by four (I–IV) different Hfrs. (A) The difference in the order does not affect the established distances (linkage) between the genes, unless they get separated by the insertion of the F factor. (B) The characteristic orientation in each can be explained by assuming that *E. coli* chromosome is circular. (After Broda 1967).

the site of integration in such a way that the chromosome is broken and on the basis of the orientation of the F factor, one side provides the origin (*oriT*) while the rest comes at the end which is rarely transferred. The different order and orientation in which markers are transferred by different Hfrs appears to be determined by the site of integration of the F factor in the chromosome. Implicit in this model is the concept that chromosome transfer is mediated by the sex factor through its capability to recombine with the bacterial chromosome.

6. Insertion of F into the Chromosome

From the behaviour of the F^+ and Hfr strains of *E. coli*, it is clear that in F^+ cells the F factor exists in an autonomous state independent of the bacterial chromosome. But in the Hfr it is integrated at a site which is characteristic for each Hfr strain.

On the basis of a fluctuation test Jacob and Wollman (1961) suggested that Hfrs arise in an F^+ population with random probability with regard to time and are inherited in a stable manner. Thus, the probability of an Hfr generation is ten per cell per generation. Hfr cells may also revert to the F^+ state with characteristic probabilities which are generally about the same order of magnitude as the probability of their formation.

In 1967, Broda concluded from his experiments on Hfr generation that they arise by nonrandom integration of F into the chromosome at limited but specific sites.

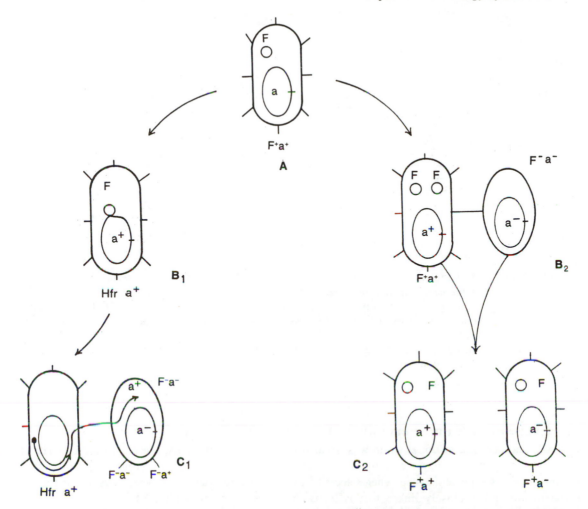

Fig. 10a. Conjugational cycle of *E. coli*. The nature of donor strain (A, B$_1$) affects the outcome of conjugation. F$^+$ × F$^-$ mating (B$_2$) leads to conversion of F$^-$ to F$^+$ (see also Fig. 3A). The other event which leads to the insertion of an F factor into the bacterial chromosome gives rise to a new donor type, Hfr (high frequency recombination male) (B$_1$). Hfr shows a high level of chromosome transfer and no conversion of F$^-$. If the two strains differ in genetic markers (e.g., a^+/a^-), the first event will generate two types of exconjugants, all F$^+$ but 50% a$^+$ and 50% a$^-$ (C$_2$). In the second event, the two types of exconjugants will also show a similar genetic segregation but will all be F$^-$ (C$_1$) (after Lederberg et al. 1952; Hayes 1953a, b).

The insertion of F factor can be considered analogous to the integration of certain temperate phages creating a prophage that is carried as a chromosomal locus. Such a model was first proposed by Campbell (1962) and could be applied to the sex factor as well. The model presumed that both the sex factor and the chromosome are circular and a single crossover taking place between the homologous sites on the two is sufficient for integration. Moreover, it also implies that a reversal of the same event can lead to the release of the F factor (or prophage) from its integrated state (Fig. 10b).

Evidence to support this model is considerable now and includes physical characterization of the structure as well as the genetic analyses. Jacob et al. (1960) coined the term 'episome' to describe such genetic

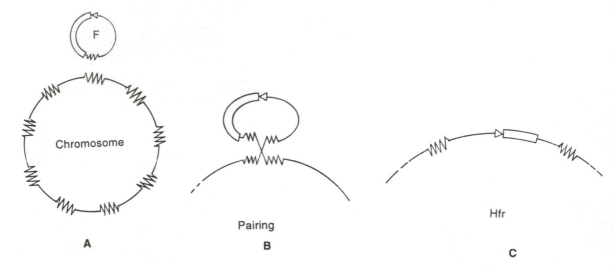

Fig. 10b. (A, C) Model for the insertion of the F factor into a chromosome. (A) Bacterial chromosome and F factor carry homologous regions (*MMM*) that can lead to their pairing (B). A subsequent crossing over results in the insertion of a F factor and formation of Hfr (B, C). If both the structures are circular, a single cross-over is sufficient for such an insertion (after Campbell 1969).

elements which could exist both in free autonomous or integrated states. The term plasmid, on the other hand, was used for such elements that could exist only in an autonomous form. These terms are no longer used with such precise meaning because mutants can be isolated that render episomes unable to integrate. So the plasmid is more popularly used and F factor can be rechristened as F plasmid.

7. Excision of F and Formation of F-Prime (F′) Factor

In 1960, another type of donor strain was isolated by Adelberg and Burns. They used the P4x Hfr strain which transferred *pro*, *leu*, *thr* as early markers and *lac* as a late marker. The progenies of P4x Hfr × F⁻ mating are F⁻ as expected, except when selected for *lac* which are usually Hfr type. Thus, Hfr is transferred distal to *lac* and is closely linked to this locus. A subclone of this strain, P4x–1 was unusual as the recombinants gaining early markers also acquired the capacity to act as genetic donors, and this property was perpetuated. Such a gene transfer phenomenon is also referred as sexduction. When used as donors, these recombinants transferred the host chromosome with the same orientation as the parent Hfr although at somewhat lower frequency. So this strain appeared to carry a modified sex factor, called F′ (F-prime), with characteristics intermediate between Hfr and F⁺. While chromosome transfer has an Hfr-like orientation, the sex factor transfer itself is infectious. A model to account for these features supposes that an illegitimate genetic exchange took place when the F factor of the P4x was excised from the chromosome (Fig. 11). Instead of reversing integration by recombination between the original sites, this genetic exchange involved one site located on the F factor and another on the chromosome, generating an F′ like P4x–1.

This type of exchange also known as type I exchange, generates a sex factor that can carry certain flanking markers, either proximal or distal, and some material of its own. Alternatively, the chromosome gains some portion of the sex factor but loses certain genes in the bargain. There is another way that an illegitimate exchange could generate F′. This takes place between two chromosomal sites flanking the inserted F factor. The resultant F′ factor picks up both proximal and distal markers but does not leave any of its own material in the chromosome (type II exchange).

Upon transfer to an F⁻ cell, the F′ integrates preferentially by homologous pairing between the genes

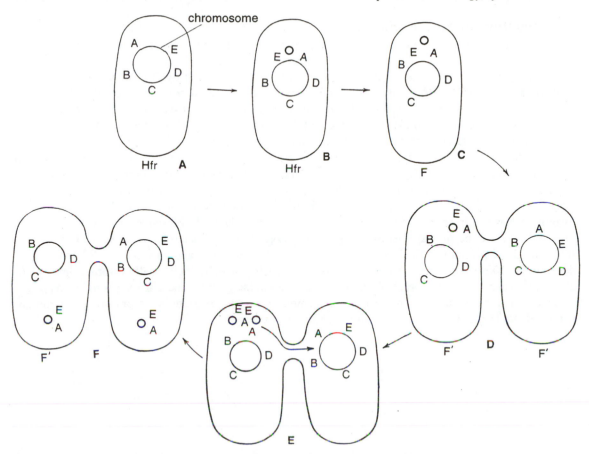

Fig. 11. (A, F) Generation of a new class of donor strain, F′ and the subsequent sexduction of F⁻. The F factor may be excised from its integrated state in Hfr (A), and if this excision is imprecise it can pick up one or more bacterial genes (B, C). When such an F′ mates with an F⁻ (D), the F⁻ becomes a partial diploid or merozygote for the genes carried by F′ (E,F) (after Adelberg and Burns 1960).

it carries and the corresponding region of the bacterial chromosome. F′ generated by type II exchange can also follow the same strategy in an F⁻ cell, but in the original cell it fails to integrate at the specific site because that site is deleted from the chromosome.

Several F′ factors have now been isolated. The primary strain in which an F′ factor is generated remains haploid for the gene(s) which are transferred from the chromosome to the F factor. When a secondary strain is generated by introducing the F′ factor into a recipient carrying a copy of similar chromosomal genes, the bacterium becomes a partial diploid for this region. If the factor and the chromosome carry two different alleles they can be described as heterozygotes or heterogenotes. These are, however, not very stable as recombination to exchange the markers or conversion into homogenotes are quite frequent. Such diploids can be maintained only in *recA* cells as recombination function is involved in all these events. F′ are thus commonly sought by the microbial geneticists for the purpose of understanding the allelic relationship and control of gene expression (Low 1972).

8.   Structure of Sex (F) Factor

From the initial integration model which could also explain the generation of F′, it was assumed that the sex factor or F plasmid was circular. Subsequently, its circular nature was also revealed by the characteristic sedimentation properties under appropriate conditions such as centrifugation in alkali or equilibrium centrifugation in CsCl-EtBr density gradients. However, the major evidence came from direct visualization under electron microscope of heteroduplexes created from covalently closed circles of F and F′ molecules. This procedure even allowed the measurement of the length more accurately (F measures approximately 100 kb) and understanding of the structure in relation to function.

The F factor carries a *tra* operon, its genes are responsible for the synthesis of pilin and the assembly of functional pilus, pairing, triggering transfer and mediating actual transfer, nicking at the transfer origin, *oriT*, DNA replication from normal (non-transfer) replication origin (*oriV/oriS*), and surface exclusion. Genes that control plasmid incompatibility and fertility inhibition are also located on F plasmid. A site *oriT* (origin of transfer), responsible for F transfer is also present. The F DNA sequences located between the *oriT* and *oriV* are the first to enter the recipient and have therefore been referred to as the leading region. Interestingly, F plasmid carries two copies of insertion sequence IS3 and one each of IS2 and γδ(gamma delta or Tn 1000). They are supposed to provide the sites of homologies with several copies of these elements on chromosome, when F plasmid engages itself in integration (and also excision) events (Fig. 12). The location and orientation of these insertion elements vis-a-vis the bacterial chromosome determines the nature of the Hfr and the orientation of the chromosome transfer.

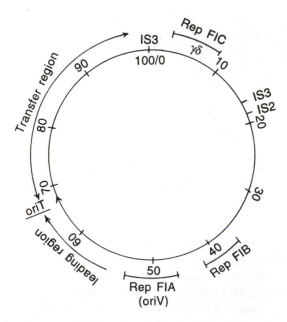

Fig. 12.   A map of the F plasmid. Map positions are δβ indicated in kilobases. The insertion sequences present on the plasmid are (gamma, delta), IS2 and IS3. Also shown on the map are location of *oriT*, two origins of replication Rep FIA (*oriV*) and RepFIB and third non-functional origin of replication, RepFIC, the leading region and a large transfer region which carry the *tra* genes (after Willets and Skurray 1987; Firth et al. 1996).

9.   Surface Properties

The ability of the Hfr or F⁺ to act as donors depends upon specific properties acquired by the cell as a result of the presence of F plasmid, in a free or integrated state. One early observation on the difference between donor and recipient cells was made by Loeb (1960) that certain bacteriophages lyse only those *E. coli* cells that had the ability to act as donors. Phages with this male specific selectivity will infect only donor strains. These include RNA phages f2, R17, MS2, M12, Qβ, and single stranded DNA phages f1, fd, and M13.

This characteristic surface property of the donor has been identified as the possession of F pili or sex fimbriae. While all *E. coli* cells are surrounded by hairy appendages called type I pilus, male cells are endowed with an additional type of pilus, the F pilus. The male-specific phages adsorb either to the length (usually RNA phages) or to the tip of the F pilus (DNA phages). The F pili are 1–2 μm long and grow perpendicular to the cell surface. Their number varies with the growth conditions (generally one to five), and they are essentially hollow rods of an external diameter of 85 Å with an axial hole of 20–25 Å diameter. The F pilus is assembled from a single type of protein subunit, called F pilin which is approximately 11.8 kDa in size and has an isoelectric point of pH 4.5. The production, organization, and assembly of functional sex pili is controlled by over a dozen genes located on the F plasmid.

Structures similar to the F pilus but not identical, are also produced by some R factors. They serve the same function and if F and such R factors coexist and express in a cell they can produce mixed pili assembled from both types of subunits. On the other hand, certain other plasmids (I-type) produce I-pili with the same role as the F pilus, but there is no intermixing between I and F pili subunits.

The mechanical removal of F pili or addition of anti-F pilus antibodies prevent donor activity which confirms their essential role in conjugation. Electron microscope studies have also shown donor and recipient pairing with the help of a sex pilus (Fig. 13). It has been suggested by some that donor DNA travels through the axial hole of the pilus to reach the recipient cytoplasm, but no evidence exists to confirm this. Several other authors have, however, suggested more rigid structures or wall-to-wall pairing, with the pilus only providing the initial contacting role and triggering the subsequent conjugation-specific events, a process akin to signal transduction. Some investigators have described cell aggregate formation during mating. It is also known that pili are depolymerized shortly after the contact, with their subunits returning to the donor cell membrane. This might be responsible for drawing the mating cells together into aggregates. It has also been shown that dissociation of pili by detergents such as SDS after mating pair stabilization but before DNA transfer does not inhibit transfer. Thus, the pilus base which survives the detergent treatment, may provide a channel sufficient for DNA transit. Indirect evidence also suggests that the pilus helps in establishing the actual contact with the cytoplasmic membrane of the recipient. The *mob* factor then may anchor the plasmid to the cytoplasmic membrane, which will trigger its transfer (Heinemann 1992).

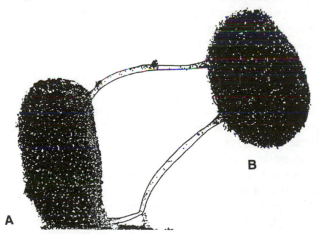

Fig. 13. **Electron micrograph of conjugating** *E. coli* **cells A and B (x 34300). The cells are connected by long appendages (sex pili) arising from the male (Hfr) cells, absent in the female (F⁻) cells. The pili have been visualized by the addition of male specific phages (after Loeb 1960).**

The surface proteins produced by the recipient cell also play an important role. The *con* mutants fail to form mating pairs since they contain much reduced amounts of the two of the outer membrane proteins.

Although both Hfr and F⁺ bacteria may possess sex pili, they can be easily distinguished by the effect of treatment with agents like acridine orange or acriflavin. Hirota (1960) reported that an F⁺ strain can be converted to the F⁻ state by treatment with acridine orange and this recipient status is stably inherited by their progeny. This compound has no effect on Hfr strains suggesting that it can lead to the loss of free autonomous F factor but not of integrated one. This phenomenon is called as plasmid curing and is based on prevention of replication of autonomous sex factor. Use of acridine orange has become a standard test to determine whether a cell carries F factor in free, autonomous (susceptible to curing) or integrated (can not be cured) state.

10.   Chromosome Transfer and Recombination

As mentioned earlier, the transfer of the bacterial chromosome from donor to recipient is under the control of the integrated sex factor and commences at the *oriT* site, located within the factor. One requirement is that a break must be made in the circular chromosome so that it can be transferred linearly (Firth et al. 1996).

In 1963, Jacob, Brenner, and Cuzin formulated two classes of model for chromosome transfer. One is to suppose that the chromosome is replicated prior to its transfer and only a copy is transferred to the recipient. This model, however, could not explain the nature of the signal required to initiate the replication.

An alternative was to suppose that transfer requires concomitant replication in the donor. Although in its original form this model was based on the transfer of one of the replicas of the donor chromosome, subsequently it could also be applied to the idea that only one of the strands of DNA is freed by replication for its passage to the recipient. Under both the situations explainable by this model, the replication is under the control of the sex factor which is presumably activated by the contact between donor and recipient.

Various pieces of genetic, biochemical, and biophysical evidence clearly point out that only one donor strand is transferred to the recipient by all the three donor types, F⁺, Hfr, and F′. Such a transfer was explained on the basis of rolling circle replication by Gilbert and Dressler (1968). They argued that bacterial DNA is nicked in one strand, which will allow the generation of a free 3′OH end to be used as a primer for the synthesis of a new strand. As the new strand grows in 5′–3′ direction, it displaces the original parental strand on the other side of the nick. This freed single stranded DNA with a 5′ terminus then migrates towards the recipient (Willets and Skurray 1987, Firth et al. 1996). It is reasonable to suppose that the same origin of transfer is used by all donor types, free or integrated. In the recipient cell, the single strand of transferred Hfr DNA provides a template for lagging strand synthesis. During the transfer the leading 5' end is probably attached to DNA helicase I at the site of DNA transfer, so that a growing loop of partially duplex DNA is presented to the recipient (Lloyd and Buckman 1995).

10.1   Integration of Donor DNA

When mating terminates, the transferred DNA is released as a linear fragment consisting of a segment of F plasmid at the leading end and a single strand overhang of variable length at the distal 3' end. Recombinants arise from exchange between this fragment and the circular recipient chromosome.

In order to produce a viable recombinant chromosome at least two crossovers are needed to integrate the Hfr DNA (Fig. 14). The various enzymatic steps involved in this process have been elucidated with the help of mutants that affected recombination (Smith 1985, Lloyd and Low 1996). Amongst these, the most central role is played by *recA* genes product. RecA protein coats the single stranded DNA and forms a nucleoprotein complex. In presence of ATP this complex interacts with duplex recipient DNA. After the initial sequence-independent interaction the two molecules move in conjunction to align by homologous sequence pairing (synapsis). From one end now *recA* actively promotes the displacement of one recipient strand by unwinding (helicase activity) and assimilation of donor strand (strand transfer), creating a Holliday joint and subsequently a heteroduplex. The generation of heteroduplex is an important step in homologous recombination in all sexually reproducing organisms, and strong genetic evidence for same exists. Another enzyme which plays an important role is a multifunctional protein complex, Rec BCD, which has DNA helicase and nuclease activities. It can initiate crossovers near each end of the Hfr DNA that leads to its integration. Recombination could also be initiated by making use of transient single-stranded gap in the loop of Hfr DNA (Lloyd and Buckman 1995). The RuvAB and RecG proteins catalyze branch migration and RuvC resolves the Holliday junction by endonuclease cutting (Smith 1985, Lloyd and Low 1996).

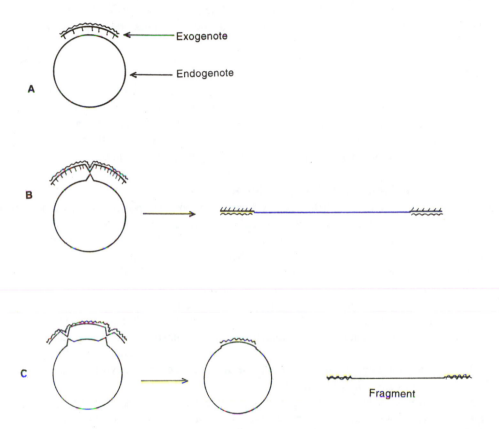

Fig. 14   **A-C Sexual reproduction leads to the transfer of a small part of the male or the donor chromosome to the female or the recipient; the former is designated 'exogenote' and the latter 'endogenote'. This has two consequences: (1) first, a complete diploid zygote is never formed in bacteria. The structure at best can be referred to as partial diploid or mero-zygote (A), (2) in order to produce the recombinants the fragment must undertake at least two cross-over joints (C). If only one cross over occurs, the circular chromosome opens out into a linear form (B after Hayes 1968).**

11.   Conjugation in Other Gram-Negative Bacteria

The F plasmid can replicate only in enteric bacteria, so the type of conjugation described above can occur only in this group. But a large number of plasmids called conjugative plasmids, capable of self-transfer exist in other gram-negative bacteria. However, all conjugative plasmids cannot readily mobilize the transfer of the chromosomal DNA. Some of such plasmids can acquire this property by gaining small bits of DNA such as insertion sequence. An example of this is known in plasmid R68 that itself transfers very efficiently, but mobilization of another plasmid or chromosomal DNA takes place at a very low frequency.

Derivatives of R68, such as R68.45, have been isolated which mobilize both chromosomal and plasmid DNA in *Pseudomonas aeruginosa* and a variety of other gram-negative bacteria at a high frequency. The difference between R68 and R68.45 is that the latter contains a 2.10 kb insertion sequence designated IS21.

As described earlier, even the F plasmid owes its chromosomal mobilizing activity to its complement of insertion sequences. However, the mechanism of F-mediated and R68.45-mediated mobilizations appear to differ. The insertion sequence on F also occurs on the chromosome and mobilization requires homologous recombination between these two regions, while the IS21 sequence may not occur on the chromosome of those bacteria where R68.45 can function. Mobilization in such cases may take place as a consequence of transposition.

Hfr-like strains have also been described in *Pseudomonas aeruginosa* that carry plasmids called FP. They can carry out similar function in this bacterial species.

In *Agrobacterium tumefaciens*, the causative agent of plant crown-gall disease, the pathogenicity is mediated by the transmission of plasmid DNA (Ti) from the bacterial cell to the plant host cell. The transferred sequences cannot be replicated autonomously in the plant cells; instead a part thereof (T-DNA) is integrated into the chromosomes. Ti plasmid can transmit itself entirely to other strains of Agrobacteria conjugationally. This interbacterial transmission is provoked by plant exudates.

12. Conjugation in Gram-Positive Bacteria

Conjugation as a mode of genetic exchange is less documented in gram-positive bacteria. In *Streptococcus (Enterococcus) faecalis*, a conjugative system has been described. Here pili do not play a role, but rather the plasmid-carrying cells form clumps with cells that lack them, and plasmid transfer occurs within these clumps. The clumping results from the interaction between an aggregation substance (adhesin) on the surface of the plasmid-containing donor cell and a binding substance on the surface of the plasmid-free recipient. The binding substance is present on both plasmid-containing and plasmid-free cells, but the aggregation substance is produced by the donor only when it comes into contact with the recipient. The recipient, in such a situation, produces a chromosomally encoded small hormone molecule, a pheromone, that diffuses out of the cell, enters the donor and induces a plasmid encoded gene to synthesize the aggregation substance. This circuit is broken once the two cells separate and the synthesis of aggregation substance stops. Under non-conjugating conditions, the gene for pheromone synthesis in donor cells remain repressed by a plasmid gene *Ica* (Clewell 1993).

In *E. faecalis, Streptococcus pneumoniae*, and *S. sanguis* a conjugal system not based on plasmids but on conjugative transponsons has been described. The intercellular transposition of conjugative transposon, such as, Tn 916, Tn 918, Tn 920, and Tn 1545 is resistant to deoxyribonuclease and requires cell-to-cell contact. Tn 1545 codes for two proteins, Xis-Tn and Int-Tn that like phage lambda's Xis and Int promote site-specific integration of DNA with minimum of homology requirement.

In *Bacillus subtilis*, chromosomal DNA is transmitted without the intervention of an autonomous plasmid. This transfer requires cell contact, initiates at a single site (origin of transfer), and correlates with chromosomal replication, as in Hfr-mediated transfer in *E.coli*. A possible difference in the two systems is that in *B. subtilis* such a transfer is limited to the first replication of the chromosome of a germinating spore.

In mycoplasma, the *Spiroplasma citri* chromosomal transmission system resembles that of *B. subtilis*. Consistent with the presumed lineage of mycoplasma from gram-positive bacteria, conjugation has been demonstrated between *Mycoplasma hominis* and *E. faecalis*.

12.1 Actinomycetes

A conjugation mechanism occurring in certain members of actinomycetes differ markedly from other gram-positive bacteria. One of the best studied members is *Streptomyces coelicolor*, where crosses can be effected by growing together two genetically different strains on non-selective media and then transferring the resulting cell mass on the selective plates. This allows the identification of a recombinant class of progenies. Three different fertility types have been identified in *S. coelicolor*: IF (initial fertility) with relatively low fertility, NF (normal fertility) and UF (ultrafertility) with markedly improved fertility. All fertility types are self- and inter-fertile, although the level of fertility varies.

Acknowledgements: I thank Mr Satish Chand for computer typing the manuscript.

13. Glossary

Acridine dyes. A group of compounds such as acriflavin, acridine orange, proflavin that have the capability of stacking or intercalating between the bases of a DNA molecule. This action confers a strong mutagenic activity and may also impede replication.

Adhesin. An aggregation substance produced by plasmid containing donor cells of some Gram$^+$ bacteria that helps in producing cell clumps with recipient.

Antibiotic. A substance that kills cells or inhibit their growth.

Anti-F pilus antibody. Antibodies that cross reacts with the pilin molecules constituting the F pilus.

Autonomous. A free-replicating independent unit such as plasmid or a phage.

Autoradiography. A method in which radiolabelled materials are incorporated into cell structures, which are then placed on a photographic film. A pattern produced on the film helps in tracking the location of the radioactive compounds within the cell.

Auxotroph. A strain that requires additional nutrient/s not required by wild-type organisms.

Bacteriophage/phage. Viruses that parasitise bacteria.

Chromosome. In a bacterial cell, the DNA molecule that contains the majority of genes required for cellular growth and maintenance.

Circular map. Refers to a linkage map carried on a circular chromosome in bacteria.

Conjugation. A process in which genetic material from one cell is transferred to another through cell-to-cell contact.

Conjugative plasmid. A plasmid that helps making effective cell contact.

Cross over. The point at which the genetic material is exchanged between the homologous chromosomes.

Donor (male or F$^+$). In bacteria, a strain that functions as male and donate the genetic material to female or recipient.

Donor strand. Refers to the DNA of the male cell that is transferred to the recipient.

Episome. An extrachromosomal genetic element such as a plasmid or phage that may exist in a cell either in a free autonomous state or integrated in to the bacterial chromosome.

Equilibrium density gradient centrifugation. A technique in which macromolecules are centrifuged at high speed while suspended in a solution of CsCl. The CsCl ions are sedimented to form a density gradient in the centrifuge tube and the macromolecules settle at a point where its density matches with that of CsCl on the gradient.

F factor. The plasmid found in male or donor *E.coli* cell that endows it to functions as donor in conjugation.

Fluctuation test. A statistical test which demonstrated that spontaneous mutations are non-adaptive.

F pili (singular Pilus) or sex fimbriae. The long appendages produced on the donor cell alone which are involved in conjugation but do not have a locomotary function.

F-prime (F′). A donor *E.coli* cell containing an F plasmid that picks up a few chromosomal genes.

Gene. A stretch of DNA whose function is responsible for a phenotype. It is a fundamental physical and functional unit of heredity.

Genetic exchange. The process of reshuffling of genetic material between two chromosomes by breakage and reunion and results in recombination.

Helicase. The enzyme that brings about helix unwinding in DNA during different processes such as replication, transcription, etc.

Heredity. Refers to the genetic phenomenon that explains the similarity among offsprings and parents.

Hfr (High frequency recombination). A donor *E.coli* strain in which F plasmid is integrated into bacterial chromosome and sponsors highly efficient chromosome transfer and thus recombination.

Heteroduplex. A DNA duplex formed by annealing single strands from different sources. A heteroduplex can be used to determine the structural dissimilarity.

Heterogenote (Partial diploid). *E. coli* cell carrying partially diploid genome such as one complete genome and a fragment of the other.

Heterozygote. An individual with a pair of different alleles.

Hybridization (=Cross). The designed mating of two parental types of organisms in genetic analysis.

Illegitimate genetic exchange. A crossing over between nonhomologous genetic units.

Infectious transfer. A highly efficient (~100%) transfer of genetic material such as that takes of F plasmid in an $F^+ \times F^-$ cross.

Insertion sequences (IS). A type of genetically mobile element that can insert into a gene thereby in activating it. A number of these can occur on bacterial chromosome and plasmids.

Integration. A process by which an extrachromosomal genome or a part of genome is accommodated into another genome.

Interrupted mating. A technique in which the transfer of a gene can be interrupted. This has been used in mapping the order in which donor genes enter the recipient cell.

Linear map. A linkage map located on a linear chromosome. Many circular maps have been converted into linear maps as the number of genes mapped may be too high to be depicted on a circular map (e.g. *E. coli*).

Linkage. The tendency of the genes, located closely on the same chromosome, to remain together during transmission.

Male specific phage. Some phages which specifically attack the male bacterial cells.

Marker. Equivalent to gene. In a bacterial cross the genes that are employed for the selection of recombinants are known as selected markers and those which are inconsequential to the initial selection of recombinants are known as unselected markers.

Mating types. The equivalent in lower organisms of sexes in higher organisms. They differ in function but not in morphology.

Merozygote. See heterogenote.

Minimal medium. A medium that consists of essential macro- and micronutrients and allows the growth of only prototrophic organisms.

Mobilization. Refers to the process that prepares a chromosome for its transfer to another cell. Many plasmids can offer this function.

Origin of transfer (*oriT*). A site at which the Hfr chromosome is nicked for its linear transfer.

Partial diploid. See heterogenote.

Phenocopy. An environmentally-induced phenotype that resembles a gene-based phenotype.

Pheromone. A sex hormone produced by some $Gram^+$ bacteria that helps in conjugation.

Pilin. Protein that constitutes the pili.

Plasmid. An extrachromosomal autonomously replicating DNA.

Plasmid curing. A process by which a cell can be made to lose the plasmid.

Prophage. A phage genome that gets integrated into host chromosome and does not express any viral function.

Prototroph. A bacterial wild-type strain that could grow on minimal medium.

Rec (Rec ABCDE etc). The proteins involved in the pathways of recombination and repair.

Recipient (Female, F⁻). Refers to the bacterial cell in conjugation that functions as female and lacks F plasmid. Also, any cell that is used to receive the genetic material.

Recombination. A process that generates individuals with new combination of genes (recombinants) by crossing over.

Recombination frequency. The proportion of recombinant cells or individuals in a population.

R factor. A type of plasmid that contains genes for resistance to several antibiotics.

Reproduction. The process that leads to the multiplication of an organism. Asexual reproduction defines one type in which the same individual employs a mechanism to produce progenies. Sexual reproduction, on the other hand, requires the differentiation of two sexes that participate in the process.

Ruv (Ruv ABC). Proteins which play important roles in recombination along with Rec proteins.

Temporal Map. A method of gene mapping in bacteria like *E. coli* where the genes are located on the basis of time of entry into recipient cell. The markers appearing early are known as proximal in contrast to those that appear late and called distal markers.

Transposition. The movement of transposable elements from one site to another and from one genome to another.

UV-irradiation. Ultraviolet electromagnetic radiation of a wave-length of 260 nm which is strongly absorbed by nucleic acid. Strongly mutagenic.

Variability. Refers to the differences created in different characters of an organism in a population.

References

Adelberg EA, Burns SN (1960) Genetic variation in the sex factor of *Escherichia coli*. J Bacteriol 79: 321–330

Bachmann BJ, Low KB, Taylor AL (1976) Recalibrated linkage map of *Escherichia coli* K-12. Bacteriol Rev 40: 116–167

Beadle GW, Tatum EL (1941) Genetic control of biochemical reactions in *Neurospora*. Proc Natl Acad Sci, USA 27: 499–506

Broda P (1967) The formation of Hfr strains in *Escherichia coli* K-12. Gen Res 9: 35–47

Cairns J (1963) The chromosome of *Escherichia coli*. Cold Spring Harbor Symp Quant Biol 28: 43–45

Campbell AM (1969) Episomes. Harper and Row, New York

Clewell DB (1993) Bacterial sex pheromone-induced plasmid transfer. Cell 73: 9–19

Clowes RC, Rowley D (1954) Some observations on linkage effects in genetic recombination in *E. coli* K-12. J Gen Microbiol 11: 250–260

Davis BD (1950) Non-filterability of the agents of genetic recombination in *E. coli*. J Bacteriol 60: 507–509

Firth N, Ippen-Ihler K, Skurray R (1996) Structure and function of F factor and mechanism of conjugation. *In*: Neidhardt FC, Curtiss III R, Ingraham JL, Lin ECC, Low KB, Magasanik B, Reznikoff WS, Riley M, Schaechter M, Umbarger HE (eds) *Escherichia Coli* and *Salmonella*. Cellular and Molecular Biology, Vol 2, American Society of Microbiology Press, Washington DC, pp 2377–2401

Gilbert W, Dressler D (1968) DNA replication: the rolling circle model. Cold Spring Harbor Symp Quant Biol 33: 473–484

Gowen JW, Lincoln RE (1942) A test for sexual fusion in bacteria. J Bacteriol 44: 551–554

Hayes W (1952a) Recombination in *Bact coli* K-12: unidirectional transfer of genetic material. Nature 169: 118–119

Hayes W (1952b) Genetic recombination in *Bact coli* K-12: analysis of the stimulating effect of ultraviolet light. Nature 169: 1017–1018

Hayes W (1953a) Observations on a transmissible agent determining the sexual differentiation in *Bact coli*. J Gen Microbiol 8:72–88

Hayes W (1953b) The mechanism of genetic recombination in *E. coli*. Cold Spring Harbor Symp Quant Biol 18: 75–93

Hayes W (1955) A new approach to the study of kinetics of recombination in *E. coli* K-12. J Gen Microbiol (Proc Soc Gen Microbiology) 13:ii

Hayes W (1957) The kinetics of the mating process in *E. coli*. J Gen Microbiol 16: 107–119

Hayes W (1968) The Genetics of Bacteria and their Viruses. 2nd Edn Blackwell, Oxford (UK)

Heinemann JA (1992) Conjugation, Genetics. *In*: Lederberg J (ed.) Encylopaedia of Microbiology Vol I. Academic Press, New York, pp 547–558

Hirota Y (1960) The effect of acridine dyes on mating type factors in *Escherichia coli*. Proc Natl Acad Sci, USA 46: 57–60

Jacob F, Wollman EL (1961) Sexuality and the Genetics of Bacteria. Academic Press, New York

Jacob F, Schaeffer P, Wollman EL (1960) Episomic elements in bacteria. Symp Soc Gen Microbiology 10: 67–91

Jacob F, Brenner S, Cuzin F (1963) On the regulation of DNA replication in bacteria. Cold Spring Harbor Symp Quant Biol 28: 329–348

Lederberg J (1947) Gene recombination and linked segregation in *E. coli*. Genetics 32: 505–525

Lederberg J (1949) Aberrant heterozygotes in *Escherichia coli*. Proc Natl Acad Sci, USA 35: 178–184

Lederberg J (1950) The selection of genetic recombinations with bacterial growth inhibitors. J Bacteriol 59: 211–215

Lederberg J (1951) Prevalence of *E. coli* strains exhibiting genetic recombination. Science 114: 68–69

Lederberg J (1955a) Recombination mechanisms in bacteria. J Cellular Comp Physiol 45: 75–108

Lederberg J (1955b) Genetic recombination in bacteria. Science 122: 920

Lederberg J (1956) Conjugal pairing in *E. coli*. J Bacteriol 71: 497–498

Lederberg J, Tatum EL (1946a) Novel genotypes in mixed cultures of biochemical mutants of bacteria. Cold Spring Harbor Symp Quant Biol 11: 113–114

Lederberg J, Tatum EL (1946b) Gene recombination in *E. coli*. Nature 158: 558

Lederberg J, Tatum EL (1953) Sex in bacteria: genetic studies, 1945–1952. Science 118: 169–175

Lederberg J, Cavalli LL, Lederberg EM (1952) Sex compatibility in *E. coli*. Genetics 37: 720–730

Lloyd RG, Buckmann C (1995) Conjugational recombination in *Escherichai coli*. Genetic analysis of recombinant formation in Hfr × F⁻ crosses. Genetics 139: 1123–1148

Lloyd RG, Low KB (1996) Homologous recombination. *In*: Neidhrdt FC, Curtiss III R, Ingraham JL, Lin ECC, Low KB, Magasanik B, Reznikoff WS, Riley M, Schaechter M, Umbarger HE (eds) *Escherichia Coli* and *Salmonella*. Cellular and Molecular Biology Vol 2, American Society of Microbiology Press, Washington DC, pp 2236–2255

Loeb T (1960) Isolation of a bacteriophage for the F⁺ and Hfr mating types of *Escherichai coli* K-12. Science 131: 932

Low KB (1972) *E. coli* K-12 F' factors, old and new. Bacteriol Rev 36: 587–600

Sherman JM, Wing HU (1937) Attempts to reveal sex in bacteria with some light on fermentative variability in the *coli-aerogenes* group. J Bacteriol 33: 315–321

Sinha U, Srivastava Sheela (1983) An introduction to bacteria. Vikas, New Delhi

Smith GR (1985) Homolgous recombination: the roles of *chi* sites and RecBC enzyme. *In*: Scaife J, Leach D, Galizzi A (eds) Genetics of bacteria. Academic Press, New York, pp 239–254

Tatum EL, Lederberg J (1947) Genetic recombination in the bacterium *Escherichia coli*. J Bacteriol 53: 673–684

Taylor AL, Adelberg EA (1961) Linkage analysis with very high frequency males of *Escherichia coli*. Genetics 45: 1233–1243

Willets N, Skurray R (1987) Structure and function of F factor and mechanism of conjugation. *In*: Neidhardt FC, Ingraham JL, Low KB, Magasanik B, Schaechter M, Umbarger HE (eds) *Escherichia coli* and *Salmonella typhimurium*. Cellular and Molecular Biology. Vol 2. American Society of Microbiology Press, Washington DC, pp 1110–1133

Wollman EL, Jacob F, Hayes W (1956) Conjugation and genetic recombination in *Escherichia coli*. Cold Spring Harbor Symp Quant Biol 21: 141–162

Zelle MR, Lederberg J (1951) Single cell isolations of diploid heterozygous *E. coli*. J Bacteriol 61: 351–355

4

Reproductive Biology of Cyanophycota

V. Krishnamurthy

1. Introduction

The Cyanophycota (the blue-green algae) are prokaryotic, photoautotrophic organisms which show considerable affinity with the bacteria. Hence, these blue-green algae are frequently described as Cyanobacteria.

This division comprises the single class, Cyanophyceae. The Cyanophyceae may be free-living, unicellular organisms, or may form definite or indefinite colonies of single cells, or may be unbranched or branched filaments. In filamentous forms, the cells may be uniform in structure or may include differentiated cells called 'heterocysts'.

In all Cyanophyceae the cell is surrounded by an inner investment and the entire plant body is enclosed in an outer sheath. In filamentous forms, the cells are serially arranged to form a trichome which is enclosed in a sheath; trichome and sheath together constitute the filament. The trichome of certain genera like *Oscillatoria* exhibit a continuous spirally oscillating movement, while in others such movement is absent.

It is necessary to keep these structural features in mind when considering the reproductive mechanisms in these algae.

2. Vegetative Propagation

Vegetative propagation of the Cyanophycota takes place by simple binary fission in unicellular forms, and by hormogonia and pseudohormogonia in filamentous forms.

2.1 Fission

A simple division of the cell occurs in unicellular blue-green algae like *Chroococcus* and *Gloeocapsa*. In these genera, divisions occur in quick succession and the new cells formed remain together in a packet inside a mucilagenous sheath, for some period. Therefore, one most commonly sees these algae in the form of packets of four or eight cells (Fig. 1A-C). A short or long resting period follows, and after that the gelatinous sheath dissolves, liberating the individual cells, which resume division.

In several colonial members, such as *Gleothece* and *Aphanothece*, the divisions may take place along a single plane perpendicular to the long axis of the cell and at right angles to each other. In *Merismopedia* the divisions occur in two planes (Fig. 1D) and in *Eucapsis* in three planes (Fig.1E). In *Microcystis*, the division occurs in many planes. In all these algae, the cells, after division, remain enclosed in a mucilagenous sheath to form colonies of various complexity. In several blue-green algae forming indefinite colonies, the colonies may dissociate into smaller colonies (Fig. 1G).

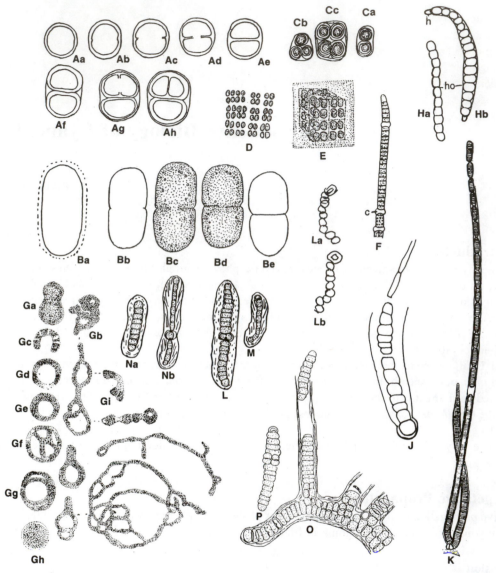

Fig. 1. (A-P Aa-Ah) *Chroococcus turgidus*, stages in cell division and colony formation. Ba–Be *Synechococcus major*, successive stages of cell-division. Ca–Cc *Gloeocapsa* sp., divisions of cells show two-to three-and four-celled colony formation. D *Merismopedia puncatata* E *Eucapsis alpina*. F *Oscillatoria brevis*, filament with concave (c) cells. Ga–Go *Microcystis aorunginosa*, colonies show various shapes and fission products. Ha–Hb *Nostoc punctiformo*, hormogonium and its germination product. Ia–Ib *Nostoc* sp., formation of hormogonium after discarding the 'hair'. J Pseudohormogonia. K *Calothrix confervicola*, hormogonia formed in a series. L *Camptylonema indicum*, germination of pseudohormogonium. M *Scytonema geitleri*, psedohormogonia. Na–Nb *Fischerella* sp., liberated homogonia developing heteroysts. O,P *Stigonema turfaceum*, formation and germination of hormogonium. (A, B after Geitler 1932, C,I,J original, D after Smith 1950, E after Clements and Shantz 1909, F after Gomont 1892, G after Huber-Pestalozzi 1938, H after Harder 1918, K after Bornet and Thuret 1876, L after Desikachary 1946a M, N after Bharadwaja 1933, O, P after Borzi 1914).

2.2 Hormogonia

Hormogonia are small pieces of trichome with one to many cells of uniform structure; the end cell of each hormogonium is round. The hormogonium is motile within the delicate mucilagenous sheath of the filament.

Hormogonia commonnly occur in members of the Nostocales and Stigonematales. In some members of these orders, this is the only means of propagation.

Hormogonia are formed by the death of one or more intercalary cells of a trichome, or by the formation of a separation disc (Fig. 1F), which is a biconcave cell.

In the Oscillatoriaceae and the Nostocaceae (Fig. 1F, H), any portion of the trichome may become differentiated as a hormogonium. In the Scytonemataceae and some Stigonemataceae, they are generally formed on special branches, as in *Stigonema* (Figs. 1O, P and 2F) and *Fischerellopsis* (Fig. 2G, H). In the Rivulariaceae, the hair-like apical portion is thrown off and then the intercalary portion of the trichome becomes a hormogonium (Fig. 1J). Sometimes, as in *Calothrix confervicola*, hormogonia are formed in a series (Fig. 1K). Similar hormogonia are also reported in *Brachytrichia* and *Mastigocladus*.

Liberation of hormogonia takes place by their movement out of the parent sheath (Fig. 2A-C). The liberated hormogonia develop into new individuals. In heterocystous forms, a heterocyst (terminal or intercalary) differentiates as soon as the hormogonium is liberated (Fig. 1H, I). In some genera, intercalary hormogonia are formed and they germinate in situ (Fig. 2D, E).

A different mode of formation of hormogonia is reported in *Aulosira implexa* var. *crassa* (Gonzalves and Kamat 1954). In this alga, a break in the sheath develops on one side and is followed by a fragmentation of the trichome. The cells near the deflected portion of the sheath die out.

The deflections then encase small pieces of the trichome of two or more cells and these pieces serve as hormogonia. The deflected portion and the outer sheath of the filament gelatinize and liberate the hormogonia.

The 'planococci' described for *Desmisiphon* (Borzi 1916), and probably also for *Stauromatonema* (Fremy 1929) are unicellular hormogonia which show a creeping movement.

2.3 Pseudohormogonia

Pseudohormogonia, (hormocysts or hormospores) are homologous to the hormogonia and are similarly formed. However, these are not motile, but enclose themselves in a thick lamellated and pigmented sheath. These occur in the Scytonemataceae and the Stigonemataceae.

On germination, the two ends grow out and an intercalary heterocyst develops. It is not known whether there is a resting period before germination. But in situ germination of pseudohormogonia has been reported (Fremy 1929, 1930; Desikachary 1948).

Pseudohormogonia are known for *Westiella* (Fig. 2A-C), *Camptylonema* (Fig. 1L), *Scytonema* (Fig. 1M), *Fischerella* (Fig. 1N), *Leptopogon, Handeliella* and *Spalaeapogon*.

Janet (1941) described 'pseudohormocysts' in *Westiellopsis* but these are likely to be endospores.

3. Asexual Reproduction

Asexual reproduction in the Cyanophyceae is by production of three kinds of spores: endospores, exospores and akinetes.

3.1 Endospores

Endopores are formed within a cell by a quick succession of divisions in three planes (e.g. *Dermacarpa*). These are commonly formed in certain unicellular members which do not have any other mode of reproduction. The number of endospores formed by a cell varies from one, as in *Dermacarpella*, or two, as in *Endonema* (Fig. 3H), to many as in *Dermocarpa* spp. (Fig. 3C-E). In *Dermacarpa coccoides, Dermacarpa prasina*,

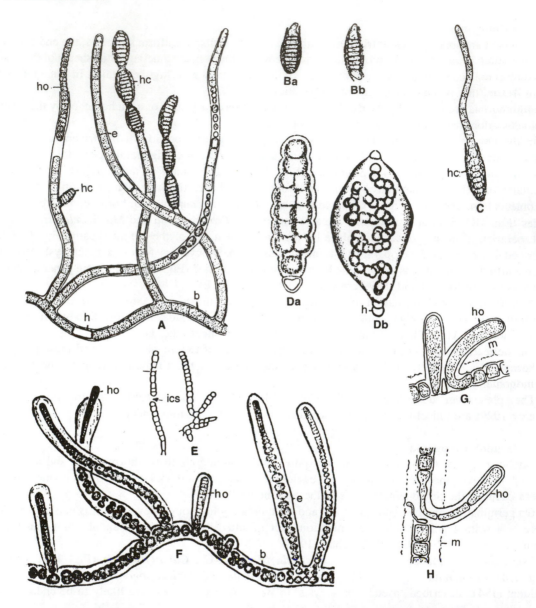

Fig. 2. A-H. *Westiella lanosa.* A Hormogonia and hormocysts. Ba–Bb, C Germinated hormocysts. Da, Db *Nostoc commune,* germlings. E *Nostochopsis lobatus,* formation and germination of hormogonia. F *Stigonema ocellum,* hormogonia. G.H *Fischerellopsis harrisii,* hormogonia. *b* basal filament, *e* erect filament, *hc* heterocyst, *ho* hormocyst, *ics* intercalary separating cell, *m* mucilage sheath. (A-C, F after Fremy 1930, D after Geitler 1932, E after Desikachary 1946b, G.H after Fritsch 1945)

and others, the entire cell becomes converted into a sporangium. But in others, e.g. *D. clavata,* the first division of the protoplast results in an upper cell which forms a number of endospores. After the libertion of these spores the basal cell enlarges and a similar endospore formation takes place. In certain other taxa (e.g. *D. suffulta*), the basal cell does not form endospores.

Endospores superficially resemble alpanospores of the green algae and are generally naked, but walled endospores are also known (Geitler 1921).

Endospores are common in the Dermacarpaceae but have been reported in *Siphonema*, in some Oscillatoriaceae, Nostocaceae, *Herpyzonema* and *Gomphosphaeria*.

In *Westiellopsis*, Janet (1941) describes 'pseudohormocysts', clusters of cells formed terminally on erect lateral branches by repeated transverse and longitudinal divisions. These should be considered as endospores.

Nannocytes described in *Gloecocapsa, Aphanothece, Microcystis, Chlorogloea*, and other genera are generally naked protoplast and should also be considered as endospores (Fig. 3A, B).

3.2 Exospores
In *Chamaesiphon* (Fig. 3 I, J) and *Stichosiphon* (Fig. 3F, G.) the apical portion of the membrane surrounding the protoplasts ruptures, and from the exposed apical end, spores are abstracted by transverse divisions and are serially liberated. These spores are exospores. Rarely, exospores do not get liberated and give rise to colonies of various kinds (e.g. *Chamaesiphon fuscus*).

3.3 Akinetes
Akinetes (resting spores) are very large, spherical, oblong or cylindrical cells and have thick walls and a food reserve of cyanophycin granules. They are without photo-synthetic pigment, but may develop colouration in the outer wall. The spore has two wall layers (Fritsch 1945) corresponding to the cell sheath and the inner investment of vegetative cells, but much thicker. Centroplasm is indistinct. The mature akinetes do not have gas vacuoles or metachromatin granules (Elenkin and Danilov 1916).

Spores are always formed in specific positions in relation to heterocysts. In some genera like *Cylindrospermum* (Fig. 4A-C), the spores are formed singly on one side of the heterocyst. Rarely, however, they may be formed on both sides of the heterocyst, and in greater numbers, forming short or long chains. Spore formation in *Nostoc* spp. (Fig. 4D,E) and in some species of *Anabaena* (Figs. 4F, G and 5E) and *Anabaenopsis* (Fig. 5F) starts in between the heterocysts and progresses towards them. Very often, the entire length of the trichome becomes converted into a series of akinetes, occasionally interrupted by a vegetative cell, or sometimes by dead cells.

3.4 Germination of Akinetes
In some Cyanophyceae, akinetes germinate immediately (Fritsch 1905; Spratt 1911; Harder 1917; Bharadwaja 1933; Palik 1941; Singh 1942), while in others there is a resting period which may sometimes be long (Lipman 1941; Rose 1934). They may withstand desiccation and high temperature (Glade 1914).

Germination of akinetes has been observed in many genera. It appears that irradiance influences their germination in *Nostoc* (Harder 1917).

Germination follows a similar pattern in all cases studied. The protoplast divides into a two-celled or few-celled germling. The inner investment of the akinetes is first dissolved. The outer (Fig. 4H,I) investment either dissolves at one end only, or entirely, and the germling gets liberated (Fig.4J,K). The liberated germling generally develops a heterocyst and then gives rise to a group of filaments by hormogonium formation. In rare cases, the entire germling divides into a number of hormogonia. These then form new filaments.

4. Heterocysts
Heterocysts are special cells with thick walls, and generally homogeneous and with pale-yellow contents in some filamentous Cyanophyceae (Fig. 5A, B, E).

Germination of the heterocyst has been documented by Desikachary (1959). Fully-grown filaments have

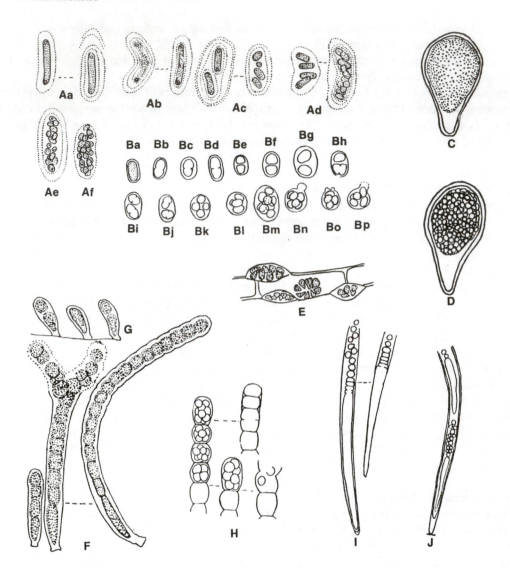

Fig. 3 A-J. Aa-Af *Aphanothece caldarium*, successive stages in the formation of nannocytes. Ba-Bp *Aphanothece bullata* var. *major*, stages in nannocyte formation. C, D *Dermacarpa olivacea* var. *gigantea*, formation of endospores. E *Dermacarpa hemispherica*, formation of endospores. F *Stichosiphon sansibaricus*, formation and liberation of exospores. G Germination of endospores. H *Endonema moniliforme*, endospore formation. I. J *Chamaesiphon curvatus*, exospore formation and liberation. In J two individuals are enclosed in a single sheath (A, I, J after Geitler 1932, B after Geitler and Ruttner 1936, C, D after Rao 1936, E after Iyengar and Desikachary 1946, F, G after Skuja 1949, H after Pascher 1929).

been obtained from the germlings only in *Nostoc commune* (Fig. 5H), *Tolypothrix elenkinii*, *Calothrix weberi* (Fig. 5I, J) and *Gloeotrichia raciborskii* (Fig. 5C, D). Only rarely has such heterocyst germination been reported in nature. Geitler (1921) and Desikachary (1946a) observed germination after a short low

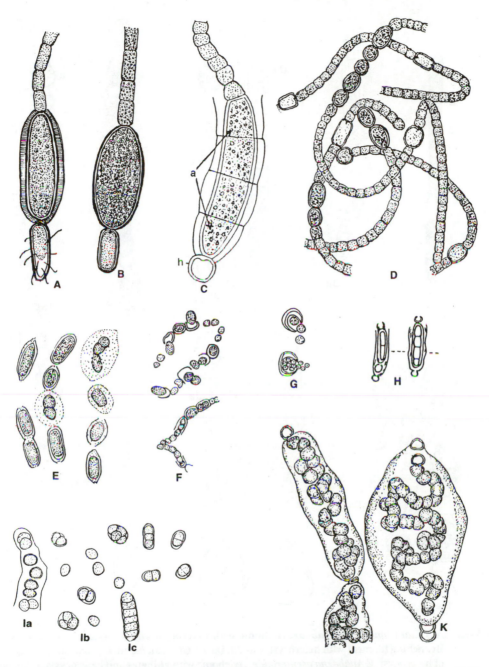

Fig. 4 A-K. A *Cylindrospermum alatosprum*, akinete. **B** *Cylindrospermum muscicola* var *longispora*. **C** *Gloeotrichia natans*. **D** *Nostoc ellipsoporum* var *violacea*. **E** *Nostoc ellipsoporum*, akinetes and their germination. **F** *Anabaena azollae*. **G** *Anabaena cycadae*, endospores. **H** *Gloeotrichia natans*, stages in the germination of akinete. **Ia-Ic** *Stigonema turfaceum*; Ia group of spores at apex of branch, Ib stages in the germination of spores, Ic a young filament. **J,K** *Nostoc commune*, germlings. (A, H after Desikachary 1959, B after Dixit 1936, C after Tilden 1910, D after Rao 1936, E after Geitler and Ruttner 1936, F after Fritsch 1945, G after Spratt 1911, I after Borzi 1878, J, K after Geitler 1932).

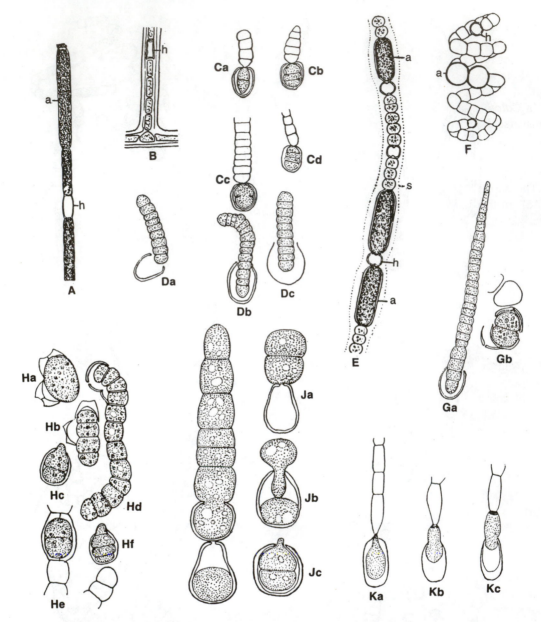

Fig. 5 A-K. A *Aphanizomenon flos-aquae*, trichome with heterocyst and an akinete. B *Hapalosiphon* sp. Branching filament with heterocyst, Ca-Cd, Da-Dc *Gloeotrichia raciborskii*, stages in germination of heterocyst. E *Anabaena oscillarioides*, trichome with akinetes and heterocysts, F *Anabaenopsis milleri*, pair of akinetes, away from the terminal heterocysts. Ga-Gb *Nostoc commune*, stages in germination of heterocyst. Ha-Hf *Nostoc commune*, various stages in germination of heterocyst. I-Ja-Jc *Calothrix weberi*, stages in germination of heterocyst. Ka-Kc *Gloeotrichia raciborskii*, stages in rejuvenation of heterocyst. (A after Lemmermann 1910, B original, C, D, K after Desikachary 1946a, E after Skuja 1949, F after Woronichin 1929, G, H after Geitler 1932, I, J after Steinecke 1932).

temperature treatment, while others have observed it in ordinary cultures. Fogg (1949) reported germination in cultures supplied with succinate and not under any other cultural condition.

Geitler (1921) observed an initial two-celled germling which became four-called and then often came out of the hetrocyst by rupture of the heterocyst wall, or by the gradual dissolution and widening up of the pore.

In *Calothrix weberi* (Fig. 5I, J) the contents of the heterocyst first divide into two daughter cells by a transverse division. The upper cell squeezes out of the heterocyst through the pore, and later on develops into a new filament. The lower cell degenerates inside the old heterocyst wall. During germination the inner cellulose layer disappears (Geitler 1921, Desikachary 1946), and it is presumed that it serves as a reserve cellulose.

The entire contents may develop into a new individual, as in *Gloeotrichia raciborskii* (Desikachary 1946). In the same species, what is described as 'rejuvenation' of the heterocyst (Fig. 5K) takes place (Desikkachary 1946). Sometimes, endospores have been observed to form inside the heterocysts of *Nostoc commune, N. microscopicum* and *Anbaena cycadeae* (Spratt 1911).

4.1 Nature of Heterocysts

Geitler (1921) considers the heterocysts as archaic reproductive bodies which have lost their functions. This view is supported by the occasional germination of their contents. Steinecke (1932) observed germination in 80% of heterocysts in cultures in *Calothrix weberi*. He therefore concluded that heterocysts function as secondary reproductive organs.

4.2 Relation Between Heterocysts and Spore Formation

There appears to be a relation between the heterocysts and spore formation in a number of genera. In many taxa the spores are restricted to the immediate neighbourhood of the heterocyst. In others, spore formation and a gradual depletion of the contents of neighbouring heterocysts has been observed. This is considered as evidence of the heterocyst supplying food material to the developing spores (Bharadwaja 1933; R.N. Singh 1942). In *Gloeotrichia*, spores are formed by cells next to the heterocyst. The protoplasmic connection between the spores and subtending heterocyst, persist till the spores are completely formed. While the spore is being formed, the contents of the heterocyst gradually becomes vacuolated and finally disappears. Often, the empty heterocysts get crushed but still remain attached to the mature spores.

Fritsch (1905) suggested that heterocysts seem to secrete a substance for promoting spore formation. In non-heterocystous spore-producing forms, this substance is probably produced by the entire trichome. In heterocystous forms this capacity appears to have become restricted to the heterocyst. This is probably the only explantation when spore formation takes place away from the heterocyst. However, the contents of the heterocysts become vacuolated in some species of *Scytonema* even without any spore formation.

Since the heterocysts have now been conclusively established as the site for nitrogen fixation in heterocystous Cyanophyceae, one may have to discard any theory as to their role in spore formation. The depletion of the contents and vacuolation of the heterocysts alongside spore production may simply be a nutritive relationship.

5. Evidence of Gene Transfer in the Cyanophycota

In the Cyanophyceae, sexual reproduction does not occur, although gene transfer between individuals has been reported. An apparent genetic recombination was reported by Kumar (1962), between streptomycin- and penicillin-resistant mutants of *Anacystis nidulans*. Subsequently, Bazin (1968) pointed out that the strains of *Anacystis nidulans* resistant to either polymyxin or streptomycin could produce a double resistant strain, when grown together. The mechanism involved in diffusion of DNA into the medium and its absorption by recipient cells. Gene transfer has also been reported by Steward and Singh (1975) and Padhy and Singh (1978). Experiments by Devilli and Houghton (1977) and Stevens and Porter (1981), who incubated members of the Cyanophyceae with extracted DNA, have confirmed donation of functional genes.

Cyanophages have been studied extensively but there is no conclusive evidence of transduction in the Cyanophyceae, although R.N. Singh and P.K. Singh (1972) reported phage-mediated streptomycin resistance. Conjugation between cells is not known. However, in *Nostoc muscorum*, Delaney et al. (1976) have reported filament anastomosis and consequent formation of different clones.

6.   Future Studies in the Twenty-First Century

The foregoing account is a summary of all that is known about the reproductive mechanisms in the Cyanophycota. These are mainly simple processes such as binary fission or formation of fragments of the trichome or formation of spores. Although heterocyst germination has been documented, it is still not considered as its primary function. Except for one report of succinate requirement for germination of a heretocyst, there is no further information on the conditions leading to this phenomenon.

Gaps do exist in our present knowledge regarding reproduction in this group of algae. It has been pointed out erlier that Delaney et al. (1976) have reported anastomosing filaments in *Nostoc muscorum* as leading to formation of different clones. Investigation into such events would reveal whether there is a mechanism of conjugation between cells and consequent sexual reproduction. We still need further data on the mechanism of gene transfer.

Future work on the Cyanobacteria will have to address the genetics of these organisms on the lines adopted for bacterial genetics. Some effort has already been made in the use of molecular, biological and genetic engineering techniques to identify the genes which encode several cyanophycean enzymes. Gene cloning and generation of mutants have also been attempted. It is envisaged that in the early years of the twenty-first century, tailor-made Cyanophycota for various uses may be created for the benefit of mankind.

Glossary

Akinete. A resting spore formed by the conversion of a vegetative cell, often thickwalled and adapted to resist unfavourable environmental conditions

Centroplasm. Central portion of the protoplast of a cyanophyte, containing genetic material

Cyanophage. Virus attacking cells of cyanophytes

Endospore. Spores formed within a cell by successive division of its contents in three planes (Cyanophycota)

Exospores. Spores serially abstricted from the open end of a sporangium by transverse division of its contents (Cyanophycota)

Filament. A long series of cells placed end to end (a trichome) and surrounded by a mucilagenous sheath (Cyanophycota)

Heterocyst. Specialised, thick-walled cell, with homogenious, generally pale yellow contents (Cyanophycota)

Hormogonium. A piece of trichome with one to many uniform cells, provided with a delicate gelatinous sheath, motile within the parent sheath (Cyanophycota)

Nannocytes. Naked protoplasts formed by successive divisions of the contents of a cell but without causing cell enlargement (Cyanophycota)

Planococci. Single-celled hormogonia (Cyanophycota)

Pseudohormogonium. Small piece of trichome, enclosed in a thick lamellated, pigmented sheath. Homologous with hormogonium (Cyanophycota)

Pseudohormocysts. Clusters of cells formed terminally on erect lateral branches by repeated divisions, contents of each cell forming a single gonidium (Cyanophycota)

Trichome. A long linear row of cells placed end to end within a mucilagenous sheath. Term applied to the cellular portion, excluding the sheath (Cyanophycota)

Transduction. Gene transfer through a vector such as a virus or phage

References

Bazin MJ (1968) Sexuality in a blue-green algae: genetic recombination in *Anabaena nidulans*. Nature 218: 282–283

Bharadwaja Y (1933) Contribution to our knowledge of the Myxophyceae of India. Ann Bot (Lond) 47: 117–143

Bornet E, Flahault C (1986) Revision des Nostocacées heterocystes 1, 2. Ann Sci Nat Bot Ser 7, 3: 323–381, 4: 347–373, 5: 51–129, 7: 177–262

Bornet E, Thuret G (1876) Notes Algologiques 1, pp I-XX + 1–72. G. Masson, Paris

Bornet E, Thuret G (1880) Notes Algologiques 2, pp 73–196. G. Masson, Paris

Borzi A (1878) Nachtrage zur Morphologie und Biologie der Nostocaceen. Flora 61: 465–471

Borzi A (1914) Studi sulle Mixoficee 1. Nuovo Bot Ital 21: 307–360

Borzi A (1916) Studi sulle Mixoficee 2. Nuovo G Bor Ital 23: 559–588

Borzi A (1917) Studi sulle Mixoficee 3. Nuovo G Bot Ital 24: 17–112

Bristol BM (1919) On the retention of vitality by algae from old stored soils. New Phytol 18: 92–107

Clements E, Shantz HL (1909) A new genus of blue-green algae. Minn Bot Stud 4: 133–135

Delaney SF, Herdman M, Carr NG (1976) Genetics of blue-green algae. In: Lewin RA (ed) The genetics of algae. Blackwell, Oxford, pp 7–28

Desikachary TV (1946a) Germination of the heterocysts in two members of the Rivulariaceae. J Indian Bot Soc 25: 11–17

Desikachary TV (1946b) On the genus *Nostochopsis* Wood. J Indian Bot Soc MOP Iyengar Comm Vol 225–238

Desikachary TV (1948) On *Camptylonema indicum* Schmidle and *Camptylomemopsis* gen. nov. Proc Indian Acad Sci B 28: 35–50

Desikachary TV (1959) Cyanophyta. 685 pp. Indian Council Agric Res, New Delhi

Devilli Cl, Houghton GA (1977) A study of genetic transformation in *Glaeocapsa calcicola*. J Gen Microbiol 98: 277–280

Dixit SC (1936) The Myxophyceae of the Bombay Presidency, India. Proc Indian Acad Sci B 3: 93–106

Elenkin AA, Danilov AN (1916) Recherches Cytologiques sur les cristeau et les granis de secretion dans les cellules du *Symploca muscorum* (Ag) Gom., quelques autres Cyanophycees. Bull Jard Imp Bot Pierre Grand 16: 40–100

Fogg GE (1949) Growth and heterocyst production in *Anabaena cylindrica* Lemm. in relation to carbon and nitrogen metabolism. Ann Bot (Lond) 13: 241–259

Fremy P (1929) Less Myxophycées de I'Afrique equatorial Francaise. Arch Bot Caen 3: Mem 2: 1–507

Fremy P (1930) Les Stigonémacées de la France. Rev Algol 5: 147–214

Fritsch FE (1905) Structure of the investment and spore development in some Cyanophyceae. Beih Bot Zbl 18: 194–214

Fritsch FE (1945) The structure and reproduction of the algae vol 2. Cambridge University Press, Cambridge (UK), 939 pp

Geitler L (1921) Versuch einer Losung des Heterocystenproblems. Sitz ungsber Akad Wiss Math-Nat K1 130: 223–245

Geitler L (1932) Cyanophyceae in Rabenhorst's Kryptogamenflora. Akademische Verlags gesellschaft Leipzig 14: 1–1196

Geitler L, Ruttner I (1936) Die Cyanophyceen der Deutschen Limnologischen Sunda Epedition, ihre Morophologie, Systematik und Ökologie. Arch Hydrobiol Suppl 14. Trop Binnenge wasser 6: 308–483

Glade R (1911) Zur Kenntnis der Gattung *Cylindrospermum*. Beih Z Biol Pflanz 70: 1–11

Gomont M (1892) Monographie des Oscillarées 1, 2, Ann Sci Nat Bot Ser 7, 15: 263–368, 16: 91–264

Gonzalves E A, Kamat ND (1954) A note on hormogone formation in *Aulosira implexa* Born et Flash var *crassa* Dixit. J Indian Bot Soc 33: 351–353

Harder R (1917) Ueber die Beziehung des Lichtes zur Keimung von Cyanophyceensporen. Jahrb Wiss Bot 58: 287–294

Harder R (1918) Ueber die Bewegung der Nostocaceen. Z Bot 10: 117–244

Huber-Pestalozzi G (1918) Das Phytoplankton des Süsswasser. *In:* Theienmann A (ed) Binnenge Wasser, Vol 16: Stuttgart 342 pp

Iyengar MOP, Desikachary TV (1946) A systematic account of some marine Myxophyceae of the South Indian Coast. J Madras Univ B 3: 93–106

Janet M (1941) *Westiellopsis prolifica* gen. et sp nov,. A new member of the Stigonemataceae. Ann Bot (Lond) 5: 167–170

Kumar HD (1962) Apparent genetic recombination in a blue-green alga. Nature 196: 1121–1122

Lemmermann E (1910) Algen in Kryptogamenflora der Mark Brandenburg. Leipzig, 256 pp

Lipman CB (1941) The successful revival of *Nostoc commune* from a Herbarium specimen 87 years old. Bull Torrey Bot Club 68: 664–666

Pascher A (1929) Über die Teilungsvorgange bei einer neuen Blaualge *Endonema*. Jahrb Wiss Bot 70: 329–347

Padhy RN, Singh PK (1978) Genetical studies on the heterocyst and nitrogen fixation of the blue-green algae. Mol Gen Genet 162: 203–211

Palik P (1941) *Gloeotrichia tuzonii* eine neue Blaualge. Arch Protistenka 95: 45–51

Rao CB (1936) The Myxophyceae of the United Provinces, India 2. Proc Indian Acad Sci B 3: 165–174

Rose E (1934) Notes on the life history of *Apanizomenon flos-aquae*. Univ Iowa Stud Nat Hist 16: 129–140

Singh RN (1942) *Wollea bharadwajae* sp. nov. and its autecology. Ann Bot (Lond) 6: 593–666

Singh RN, Singh PK (1972) Transduction and lysogeny in blue-green algae. *In:* Desikachary TV (ed) Taxonomy and biology of blue-green algae. University of Madras pp 258–261

Skuja H (1949) Zur Süsswasseralgenflora Burmas. Nova Act Reg Soc Sci Upps Madras Ser 4, 14: 1–189

Spratt ER (1911) Some observations on the life history of *Anabaena cycadae*. Ann Bot (Lond) 25: 369–380

Smith GM (1950) Freshwater algae of the United States of America, 2nd edn. McGraw-Hill, New York 79 pp

Steinecke F (1932) Über Beziehungen wischen. Färlung und Assimilation bei eimiger Süsswasseralgen. Bot Arch 4: 317–327

Stevens SE, Porter RD (1981) Transformation in *Agmenellum quadruplicatum*. Proc Natl Acad Sci USA 77: 6052–6056

Steward WDP, Singh HN (1975) Transfer of nitrogen-fixing (nif) genes in the blue-green alga *Nostoc muscorum*. Biochem Biophys Res Commun 52: 62–69

Tilden JE (1910) Minnesota Algae I. University of Minnesota, Minneapolis, 550 pp

Woronichin NN (1929) Materialy K izu cheniyu algologiches koj restitel' noste ozur kunundinskoj stepi. Izvestia Glaonogo Botanicheskogo Sada SSSR 28: 12–40

5

Reproductive Biology of Eukaryotic Algae

V. Krishnamurthy

1. Introduction

The algae comprise a large heterogeneous assemblage of photoautotrophic organisms which vary vastly in size, shape and mode of life. The diverse phyla of algae show morphological, cytological, physiological and biochemical differences. The reproductive systems of these algae also show similar diversity. Reproduction in the algae may take place by vegetative or asexual or sexual methods.

A more recent classification of the algae differentiates prokaryotic and eukaryotic algae. The divisions Cyanophycota and Prochlorophycota are included in the Prokaryotae, and four divisions—Rhodophycota, Chromophycota, Euglenophycota and Chlorophycota with 15 classes of algae in the Eukaryotae.

Among the Eukaryotae, the Chlorophycota come closest to the higher plants in their plastid pigment composition. Three classes are included, Chlorophyceae, Charophyceae and Prasinophyceae. Of these, the Charophyceae are considered to have greater affinity to higher plants than the other two classes. Vegetative multiplication occurs by fission (cell division) in unicellular algae. In some algae this is the only means of multiplication. In multicellular algae multiplication may occur by fragmentation. In some algae special vegetative propagula or gemmae are formed which, when severed from the parent plant, develop into new thalli. Autocolony formation in certain coenobial forms represents a unique mode of vegetative multiplication.

Asexual reproduction takes place by spores of different types formed in specialized cells called sporangia. These spores are formed either after a series of mitotic divisions, or after an initial meiotic division.

The spores vary in their morphology and function. In many algae the spores have flagella and are motile, the zoospores. In many other algae the spores are non-flagellate, and are called aplanospores. Both zoospores and aplanospores germinate on a suitable substratum, without any resting phase. Most of these do not have distinct cell walls; some may have a wall of rigid, non-living matter.

Besides these, of common occurrence are the resting spores or akinetes. These are heavy-walled, usually formed singly in a cell of the alga, and have a short or long dormant period before germination. Other forms of resting spores are cysts, statospores and hypnospores.

Sexual reproduction takes place by the fusion of gametes, produced in specialized or unspecialized cells of the alga called gametangia. The gametes fuse in pairs to form a zygote which may or may not undergo a resting stage before germination.

The fusion between the nuclei of the two gametes is an essential feature of sexual reproduction and is termed karyogamy. The fusion of the cytoplasm of the two gametes is termed plasmogamy. Karyogamy brings together genetic material from different sources and is a most significant feature of gametic fusion.

On the basis of relative size and behaviour of the fusing gametes, we can differentiate three modes of sexual reproduction; isogamy, heterogamy and oogamy.

Although many algae may show both asexual and sexual reproduction in the same thallus, quite often the two processes are involved in a cylic succession of two vegetative phases which together form the life cycle of the alga.

2.   Vegetative Propagation

2.1   Cell Division

In very simple, unicellular algae, vegetative propagation is achieved by mere fission or division of the cell into two. If conditions are favourable, this process can be repeated in successive progenies resulting in a large number of individuals from a single parent cell. This is the only mode of reproduction in Euglenophycota. This process may occur in unicellular algae of other divisions too, for instance, the Cyanophyceae.

In *Euglena*, the divisions of the cell are longitudinal and the single flagellum is retained by one of the daughter cells while the other forms a new flagellum. The cell membrane is shared by the two daughter cells (Fig. 1A).

Only when the cell membrane is involved in cell division, should the process be considered as true fission. Such true fission occurs in many unicellular and colonial Cyanophyceae.

In several unicellular members of the Volvocales (Fig. 1Ba-Bd), the cell wall is not involved in cell division (e.g. *Chlamydomonas* spp.) although the flagella may be shared by the daughter cells. In some members both flagella are retained by one of the daughter cells. In either case new flagella are formed by the progeny.

In many unicellular Volvocales, a palmella stage develops wherein the division products of a cell do not separate but remain together within mucilage secreted by them. The cells lose their flagella and may undergo repeated divisions giving rise to an extensive mucilaginous matrix containing immobile cells. Under favourable conditions, this stage may break down, the individual cells develop flagella and swim away.

2.2   Autocolony Formation

Autocolony formation is another form of vegetative reproduction common in coenobial Volvocales such as *Eudorina, Pleodorina* and *Volvox*.

In *Eudorina indica* (Doraiswamy 1940), all except the anterior 12 cells of the 64-celled colony divide and form daughter colonies. At the time of division, the protoplast of the cell withdraws slightly from the cell wall and somewhat gelatinizes and forms a mucilaginous vesicle. First the protoplast undergoes a longitudinal division followed by another longitudinal division in the daughter cells, but at right angles to the first wall. These four cells divide further to form a plakea of 8 and then of 16 cells. With further division, a curved plate is formed. During further division of the cells, the plakea reverses its curvature by a process of inversion (Fig. 1Ca-Ch). Before inversion, the anterior ends of the cells face inwards but after inversion they face outwards as in the parent colony. A similar but more pronounced inversion takes place in *Pleodorina sphaerica* (lyengar and Ramanathan 1961) and in species of *Volvox*. (Fig. 1 Da-Di).

2.3   Cell Division in Desmids

Among the Chlorophycota, members of the Desmidiaceae show a characteristic mode of cell division. The cell is deeply constricted in the middle to mark two halves of the cell joined by a narrow isthmus. The cell wall is marked with characteristic ornamental markings. The cell wall consists of two units called semicells which enclose the protoplast and these have overlapping edges at the isthmus. At the time of division, the semicells separate at the isthmus while the cell undergoes nuclear division followed by transverse cleavage of the protoplast (Fig. 1 Ea-Ec). The two new protoplasts bulge out at the isthmus and soon assume the

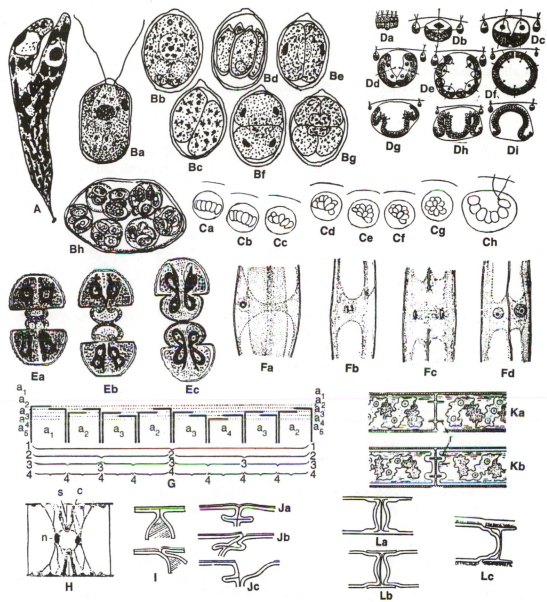

Fig. 1 A-L   Vegetative propagation of algae by cell division and fragmentation. A Cell division in *Euglena*. B
Chlamydomonas angulosa. Ba Single Cell. Bb Initation of cell division. Bc-Bg Stages in cell division
involving rotation of protoplast after first longitudinal division. Bh Palmella stage. C *Eudorina indica*.
Ca-Ch Stages in autocolony formation. D *Volvox aureus*. Da-Di Stages in autocolony formation. E
Cosmarium botrytis. Ea-Ec Stages in cell division and new semicell formation. F *Navicula halophila*. Fa-
Fd Stages in cell division. G *Eunotia* sp. Sequential vegetative divisions showing reduction in size during
four cell generations (diagrammatic). H *Spirogyra* sp. cell division. I *Mougeotia* sp. Ia-Ib Mechanism of
cell disjunction. J *Spirogyra weberi*. Ja-Jc Stages in formation of replicate end walls and disjunction. K
Spirogyra sp. Ka, Kb Stages in formation of replicate septa. L *Spirogyra colligata*. La-Lc Structure of
septum and replication (diagrammatic). Note H-piece in Lb (A Original, C after Doraiswami 1940, F
after Subrahmanyam 1947, Rest after Fritsch, 1935).

shape of the parent cell. A new semicell is formed around each protoplast half. The new wall is at first smooth and soon develops the markings characteristic of the species.

2.4 Cell Divisions in Diatoms

In the diatoms the cell wall is composed of two valves and two connecting bands which overlap each other and form a girdle. At the time of cell division, the protoplast undergoes division in a plane parallel to the valve and a new valve and a connecting band are formed on the free surface of the daughter protoplast, while the parent valves and their connecting bands separate (Fig. 1Fa-Fd). The new valves and connecting bands formed on the daughter cells fit into the connecting bands of the parent cell. Thus, while one of the daughter cells will remain the same size as the parent cell, the other daughter cell will be smaller. Repeated divisions in the daughter cells ultimately result in a clone of cells showing diminshing size (Fig. 1G).

2.5 Fragmentation

In filamentous algae fragmentation either by accident or design may lead to multiplication of the thallus. This may happen in uniseriate filamentous thalli or even in multiseriate filaments including pseudoparenchymatous and parenchymatous forms.

Of special interest is the mode of fragmentation in uniseriate filamentous genera of the Zygnematales. The cross walls between cells in some genera show inner wall layers which are invaginated (Fig. 1H-I). In others the cells exhibit "replicate end-walls" (Fig. I,K). Differential turgor in adjacent cells exerts pressure on the cross wall and leads to its rupture (Fig. I, J, L).

In *Caulerpa*, a member of the Siphonales, the older parts of the thallus decay, liberating the branches as independent plantlets.

In many genera of the Gigartinales, among the Rhodophycota, branches become constricted at the base and these may sever from the main axis at these constrictions.

In many marine algae, fragments of the thallus can survive and continue growth as individual plants. This principle is adopted in the cultivation of several marine algae of economic importance, such as *Gracilaria*, *Eucheuma* and others.

2.6 Special Propagula

In many algae modified branches may function as special propagula for vegetative multiplication. Such propagula are present in the Charophyceae (Fig. 2A-F), in Sphacelariales (Fig. 2I-K) among the Phaeophyceae and in various species of *Hypnea* (Fig. 2G, H), *Mychodea*, *Bonnemaissonia* and others.

3. Asexual Reproduction

Asexual reporduction in the algae is brought about by various types of spores which may be flagellated, the zoospores, or non-flagellated, the aplanospores. In all cases of spore formation, one of the cells of the thallus functions as sporangium. This may be an unmodified cell of the thallus as in *Ulothrix* (Fig. 3A, C, F), or a specially differentiated cell. In either case the contents of the cell withdraws from the cell wall and then undergoes division into a definite number of units, each develops into a zoospore or an aplanospore.

Usually, spore formation involves a primary meiotic division followed by one to several mitoses, resulting in a multiple of four spores. However, there are many instances of spore formation without a primary meiosis. Sometimes these two types of spores are designated as meiospores and mitospores; however, this terminology is not in common usage.

3.1 Zoospore Formation

The phenomenon of zoospore formation consists of the division of the contents of the zoosporangium into

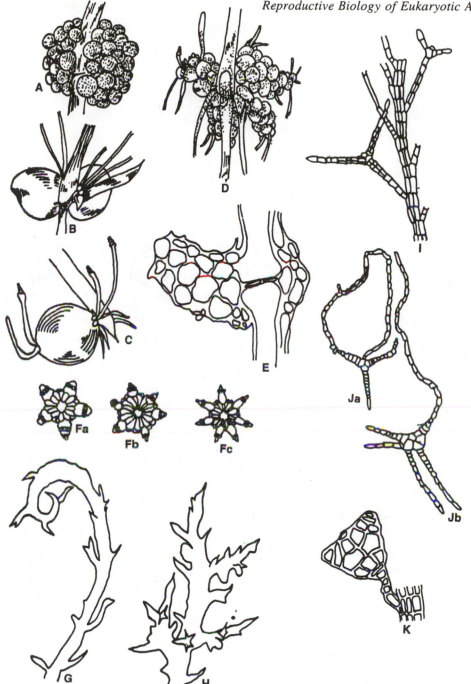

Fig. 2 A-K.   Vegetative propagation in the algae by special propagule. A *Chara fragifera* **root bulbils. B-C** *Chara aspera,* **spherical root bulbs. D** *Chara baltica,* **multicellular root bulbils. E Unilateral stem bulbilils. F** *Nitellopsis obtusa.* **Fa-Fc Star-shaped bulbils. G** *Hypnea musciformis,* **tendrillar branches. H** *Hypnea valentiae,* **stellate bulbils. I, J** *Sphacelaria furcigera***: I A branch with propagule and J A germinated propagule. K** *Sphacelaria tribuloides***, propagule (A-F after Bal et al. 1962, G, H after Ramarao 1992. I-K Original).**

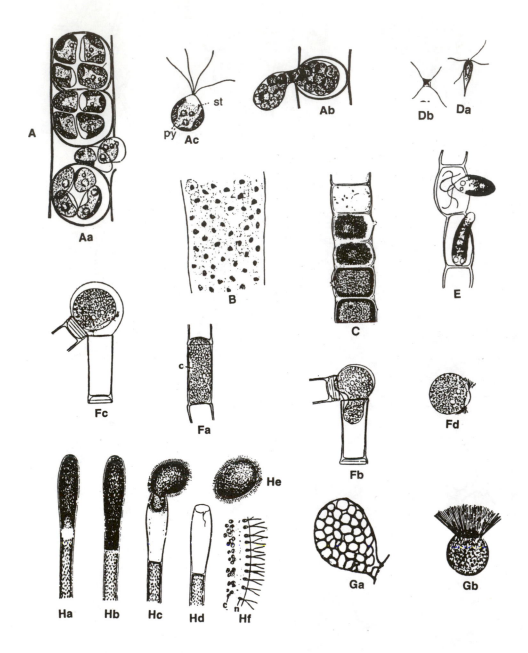

Fig. 3. A-H.   Zoospore formation in algae. A *Ulothrix zonata*. Aa Cell forms zoospores. Ab Liberation of swarmers in a vescicle. Ac Quadriflagellate Macrozoospores. B *Cladophora glomerata*, delimitation of zoospores. C *Chaetomorpha aerea*, filament with zoospore-forming cells. D *Urospora penicelliformis*. Da Zoospore in apical view. Db Zoospore in lateral view. E *Stigeoclonium tenue*, single quadriflagellate zoospore escapes from zoosporangium. F *Oedogonium concatenatum*. Fa-Fc Stages in formation and liberation of zoospore. Fd Liberated zoospore. G *Derbesia marina*. Ga Zoosporangium. Gb Zoospore. H *Vaucheria repens*. Ha-He Stages in the formation of zoospore and its liberation. Hf Periphery of zoospore. c crown of flagella, *npy* pyrenoid, *st* eye spot (after Fritsch 1935).

a definite number of units, without involving the cell wall of the zoosporangium. This aspect has been studied throughly in the unicellular green alga *Chlamydomonas*.

At the time of division, the protoplast separates from the cell wall and undergoes longitudinal division. In some species the division is slightly transverse. In such cases the protoplast withdraws from the cell wall and rotates by 90° before division. The apparent transverse division, therefore, is really a longitudinal division. Cleavage starts by furrowing of the protoplast from both ends, but usually the rate of furrowing is greater at the anterior end. During this process, the pyrenoid, the chloroplast and the nucleus divide while the flagella and contractile vacuoles are either shared between the two daughter protoplasts, or retained in one of them. These organelles are formed de novo in the other protoplast. The first division is followed by a rotation of the two protoplasts over 90° and then they cleave again. In this manner 4, 8 or 16 protoplasts may be formed, each with a plastid, an eye spot, two contractile vacuoles, and two flagella. These then escape as 'naked' zoospores.

In *Ulothrix zonata*, an unmodified cell of the filamentous thallus produces four to eight zoospores by division of the entire contents of the cell (Fig. 3A) The zoospores are usually of two kinds: (1) macrozoospores (Fig. 3Ac) which are larger and produced four per cell, and (2) microzoospores which are smaller and may be eight or more per cell (Fig. 3Ab). The zoospores are generally quadriflagellate. At the time of liberation, the cell wall of the zoosporangium ruptures laterally and the zoospores escape. However, sometimes the zoospores are enclosed in a hyaline vesicle and the packet of zoospores escapes from the zoosporangium. Then the vesicle dissolves and the zoospores are liberated.

In the Cladophorales, zoospores are produced (Fig. 3B) in large numbers inside the long, cylindrical, unmodified cells of the thallus which may be terminal or intercalary in position. The zoospores may be biflagellate as in some species of *Cladophora* or *Rhizoclonium* or may be pear-shaped and quadriflagellate as in *Urospora* (Fig. 3D). The zoospore of the latter genus is obovate and produced to a point at the posterior end and appears quadrangular when seen from the anterior end. In all Cladophorales, liberation of zoospores is through a lateral opening which may be at the anterior end of the cell.

In the Chaetophorales, the zoospores are quadriflagellate and may be either macrozoospores formed singly (Fig. 3E) in a cell or microzoospores, up to four in a cell.

In members of the Oedogoniales the contents of the zoosporangium form a single zoospore with a crown of flagella around a hyaline anterior end (Fig. 3Fa-Fd). The flagella are short but are of equal length. The liberation of the zoospore is by a transverse fracture of the cell wall of the zoosporangium close to its apical end. A small amount of mucilage oozes out of the fracture and the zoospore moves into it. After some time the mucilage dissolves and the zoospore is liberated.

In nannandrous species of *Oedogonium*, a special type of zoosporangium called an androsporangium (Fig. 4B) is formed. This is formed either singly or in series from a vegetative cell, which is designated the androsporangium mother cell. A separation ring appears close to the apex of the cell (as during vegetative cell division). The ring is thin and delicate. The nucleus divides and one of the daughter nuclei migrates to a position close to the separation ring. The cell wall splits horizontally at the separation ring and the ring itself stretches, separating the upper, smaller fragment of the cell wall and the lower larger fragment. A cross wall is formed. It separates the two fragments, and a distal discoid androsporangium and a subtending cylindrical vegetative cell differentiate. The subtending cell may again produce a second androsporangium in a similar fashion. Sometimes the androsporangium itself may divide transversely to give rise to more than one androsporangium.

In *Derbesia* zoosporangia are produced as more or less club-shaped or orbicular cells (Fig. 3Ga) produced laterally on the branches of the thallus. Very often the contents of the zoosporangium gets separated from the thallus branch by a plug. The contents then undergoes divisions and produces a large number of zoospores. Each zoospore when fully formed is spherical with a crown of flagella (Fig. 3Gb) around the apical end and resembles those of *Oedogonium*.

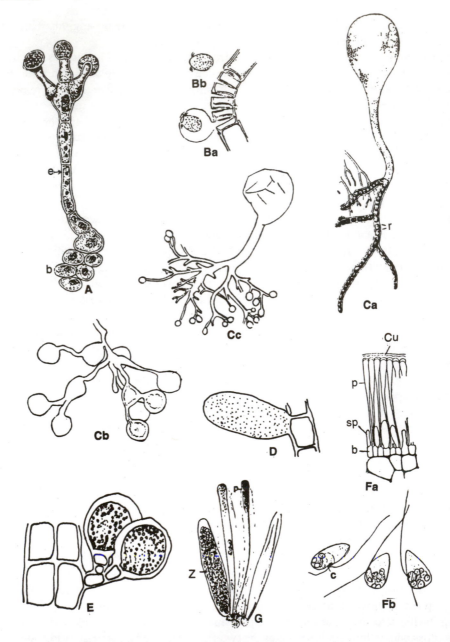

Fig. 4 A-G. **Zoosporangia and zoospores in different groups of algae. A** *Cephaleuros virescens*, **an aerial branch with stalked sporangia. B** *Oedogonium* **sp. Ba Androsporangia with one escaped androspore. Bb Single androspore. C** *Botrydium* **spp. Ca** *B. tuberosum with cysts* **at the ends of rhizoidal branches as well as intermittently in the rhizoids. Cb** *B granulatum*, **cysts occupying entire rhizoidal system. D** *Giffrodia mitchelliae*, **single unilocular sporangium. E** *Sphacelaria furcigera*, **young unicellular sporangia. F** *Laminaria saccharina*. **Fa Sorus of young unilocular sporangia. Fb Zoospores. G** *Sacchorhiza bulbosa*, **unilocular sporangia and paraphyses; one sporangium has dehisced and the zoospores liberated.** *b* **basal cells,** *c* **flagellar insertion,** *cu* **cuticle,** *e* **erect filament,** *p* **paraphyses,** *r* **rhizoid,** *sp* **sporangium,** *z* **zoosporangium (Ca, Cc after Iyengar 1944, D, E original; rest after Fritsch 1935).**

A very special type of sporangium develops in *Cephaleuros* and other members of the Trentepohliaceae, called the stallked sporangia (Fig. 4A). These have a structural mechanism for getting liberated as such. These sporangia may germinate directly but often produce swarmers.

In the Prasinophyceae, the zoospores are quadriflagellate and the flagella are covered by scales. In several members of the Xanthophyceae and the Chrysophyceae, the zoospores have unequal flagella which are also structually dissimilar. One flagellum is smooth and resembles the flagellum of green algae; the other flagellum is clothed with several small fibrillar processes, the mastigonemes. In most Chrysophyceae and the Prymnesiophyceae the flagella are clothed with scales.

In the genus *Vaucheria* which is a coenocyte, the zoosporangium is delimited by a cross wall at the end of a branch (Fig. 3He, Hf). The contents of the zoosporangium retract from the cell wall and forms a single zoospore, covered all over with numerous flagella (Fig. 3Hc, Hl, Hb) which are disposed in pairs of slightly unequal length and show structural dissimilarity, as in other Xanthophyceae.

In *Botrydium*, cysts are produced in the rhizoidal system in various ways, and these appear to be capable of germination after liberation (Fig. 4C).

The androsporangia of Oedogoniales (Fig. 4B) are also considered to be asexual spores but differ from zoospores in that they produce androspores which develop into dwarf male plants.

The zoosporangia of the Phaeophyceae are always specialized cells. Two types of zoosporangia are known: unilocular and plurilocular. The unilocular sporangia (Fig. 4 D-G) produce several, usually a multiple of four, zoospores by cleavage of the contents into a number of uninucleate parts, each developing into a zoospore. In such sporangia the first division of the nucleus is meiotic and ultimately leads to the formation of haploid zoospores. The plurilocular sporangium shows a division of the sporangial contents with formation of septa separating the daughters and ultimately produces numerous uninucleate protoplasts separated by septa, the septa arising in regular longitudinal and transverse directions. The sporangium thus assumes a multichambered appearance. Each chamber or loculus, produces a single zoospore. All divisions in the plurilocular sporangium are mitotic.

The plurilocular sporangium may be formed on diploid as well as haploid thalli as in primitive orders like the Ectocarpales. In *Ectocarpus*, plurilocular sporangia appearing on diploid thalli serve as accessory asexual reproductive organs, serving to perennate the diploid condition. Similarly, pluricular sporangia occurring in haploid thalli also perpetuate the haploid condition. Unilocular sporangia occur on diploid thalli, and the meiotic division in this sporangium reinstates the haploid thalli.

In more specialized orders of the Phaeophyceae, unilocular sporangia are formed on diploid sporophytes, which regularly alternate with the haploid gametophytes which produce gametangia of various types (e.g. Laminariales).

Zoospores of the Phaeophyceae (Fig. 4H) show a haterokont condition with one long, usually forward-projecting flagellum and the other shorter, backward-projecting flagellum. The flagella are usually inserted laterally. While the anterior flagellum is covered with fine hairs, the posterior flagellum is usually smooth.

3.2 Morphology of the Zoospores

The zoospore is unicellular, flagellated, oval to oblong to spherical in shape and is usually formed within a zoosporangium which may be an unmodified cell of the thallus or a specialized unicellular organ. The zoospore also possesses one or more plastids with pigmentation characteristic of the group, with or without pyrenoids, but with an eye-spot and probably contractile vacuoles. It may be walled or naked.

Considerable attention has been paid to the study of the flagella. These may be single (e.g. *Euglena*), two (*Chlamydomonas*), four (*Carteria, Urospora*), or more (*Oedogonium*). They may be inserted at the anterior end, sometimes in a cavity, or may be laterally inserted. Where the flagella are numerous, these may form a crown around the anterior end. When there are two or four flagella, they are usually of equal length and of similar structure. these are the types of flagella in the majority of the Chlorophycota. In some Chlorophyceae, like the Oedogoniales and some members of the Bryopsidales, a crown of flagella surrounds the apex and these algae are discribed as stephanokontan. These flagella are also of equal length and similar structure.

In most of the Chromophycota, the paired flagella are unequal in length and also show differences in fine structure. One flagellum is usually short and smooth, while the other is long and covered with fine hairs or fibrils. An older terminlogy describes these as the 'whiplash' and 'tinsel' types, respectively. In the special type of zoospores (synzoospores) in *Vaucheria* the flagella cover the entire surface of the zoospore and are arranged in pairs, the members of a pair slightly unequal in length and may also show differences in fine structure. In many Chromophycota, the flagella may also be inserted laterally.

3.3 Fine Structure of Zoospores

The fine structure of algal flagella has been reviewed exhaustively by Moestrup (1982).

The flagellum is a tublar structure consisting of a sheath surrounding a bundle of microtubules arranged in a specific manner (Fig. 5J). In most algae, there are two distinct strands in the centre, surrounded by a ring of paired strands (doublets), the whole comprising an axioneme. This 9 + 2 arrangement of microtubules (Fig. 5Jb) is more or less univeresal. There are, however, some exceptions, as in the case of pseudocilla of various green algae, where the arrangement is 9 + 0. At the place of insertion of the flagellum on the cell surface, the flagellar structure goes through a transition zone, in which the two microtubules of the central strand and the nine doublets give place to a basal cylinder showing a stellate pattern in cross-section (Fig. 5 Jc-Jd). Lower down, the central strands are not seen, while the nine peripheral doublets alone remain (Fig. 5Je). The transition zone leads down to a basal body with the incorporation of a transitional fibre; the doublets of the peripheral tubules change to a triplet condition (Fig. 5Jb). At its lowest part, the basal body shows a cartwheel configuration (Fig. 5jg). The basal bodies are attached to the flagellar roots which may be microtubular or may be striated bands.

The flagellar roots extend into different regions of the cell (Fig. 5K), and some may even contact various organelles, including the nucleus. These flagellar roots have been assigned different functions like anchorage of flagellar apparatus, conduction of stimuli, regulation of flagellar movements or function as skeletal system.

An interesting feature in the Fucalean flagella is the presence of a swelling at the base of the long flagellum, the proboscis, well-illustrated in *Fucus*, *Ascophyllum* and *Pelvetia*. The proboscis is elastic and flat and is connected to a fibrillar flagellar root. The cross banded root of other brown algae like the Chrysophyceae is not seen in the Phaeophyceae. In most of the less specialized members of this class, the root is a broad structure consisting of 12 to 15 microtubules which travel from the base of the flagellum along the plasmolemma to the anterior end of the cell. Here the microtubules bend back and run to the opposite side. In many primitive Phaeophyceae, the root consists of six to eight microtubules.

The 9 + 2 stranded structure is common to almost all algae. However, there are other details in which flagella of different groups of algae may show differences. In some green algae, in the Prasinophyceae for instance, the flagellar sheath is clothed with scales (Fig. 5G, H). In the heterokont algae (Chrysophyceae, Phaeophyceae), the longer flagellum always bears fine haris, the mastigonemes (Fig. 5B). Detailed studies have been made of the mastigonemes (Loiseaux and West 1970) and it has been found that these show a basal swelling, and have a sheath surrounding three to several microfibrils. The sheath is not complete through the length of the mastigoneme but stops short of the apex, where the naked microfibrils can be made out as a tripartite structure. In the Cryptophyceae, both flagella have hairs but differ in arrangement; while the longer flagellum has hairs on two opposite sides and the shorter flagellum has hairs on all sides (Fig. 5A). In the Prasinophyceae, all flagella (whether one, two or four) bear hairs which are tubular and their arrangement is similar in all flagella (Fig. 5C-F).

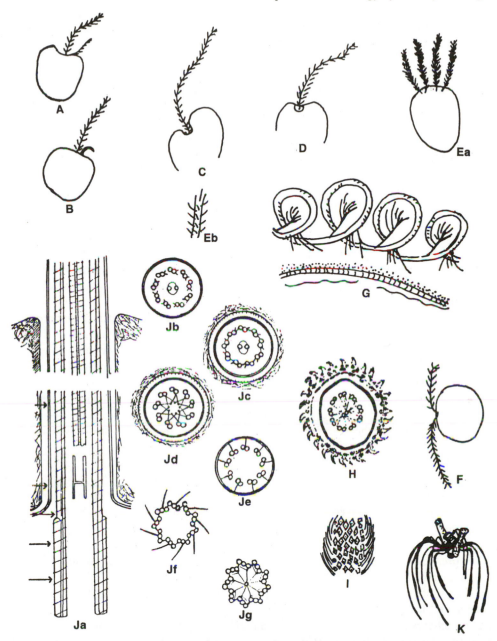

Fig. 5 A-K. Structure of flagella of algae. A-G Diagrammatic representation of flagella of different groups of algae (for explanation see text). H Cross section of flagellum covered with three layers of scales. I Longitudinal section of scaly flagellum shows inner layer of diamond-shaped scales and an outer layer of overlapping scales. J Fine structure of flagellum of *Eudorina illinoisensis* (diagrammatic). Ja Longitudinal section of flagellum. Jb-Jg Transverse sections at succeessive levels of the flagellum to show transition in arrangement of microtubules up to the basal body and flagellar root. K *Chlamydomonas* sp., diagrammatic representation of flagellar roots and microtubules. (A-I, K adapted from Moestrup 1982, J after Hobbs 1971).

In several Chrysophyceae and Dinophyceae (Fig. 5G), a paraxial rod is present within the sheath along with the axioneme. In some of these algae, the axioneme often shows helical twisting, while the paraxial rod is nearly always straight (e.g. the transverse flagellum of Dinophyceae).

In the Phaeophyceae, the flagella may bear short spines which may occur singly or in a row, along one side of flagellum. The single long flagellum of the spermatazoid of *Dictyota* bears a row of about 12 spines. The long flagella of *Himanthallia*, *Xipophora* and *Hormosira* show single spines.

3.4 Aplanospores

Aplanospores are formed in a similar way to those of zoospores, only these spores do not develop flagella (Fig. 6A, B).

In the Chlorococcales, asexual reproduction takes place by formation of either zoospores (zoosporic forms) or special types of non-motile spores called autospores (autosporic forms). The autospores are distinct from the aplanospores. Before liberation, the autospores assume the mature features of the parent cell, including the cell wall (Fig. 6C). In many Chlorococcales the autospores are not completely liberated and remain together as a colony, held together by gelatinized cell walls of the parent cells. The gelatinization may be complete or partial and the daughter cells may lie embedded in the gelatinized parent cell wall, or may be held together by fragments of the parent cell wall which persist and help to form a colony. Different types of colony formation in Chlorococcales are illustrated in Fig. 7.

More often, spore formation is limited, and four spores are common. Such tetraspores are common in the Dictyotales (Fig. 6D) among the Phaeophyceae and in the Rhodophyceae (Fig. 6 F-H). Monospores are formed in many Rhodophyceae and are a common mode of reproduction of the same phase in the life cycle. Some red algae form biospores in pairs in a sporangium (Fig. 6E). It is presumed that this results from an arrested development of a tetraspore.

Among the Rhodophyceae, one may also come across paraspores and polyspores. parasporangia contain irregular masses of monospores (Fig. 6I). Polysporangia, on the other hand, contain spores formed after regular repeated divisions of the protoplast and may represent the extended condition over the tetrasporangium (Fig. 6J).

3.5 Other Spore Types

Akinetes or resting spores are modified cells which accumulate materials and develop a thick resistant wall which may become three- to several-layered (Fig. 6K). These may be liberated by the decay of adjoining or circumjacent cells of the thallus, or may be liberated by a special mechanism as in *Pithophora* (Ramanathan 1939b). Akinetes may germinate directly, or remain dormant for a long time.

Hypnospores are formed by the contraction of the protoplast from the parent cell wall and remain dormant for a prolonged period (Fig. 6L).

In the Chrysophyceae resting spores develop in specially formed cysts which have a cell wall covered with siliceous scales and sometimes spines.

In many members of the Chrysophyceae, formation of a siliceous cyst and the migration of the protoplast into it and the subsequent plugging of the cyst, is a common occurrence. The cyst when fully formed is made up of two parts; a more or less spherical siliceous envelope with a small opening on one side and a plug-like piece closing the opening. The contents of the cyst will germinate under favourable conditions and produce one or more motile cells. In a similar manner, many members of the Dinophyceae may also go into an encysted condition under unfavourable conditions.

Diatoms also form resting spores by the contraction of the protoplast away from the parent cell wall and the formation of a new cell wall around itself. The wall also becomes ornamented with spines and warts. These are known as statospores (Fig. 6M).

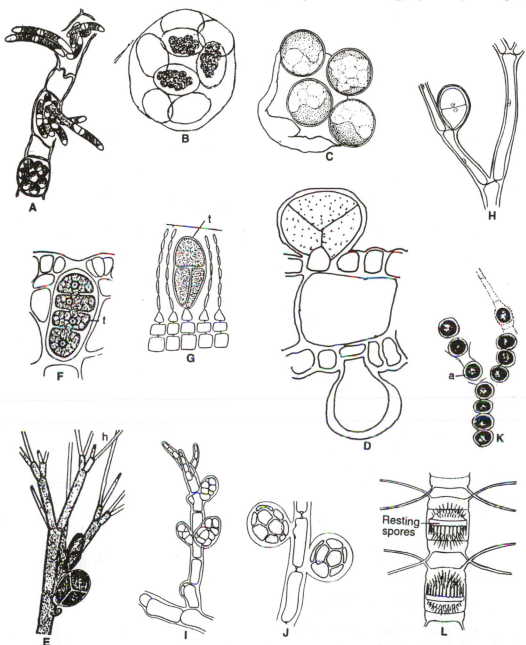

Fig. 6 A-M. Non-motile spores of algae. A *Ulothrix zonata*, aplanospores and their germination in situ. B *Halosphaera* sp., aplanospores. C *Chlorella* sp., autospores. D *Dictyota dichotoma* tetraspores. E *Callithamnion* sp., bispores. F *Cystoclonium purpureum*, zonate tetraspores. G *Peyssonelia dubyi*, cruciate tetrasporangium. H *Callithamnion corymbosum* Tetrahedral tetrasporangium. I *Callithamnion* sp. parasporangia. J *Ptilothamnion* sp. poly-sporangia. K *Pithophora polymorpha*, akinete. L *Draparnaldia glomerata*, hypnospores. M *Chaetoceras* sp., statospores. *a* akinete, *h* hair, *t* tetresporangium, (A, B, H, L, after Fritsch 1935, D original; C, J, M after South and Whittik 1967, E after Rosenvinge 1924, F, G after Kylin 1918, K after Ramanathan 1939).

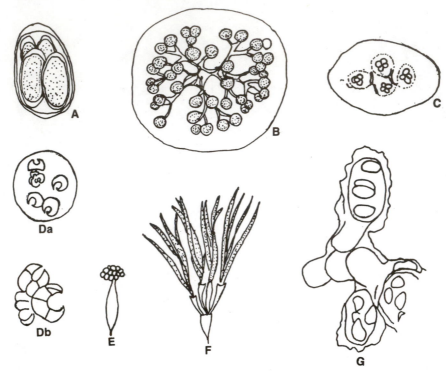

Fig. 7 A-G.   Autospores and new colony formation in Chlorococcales. A *Oocystis panduriformis.* **B** *Dictyosphaerium pulchellum.* **C** *Radiococcus nimbatus.* **D** *Kirchneriella lunaris.* **Da** Young four-celled colony, one cell has formed autospores. **Db** Mode of division of a cell prior to autospore formation. **E** *Actidesmium hookeri.* **F** *Ankistrodesmus falcatus.* **G** *Nephrocytium edysiscepanum.* **(A, C-G after Fritsch 1935, B after Iyengar and Ramanathan 1951)**

4.   Sexual Reproduction

The sexual mechanism consists of the formation of gametes which fuse in pairs to produce a zygote. It germinates and produces a new thallus, either directly or quite often by production of four to several spores which then produce new thalli. In this process fusion of gametic nuclei is an essential step and the product of fusion is a diploid nucleus. When the zygote produces spores, there is usually a meiotic division leading to the production of haploid spores.

The nuclear fusion, during sexual reproduction, brings together genetic material from two different sources, the male and female parents. It has been shown that even in the most primitive algae, there is always a difference in the genetic material of the fusing gametes, even if they are derived from the same parent cell.

In those algae where the zygote germinates and gives rise to a thallus directly, a diploid vegetative phase emerges and the thallus develops a reproductive process involving meiosis so as to restore the haploid condition in the reproductive cells. These may be of the nature of gametes or spores. If they are gametes, their fusion will restore a diploid condition. If they are spores, they will develop into haploid thalli, thus introducing an alternate haploid generation.

The gametes may be flagellated (planogametes) or non-flagellated (aplanogametes). Rarely, gametes do not fuse and develop into a new thallus. Such gametes are described as apogamous.

Depending on the morphology of the gametes and to some extent on their behaviour, sexual reproduction may be isogamous, anisogamous (or heterogamous) and oogamous. In isogamy the fusing gametes are morphologically alike and exhibit similar movements before fusion. In anisogamy, the fusing gametes are dissimilar in size and the smaller gamete is more active in its movements, the larger sluggish and often immobile. In oogamy, besides a difference in size, the larger gamete is non-motile, and in planogamous species, without flagella. The smaller gamete is highly motile and flagellated, and is called a spermatozoid.

The process of sexual reproduction may be considered as consisting of two steps; formation and liberation of gametes and gametic fusion.

4.1 Formation of Gametes

Gametes are formed from cells of the thallus that may not be very different in structure from vegetative cells, or from specialised cells of the thallus. These cells are designated gametangia. In oogamy the gametangia are described as oogonia (produce ova) and antheridia (produce antherozoids).

The number of gametes produced in a gametangium may vary from one to many. The oogonium usually produces one ovum, while other gametangia (including antheridia) produce two to many gametes. There are. however, algae which produce several ova in an oogonium (*Sphaeroplea*).

For instance, in many species of *Oedogonium* the number of antherozoids per antheridium is two. Similarly, in the Charophyceae, the cells of spermatangial filaments may produce two spermatozoids each. On the other hand, the antheridia of *Vaucheria* produce numerous antherozoids. A similar condition is reported in brown algae like *Dictyota*.

The gametangium may produce a single gamete by the transformation of the entire contents of the cell, often accompanied by a withdrawal of the protoplast from the cell wall. The protoplast may develop a thin wall or membrane around itself and then function as the single gamete.

Usually, however, a gametangium produces four to several gametes. In such cases the protoplast of the gametangium retracts from the wall and then undergoes division. At first the nucleus divides and is followed by cytoplasmic cleavage resulting in a number of uninudeate protoplasts. These may simply round off and become aplanogametes, or each may develop a pair of flagella and become planogamete.

4.2 Liberation of Gametes

This is achieved by one of two processes: (1) The gametangial wall ruptures irregularly, or by a definite transverse split, or develops a pore in the wall. (2) The wall of the gametangium gelatinizes to liberate the gametes.

Where a rupture of the gametangial wall takes place, the gametes may be liberated directly into the water, or a mucilaginous vesicle is formed first and the gametes move into the vesicle and then the gametes are liberated by dissolution of the vesicles.

4.3 Gametic Fusion

The process of gametic fusion starts with the coming together of the gametes, goes through a stage of plasmogamy or cytoplasmic fusion followed by karyogamy, or nuclear fusion, and culminates in the formation of a zygote.

In isogamous algae the gametes may be flagellated and capable of free locomotion in the surrounding water. Alternately, these may be non-flagellated and exhibit amoeboid movement within a restricted environment (e.g. members of the Zygnematales and the Desmidiaceae). Even in unicellular algae, where planogametes are involved, the gametes may be derived from cells of a single clone when the alga is said to be homothallic. Sometimes, gametes derived from cells of a single clone do not fuse. For fusion, the gametes should belong to two different clones. Such species are designated heterothallic. Such a distinction between homothallism and heterothallism is easier to notice in filamentous or thalloid forms.

4.4 Gametic Fusion in *Chlamydomonas*

The process of gametic fusion has been thoroughly studied in *Chlamydomonas*. This genus exhibits all three types of sexual reproduction: isogamy, anisogamy and oogamy. In some species, recently divided and liberated cells function as gametes. These are smaller in size than vegetative cells. In many species the fusing gametes are similar to the parent cell and are walled (e.g. *Chlamydomonas media*, *C. chlamydogama*, *C. ovata*). The cell walls are shed after the gametes establish contact and begin to fuse. In some species the gametic fusion is progressive and the zygote wall is even formed before the parental walls are discarded. (*C. moewusii*, *C. ehrenbergii*, *C. pseudopetryi*). In *C. gymnogama* the wall is discarded even before fusion. Therefore, this may be deemed to be a fusion of naked gametes. In *C. goroschankinii* and *C. braunii* typical anisogamy is reported. Extreme anisogamy bordering on oogamy occurs in *C. coccifera* and *C. heterogama*. In these taxa the larger gamete is represented by an undivided vegetative cell which loses flagella and is motionless and the smaller gametes are formed by division of the contents of the mother cell into 8, 16 or 32 units. In *C. striatum* the smaller gametes are without walls. In *C. coccifera* sexual reproduction is almost oogamous.

The process of gametic fusion has been studied with the help of both visual and electron microscopy in *C. reinhardii* (Friedmann et al. 1968) and in *C. moewusii* (Brown et al. 1968).

It is recognised that gametic fusion in *Chlamydomonas* often requires the juxtaposition of two distinct mating types which may be termed as the + and – types. In *C. moewusii*, studied by Brown et al. (1968), soon after the compatible mating types are mixed together the fusing gametes adhere initially by the tips of their flagella which is probably facilitated by an amorphous material at the tip, probably a glycoprotein, which acts as an attractant and serves to hold the gametes together (Fig. 8A, Bc). After adhesion of the fusing gametes, other gametes crowd them and form a clump. In the fusing gametes progressive agglutination of the flagella along their length takes place, but this is only transient.

In a number of species, such as *C. eugametos* and *C. moewusii*, the gametes come together in a vis-a-vis position (Fig. 8Ba, Bb) and a protoplasmic bridge is formed between them (Moewus 1933; Gerloff 1940; Lewin 1950, 1952, 1954, 1957; Mitra 1951; Lewin and Meinhart 1953; Tsubo 1956, 1957, 1961). On the other hand, in *C. reinhardii* (Friedmann et al. 1968) a fertilization tubule is formed between the fusing gametes (Fig. 8Ca, Cb) probably produced by one of the gametes (the + type). Possibly a structure resembling the fertilisation tubule is formed in a number of isogamous and even oogamous algae. Hoffman (1961) describes a delicate cytoplasmic connection between the egg and tip of the antherozoid in *Oedogonium cardiacum*.

Clumping of gametes is also recorded in a number of other algae, one classical example is that of *Ectocarpus siliculosis*.

The flagellar agglutination is probably aided by a substance which is biosynthesized during gamete formation, and the agglutination mechanism may be responsible for the species specificity of gamete contact (Wiese and Shoemaker 1970).

In *Chlamydomonas*, plasmogamy is effected either through a protoplasmic bridge as in species with vis-a-vis coming together of the gametes or through a fertilisation tubule as in *C. reinhardii*. In the latter case there is some evidence (Friedmann et al. 1968) to show that the fertilisation tubule is produced only by the + strain (Fig. 8Ca, Cb), so that in this species one can assume the existence of sex differentiation.

During plasmogamy, chloroplast fusion is initiated by the juxtaposition of the anterior margins of the two plastids, an initial breakdown of the chloroplast envelop, and subsequent fusion of the chloroplasts. About this time the gametic wall also dissolves.

During the process of plasmogamy, the nuclei of the two gametes migrate to an equatorial position and then unite. After plasmogamy and karyogamy, the primary zygote memberane is shed (Brown et al. 1968) and the secondary zygote wall is secreted.

Fig. 8 A-F. Sexual reproduction in *Chlamydomonas*. A *Chlamydomonas* sp., Clumping of gametes. B *Chlamydomonas* sp. Ba Single gamete. Bb Coming together of gametes by their anterior end. Bc Flagellar agglutination and plasmogamy. Bd Quadriflagellate zygote. C *Chlamydomonas reinhardii.* Ca production of fertilization tube. D *Chlamydomonas iyengarii*. Da Posterior alignment of gametes. Db Begining of plasmogamy. Dc, Dd Successive stages in plasmogamy. E *Chlamydomonas braunii.* Ea Gametic contact. Eb Later stage of gametic fusion. F *Chlamydomonos coccifera*, shows an advanced anisogamic condition bordering on oogamy (A after Brown et al., 1968, Ba-Bd Canter-Lund and JWG Lund 1995. Ca, Cb after Friedmann et al 1968. E, F from Fritsch 1935).

At this point it has to be stated that in some species of *Chlamydomonas*, such as *C. indica*, the gametes come together by their posterior end and fuse, giving rise to a quadriflagellate zygote (Fig. 8Da, Dd). The details of the mechanism of such fusion are scantily known and it is not known if there is any chemical attractant involved in bringing the gametes together.

While in many species of *Chlamydomonas* sexual reproduction is considered to be isogamous, there are species exhibiting an anisogamous condition bordering on oogamy, such as *C. coccifera*. This species also possesses walled gemetes in contrast to the naked gametes of other species.

5. Sexual Reproductive System in Charophyceae

In the Charophyceae, there is no asexual reproduction by spores. Sexual reproduction involves the formation of antheridia and oogonia, which are morophologically distinctive among the algae, having specialized vegetative cells surrounding the sexual cells, thus showing at least a superficial resemblance to the sex organs of archegoniate plants.

Chara spp. may be dioecious or monoecious, the antheridia and oogonia borne on the branch bearing nodes. These mostly appear on the basal node of a lateral (Fig. 9A), often of the second or even third order. One of the peripheral cells of the basal node, usually the adaxial one, gives rise to a row of three cells, the terminal cell being the antheridial primordium, the middle cell its basal node, and innermost an internodal cell. the adaxial cell of the basal node gives rise to an oogonium.

The antheridial primordium delimits a pedicel cell at the base (Fig. 9Ba) and the remaining primoridium undergoes two vertical divisions at right angles to each other and a transverse division, resulting in octants. Each octant divides periclinally into a radial row of three cells. The outemost cell expands to form a shield cell, the middle cell forms a manubrium, and the innermost, a head cell. With further development the head cells give rise to secondary headcells each of which bears two spermatogenous filments of discoid cells (Fig. 9Bc). The mature antheridium is spherical with eight shield cells which become variously ornamented and sometimes markedly pigmented. Each spermatogenous filament consists of a row of sperm mother cells. Each sperm mother cell untimately produces a single, coiled spermatozoid with two slightly-laterally inserted flagella (Fig. 9Ca-Cd). In *Chara zeylanica*, four shield cells have been reported (Groves 1931).

The oogonium in *Chara* generally arises on the basal node of the antheridium (Fig. 9Ba). The oogonial mother cell divides and produces a row of three cells; the terminal cell forms the oogonium proper. The middle cell becomes a node with five peripheral cells and these cells elongate and become wound spirally around the oogonium; these are the covering filaments. The tips of the covering filments are usually cut off by cross-walls to form five (in Chareae) or ten (in Nitelleae) crown cells or coronula (Fig. 9Bb).

The oogonial cell cuts off one (in Chareae) or three (in Nitelleae) small sterile cells, before differentiating into the single oosphere or egg cell. The mature oogonium consists of a stalk cell, five spirally wound sheath cells covering an oosphere, and one or three sterile cells.

The gross morphology of the sex organs in the Charophyceae is more or less uniform, there being minor differences in the relative positions occupied, and the mode of arrangement of the antheridia and oogonia. Such differences are used as taxonomic criteria at the generic level.

At the time of fertilization, the spiral sheathing cells elongate and separate from the apex of the oosphere, thus forming a cavity at the apex. They also separate laterally leaving small slit-like apertures leading to the cavity. The cavities as well as the apertures are filled with a colourless, viscous fluid to which the swarming spermatozoids adhere and eventually one of these sperms enters the oosphere and fertilization ensues (de Bary 1871).

The product of fertilization is the oospore, which is usually spherical to narrowly ellipsoid, differing widely in size, opacity, colour, and consistency. The mature oospore has a wall composed of four membranes. The outermost is a coloured membrane derived from the inner walls of the cover filaments of the oogonium (Groves and Bullock-Webster 1920) and is said to be composed of suberin and silicic acid. This membrane shows a number of ridges on the surface and may show various ornamental markings. This membrane encloses a closely-adhering inner coloured membrane, which is granulated. Within this is an outer colourless membrane and an inner colourless membrane, and these are quite translucent.

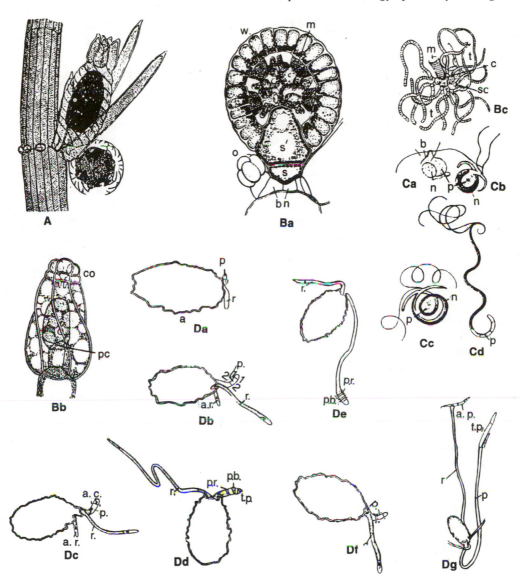

Fig. 9 A-D. Sexual reproduction in Charophyceae. A *Chara fragilis*, sex organs borne at the node of a branch. B *Nitella flexilis*. Ba Longitudinal section of an antheridium. Bb Mature oogonium. Bc Manubrium with spermatogenous filaments. C *Chara foetida*. Ca-Cc Successive changes in protoplast of spermatogenous cell. Cd Mature spermatozoid. D *Chara zeylanica*. Da-Df Germination of oospore. Dg Development of accessary protonema from primary root. *ac* apical cell, *ap* accessory protonema, *ar* accessory rhizoid, *b* beak, *bn* basal node, *C* capitulum, *Co* Corona, *m* manubrium, *n* nucleus, *O* oogonal primordium, *P* Protonema, *p* periplasm, *pb* potonemal branchlet node, *pc* cells cut off at base of oogonoium, *pr* protonemal root, *r* primary root, *rh* rhizoid, *tp* terminal process, *w* wall. (A, B after Oltmanns 1898, C Belajeff 1894, D after Sundaralingam 1954)

During germination, the contents of the oospore divide to give rise to a small cell at the apex. This cell undergoes a vertical division to form two cells and these protrude out of the oospore membrane as two knob-like structures (Sundaralingam 1954). One of these grows out into a protonema and the other into the primary root. The protonema elongates and becomes inflated towards the apex. The inflated portion cuts off an apical cell which divides to give rise to a row of three or five cells. It is from this portion that a new individual arises. The sequences of germination in *Chara zeylanica* are given in Fig. 9Da-Dg.

6. Sex Attractants

As mentioned earlier (see Section A), in *Chalmydomonas*, flagellar agglutination between fusing gametes is brought about by a glycoprotein produced at the tips of flagella, which is also the mechanism for ensuring species specificity in the fusing gametes.

Similar substances which attract compatible gametes have been reported in a number of algae, the most thoroughly studied members being the Phaeophyceae. As early as 1854, Thuret had shown that the male gametes of *Fucus* tended to be attracted towards the egg and these get attached to the egg by their flagellum. A similar observation was made by Berthold (1881) and Oltmanns (1899) in *Ectocarpus*, and by Kuckuck (1899) in *Cutleria*.

More recent studies have been carried out on the sex attractant in *Ectocarpus siliculosis* (Müller et al. 1971, Müller and Gassmann 1978), *Fucus serratus* (Müller and Seferiadis 1977) and the chemical nature of the attractants has been determined. Similar studies have also been carried out in the Laminariales (Lüning and Müller 1978). One can safely conclude that chemokinetic and chemotactic reactions are exhibited in a number of algae with oogamous sexual reproduction.

Gametogenesis in species of *Volvox* is also under pheromone control (Starr et al. 1980). A similar sex attractment has been recognized in *Oedogonium* (van den Ende 1976).

The chemical nature of marine pheromones of the Phaeophyceae has been determined. In *Ectocarpus*, *Sphacelaria* and *Adenocystis* the substance is ectocarpene, a cyclic cell olefine. Similar low molecular weight olefinic pheromones are known in *Desmerestia* (desmerestene), *Dictyota* (dictyopterene), 29 species of Laminariaceae, Alariaceae and Lessoniaceae (lamoxirene), *Cutleria* (multifidene) and *Syringoderma* (visidiene). Linear ^{8}C or ^{11}C olefinic pheromones have been identified in other Phaeophyceae: fucoserratene in *Fucus*, finaberrene in *Ascophyllum* and *Dictyosiphon*, and cystophorene in *Cystophora*. Hormosirene is a ^{11}C pheromone containing cyclopropane found in *Hormosira*, *Xipophora*, *Durvillea*, *Scytosipyhon* and *Colpomenia*.

These pheromones are functional in as low a concentration as ^{7}10 mol/l and have a range of 0.5 to 1.0 mM (Maier and Müller 1986).

7. Origin of Sexuality in Algae

Sexual reproduction is common to all groups of algae, except the Eustigmatophyceae, Raphidiophyceae, Cryptophyceae and Euglenophyceae.

The formation of gametes as fundamental units of sexual reproduction, however, is widespread among the other classes of algae. It is of interest to note that all three types of sexual reproduction (isogamy, anisogamy and oogamy) are met with even in the most primitive members of the different algal classes.

Isogamy should have originated in haploid algae as the first step in sexual reproduction. The fusion of gametes derived from a single clonal population must have been the earliest manifestation of sexuality. Subsequently, incompatability must have arisen between gametes of monoclonal origin. In such cases, the fusing gametes are always derived from two different clones, and only compatable gametes fuse.

The compatibility of gametes probably relates to definite biochemical phenomena which distinguish one clone from another. This is further amplified in heterogamous sexual reproduction involving structurally and functionally dissimilar gametes.

Genetic studies on many algae have demonstrated that gene transfer and recombination are as evident in algae (during sexual reproduction) as in higher plants. In many algae (*Cladophora, Wrangelia*) the sex chromosomes have been identified and their behaviour demonstrated. These have much resemblance to the behaviour of sex chromosomes in higher plants.

A further evolution of sexuality in algae is reported in dioecious (several Fucales, *Dictyota*) members of the Phaeophyceae and Rhodophyceae.

8. Germination of Zygote

Germination of zygote may result in direct development of the mature vegetative thallus. This is always so where the vegetative thallus is diploid and it produces gametes after a meiotic process. However, such a situation is very rare and has been recorded only for the Fucales among the Phaeophyceae. Here, again, many phycologists would like to interpret this feature as conforming to the concept of an alternation of haploid and diploid generations.

The more common condition among the algae is the production of gametes in haploid thalli. On fusion of gametes, a diploid zygote results. To re-establish the haploid thallus, the germinating zygote may adopt one of several routes.

The zygotic nucleus divides meiotically and forms four nuclei. Three of these degenerate and the surviving nucleus gives rise to a single haploid germling which then forms a new thallus (e.g. Zygnematales).

The zygote, after meiotic division, gives rise to four or more zoospores or aplanospores which give rise to new haploid thalli (e.g. *Ulothrix,* many unicellular Volvocales, many other Cholorophyceae). Sometimes the zygote undergoes a resting phase as in the Hydrodictyaceae of the Chlorococcales, where the zygote produces zoospores after meiosis and the haploid zoospore develops into a 'polyhedron'. After a prolonged resting period the polyhedron, on germination, gives rise to zoospores which from a new colony (e.g. *Hydrodictyon, Pediastrum*).

The zygote germinates and produces an alternate diploid thallus, which may be morphologically similar to the haploid phase (*Enteromorpha*), or morphologically different (*Derbesia-Halicystis, Derbesia-Bryopsis,* the Laminariales).

In most Florideophyceae the zygote undergoes amplification and produces a diploid system of filaments (carposporphyte). These filaments produce numerous diploid spores (carpospores) which produce the alternate diploid thallus (tetrasporophyte). The thalli are usually morphologically similar to the haploid phase or the gametophyte. In some genera like *Asparagopsis* and *Bonnemaissonia*, the diploid thallus is morphologically different.

In a number of algae, because of the environmental or other parameters, the gametic fusion fails, and the individual gamete assumes the shape and structure of a zygote. Then it produces a new thallus. Such parthenospores are well-known in a number of green algae, e.g. in Zygnematales.

9. Life History

The life history in sexually-reproducing algae may be haplobiontic when there is only one vegetative phase in the life cycle, or diplobiontic when there are two alternating vegetative phases, one haploid and the other diploid. It has been mentioned above as to how gametic fusion brings about a diploid state in the zygote. A restoration of the haploid stage is essential for the normal life cycle. This is due to meiotic division of the diploid nucleus, either in the zygote or in a sporangium produced on a diploid thallus. The zygotic meiosis is common in most haplobiontic algae, and the vegetative phase is always haploid. However, rarely meiosis takes place prior to gametogenesis. Here the life history is haplobiontic but the vegetative thallus is diploid. Examples which may be cited include species of *Cladophora*, the members of the Bacillariophyceae, and the Fucales among the Phaeophyceae.

In most other algae, there is an alternation of two vegetative phases, the haploid gametophyte produces gametes, and the diploid sporophyte produces spores after meiosis. The life history is diplobiontic. The alternate phases may be morphologically similar when the alternation is of isomorphic generations. On the other hand, the alternate phases may show different morphology when the alternation of generations is described as heteromorphic.

In some algae with alternate gametophytic and sporophytic phases, there may be accessory modes of reproduction in either phase, by means of spores. Often an alga reproduces asexually several times before embarking on sexual reproduction as in *Ectocarpus* spp.

The patterns of life history in different groups of algae are so varied that it is difficult to generalise under a single or a few types. The following account of the life histories in a number of selected genera will give some idea of the diversity involved.

9.1   *Chlamydomonas*

Chlamydomonas is a haplobiontic, haploid organism. The alga can reproduce itself repeatedly by zoospores, usually four from each parent cell. These zoospores may be walled or naked. In the latter the zoospores develop walls after liberation and swim away as new individuals. Under certain environmental conditions (which may be seasonal), the alga produces gametes. In isogamous species 4-16 gametes may be produced and these may be walled or naked. On liberation these gametes fuse in pairs and produce a zygote which may go through a resting period. On germination the contents of the zygote produce four zoospores after a meiotic division of the nucleus. On liberation the zoosproes become new individuals. The life history is similar in even anisogamous species which border on an oogamous condition.

9.2   Sexual Reproduction and Life History in other Chlorophycota

In the Zygnematales the gametes lack flagella and exhibit only amoeboid movement. Asexual reproduction is apparently unknown. Sometimes when the gametes fail to fuse, they round up and form parthenospores. Otherwise, sexual reproduction is the rule and is initiated by a process of conjugation between two cells, either of the same filament (Fig. 10I) or of two different filaments. The heterothallic condition is more common (Fig. 10H). The conjugating cells are connected by a conjugation tube, developed as small protrusions from each cell and their subsequent fusion (Fig. 10H, I, J). The gametes are formed from the entire contents of the cells by retraction of the protoplasts from the cell walls. In some species a small portion of the protoplast is left unused in the formation of the gamete. In isogamous species the two gametes move towards each other and meet and fuse in the conjugation tube (Fig. 10K). In an isogamous species the gamete from one cell moves across to the other cell through the conjugation tube (Fig. 10H, H, J). After gametic fusion the zygote rounds up, develops a thick wall and enters a resting stage. During germination of the zygote, meiosis takes place and of the four haploid nuclei only one survives. Then it grows out after rupturing the zygote wall. These algae, therefore, have a haplobiontic life history with zygotic meiosis.

The desmids are also unicellular algae with gametic fusion and zygotic meiosis. Each conjugated cell forms one (Fig. 10Ga-Gc) to four gametes. One to four zygotes are formed after conjugation (Fig. 10 Gd).

Zygotic meiosis is widespread among the green algae. In the oogamous *Oedogonium* there are two major types of sexual life history. Accordingly the species are classified into macrandrous (Fig. 10B) and nannandrous species (Fig. 10A). In macrandrous species the oogonia and antheridia may be produced either on the same filament or on different filaments which are morphologically alike (Fig. 10B-D, F). The oogonium arises from a cell derived after a series of vegetative divisions characteristic of the genus. The oogonial cell is therefore often crowned by a number of "cap cells" (fig. 10C). The cell bulges out, becomes spherical or globoid, its contents withdraw from the cell wall, and form the ovum. The antheridia are

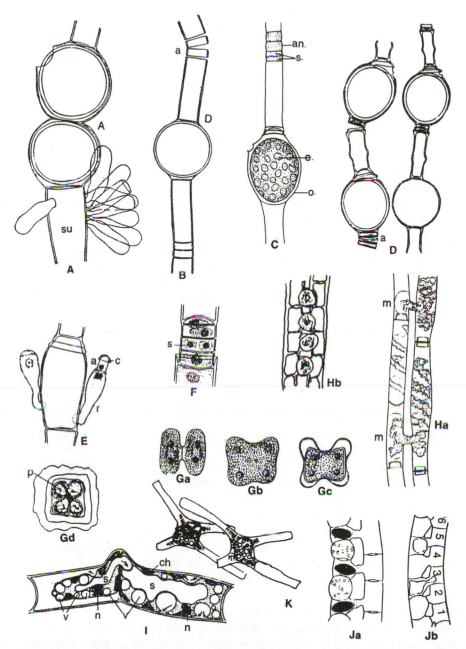

Fig. 10 A-Jb. **Life history in Chlorophyceae. A** *Oedogonium cyathigerum*, filament with two oogonia and a number of unicellular dwarf males on the supporting cell. **B** *O. zigzag* var *robustum*, monoecious thread. **C** *O. fovelatum*, Monoecous filament. **D** *O. nodulosum* var. *commune*, monoecious threads. **E** *O. concatenatum*, oogonium with two dwarf males attached, **F** *O. boschii.*, antheridia with young antherozoids. **G** *Cylinderocystis brebissonii*. **Ga-Gc** Stages in conjugation. **Gd** Young zygote with nuclei in contact, and four pyrenoids. **Ha** *Spirogyra* sp, scalariform conjugation. **I** *Spirogyra longata*, lateral conjugation. **J** *Spirogyra* sp. **Ja** Regular conjugation between opposed cells in a pair of conjugating filaments. **Jb** Formation of zygotes in alternate cells. *a* antheridium, *c* cap cell; *e* egg cell, *f* female cell, *m* male cell, *o* oogonium, *s* spermatozoid, *su* supporting cell, *r* rhizoidal cell (after Fritsch 1935)

formed by division of a vegetative cell, either once or repeatedly, and give rise to one or a series of discoid antheridia (Fig. 10F). This pattern of division is the same as that for androporangia described earlier for this genus. Each antheridium produces two antherozoids, each with a ring of flagella around a hyaline apex. The antherozoids are liberated by a transverse split of the antheridial wall (Fig. 10D) and they swim away. In the meanwhile the oogonial wall shows either a transverse median split or it develops a pore on its distal end (Fig. 10C). Mucilage is secreted around the opening and it is possible that this is chemotactic. An antherozoid enters the oogonium through the mucilage and fuses with the ovum, giving rise to a diploid oospore. The oospore germinates after a resting period and gives rise to four zoospores after a meiotic division.

In nannandrous species of *Oedogonium*, a special spore, the androspore, is produced in an androsporangium. The androsporangium may be produced on the same filament which produces the oogonium, or on a different filament. This spore is attracted to the oogonium and may settle on the oogonial wall (Fig. 10E) or on the subtending cell (Fig. 10A) and produce a short filament called a nannandrium or dwarf male with one or a few antheridia. Antherozoids produced in the nannandrium fertilize the ovum and an oospore results.

Sometimes sexual fusion fails in *Oedogonium* spp. and the ovum becomes converted into a parthenospore, and may repeat the vegetative thallus.

Cylindrocapsopsis is a unique genus in which sexual plants can be separated into one-celled dwarf males and dwarf females, analogous to the nannandria of *Oedogonium*.

The Ulotrichales, as a rule, show a life history with zygotic meiosis.

In contrast, a number of algae exhibit a life history with meiosis in the gametangium, the vegetative thallus becomes diploid and the gametes represent the only haploid stage in the life cycle. This is the condition in some members of the Chlorococcales like *Chlorochytrium* and *Apiococcus*, members of the Dasycladales, some members of the Bryopsidales and the Bacillariophyceae.

In the Dasycladales studied so far, special branches of the cell show division of the contents which become spherical, each surrounded by a thin wall, called the cysts. The cysts are really gametangia for, under certain conditions, the contents of the cyst form biflagellate gametes which escape by a tangential split of the cyst wall. The zygote formed by fusion of these gametes develops into a new diploid thallus. This thallus may be coenocytic, or as in *Acetabularia*, may have a single giant nucleus in the vegetative cell. It may undergo repeated divisions and the daughter nuclei become distributed in the thallus and form cysts.

Among the Dasycladales (Chlorophyceae), only in *Acetabularia* is the life cycle completely known. The thallus is tubular with a short, creeping, basal portion and an erect portion. The single large diploid nucleus remains in the basal portion. Sooner or later, the erect portion gives rise to horizontally-directed more or less club-shaped branches which cohere laterally to form the characteristic 'hat' (Fig. 11Ba). At the time of spore formation, the single diploid nucleus undergoes meiotic division and produces numerous haploid nuclei which then migrate into the 'hat'.

Cytoplasmic cleavage occurs in the hat portion, resulting in numerous uninucleate protoplasts which then become surrounded by distinct membranes and become cysts (Fig. 11Ba). Inside each cyst, the nucleus divides mitotically and these nuclei become incorporated into biflagellate gametes. The gametes are liberated by rupture of the cyst membrane (Fig. 11Bb), usually in a tangential plane. Liberated gametes are isogamous (Fig. 11Bc) and these produce zygotes (Fig. 11Bd) which develop into the diploid thallus.

In genera like *Codium*, some branches of the thallus (called vesicles) produce gametangia as lateral organs which produce numerous biflagellate gametes after meiosis (Fig. 11Aa-Ae). These are anisogamous and their fusion gives rise to a spherical zygote.

Occasionally, an alternation of a diploid gametophyte with gametic meiosis with a diploid sporophyte

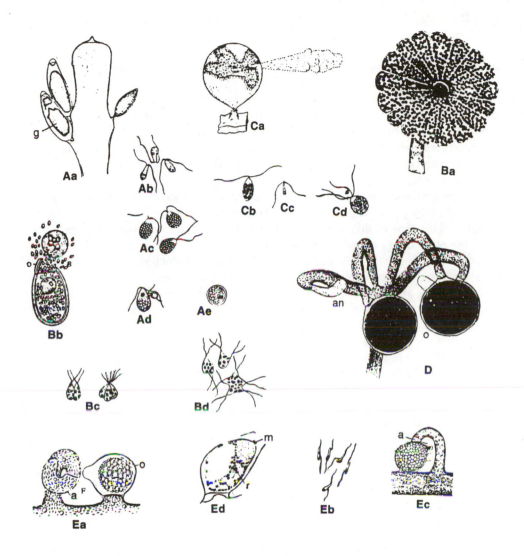

Fig. 11 A-E. Life history in Chlorophyceae and Xanthophyceae. **A** *Codium mucronatum*. **Aa** Vesicle with gametangia. **Ab-Ad** *Codium tomentosum*. **Ab, c** Anisogamous gametes. **Ad** Fusion of gametes. **Ae** zygote. **Ba** *Acetabularia wettsteninii*, view of 'hat' from above, with cysts. **Bb-Bd** *Acetabularia mediterrania*. **Bb** Cyst, liberated swarmers. **Bc** Fusion of swarmers. **Bd** Swimming zygotes. **C** *Halicystic ovalis*. **Ca** Plant, discharged swarmers. **Cb** Macrogametes. **Cc** Microgamete. **Cd** Fusion of gametes. **D** *Dichotomosiphon*, part of the thallus with two oogonia and two antheridia. **Ea, Ec, Ed** *Vaucheria sessilis*. **Ea** Sex organs. **Ec** Ogonium at fertilization. **Ed** Fertilization, **Eb** *Vaucheria synandra*, antherozoids. *an* antheridium, *g* gametangium, *O* oogonium. (After Fritsch 1935)

occurs, as in *Cladophora glomerata*. In this case, the only haploid phase is represented by the gametes, the diploid sporophyte gives rise to diploid zoospores only, by mitotic divisions. This condition, however, is rare.

9.3   Sexual Reproduction and Life History in the Chromophycota

Among the Chromophycota, in the Xanthophyceae, there is no sexual reproduction in the less-specialized members.

In *Botrydium*, isogamous sexual reproduction has been recorded. The best studied genus *Vaucheria* exhibits an oogamous reproduction, antherozoids produced in large members in more or less cylindrical, terminal antheridia and the single egg cell in an oogonium which is in proximity to the antheridium (Fig. 11Ea). Often, the oogonium is beaked. Rupture of the antheridium releases the antherozoids (Fig. 11Eb) while the wall dissolves at the beak of the oogonium (Fig. 11Ec) enabling fertilization (Fig. 11Ed).

In the Dinophyceae, most members exhibit isogamous sexual reproduction, the gametes being much smaller than the vegetative cell but with similar morphology. In some armoured Dinoflagellates, the gametes may be naked.

In the Bacillariophyceae, the cells multiply by repeated division. In the process of division, the two thecae (or valves) of the parent cell are shared by the daughter cells, the two daughter cells always fit into the parent valve. This type of division results in one daughter cell approximately the same size as the parent, while the other daughter cell is smaller than the parent cell. When division occurs in the smaller cell, again one of the progeny will be still smaller. This kind of reduction in cell size continues and may ultimately result in a clonal population of cells of varying dimensions.

When the cells have attained a particular size and under certain environmental conditions, the diatoms go through a process of conjugation. In pennate diatoms two cells come together and become approximated to each other in a common mucilage. In each cell meiotic division takes place and four haploid nuclei are formed. Depending on the genus, either two (Fig. 12Aa) or three of the haploid nuclei degenerate and gametes are formed around the remaining nuclei (Fig. 12Ab).

Thus, either two or one aplanogamete is formed in each cell. The valves of the conjugating cells separate (Fig. 12Ac) allowing the gametes to show amoeboid movement and fusion (Fig. 12Ad-Ah). The fusing garnetes are always derived from the two conjugating cells. Hence, sexual fusion is heterothallic. The zygote then grows vigorously, attaining several times the size of the parent cell and becomes an auxospore (Fig. 12Ag-Ai). The auxospore develops a primary envelope of pectic membrane, the perizonium, around it and then a wall impregnated with silica (Fig. 12Aj). The wall is in the form of two valves and connecting bands. These surround the protoplast and give rise to new cells. The auxospore may at best be described as a zygote formed by allogamic sexual fusion.

In centric diatoms there are two types of auxospore formation. In one there is an oogamous type of sexual reproduction. The contents of one cell forms the egg cell while in another cell, the contents divide and produce a number (usually a multiple of four) of uniflagellate microspores (Fig. 13Ba, Bb). The microspore (really the antherozoid) after swimming for some time, settles on a cell producing an ovum. The valves of this cell separate and the microspore enters and fertilizes the ovum. The fertilized ovum (zygote) then grows in size and becomes an auxospore.

In *Cyclotella meneghiniana*, Iyengar and Subrahmanyan (1944) report an autogamous sexual reproduction. The nucleus of the cell divides meiotically into four (Fig. 13 Aa-Ae). Of these, two degenerate and the remaining two fuse (Fig. 13 Ab-Ad) to form a diploid nucleus. The zygote develops into an auxospore (Fig. 13Ai, Aj).

In the Phaeophyceae, sexual reproduction may be isogamous, anisogamous or oogamous. Isogamy involving two similar motile, heterokont gametes is found in the less specilized members of the Ectocarples, the Cutleriales and the Sphacelariales (Fig. 14A-D). Oogamy occurs in the Dictyotales (Fig. 14 Ea-Ed), Laminariales (Fig. 14 Fa, Fb) and the Fucales. While in the Dictyotales and the Laminariales there is an alternation of sporophytic and gamoetophytic phases in the life cycle, in the Fucales, reproduction is only sexual (Fig. 14Ga, Gb).

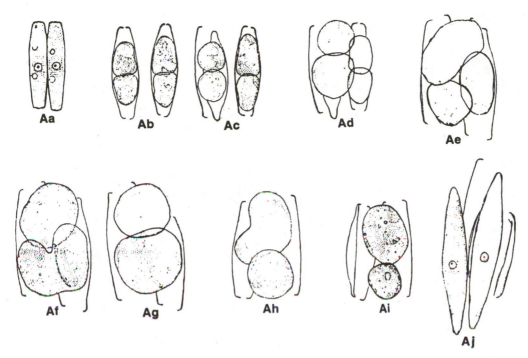

Fig. 12.   Sexual reproduction in diatoms. A *Navicula halophila*. Aa Two cells ready for conjugation. Ab Contents of cells divide to form two gametes each. Ac Separation of parent valves. Ad Gametes prepare to fuse, the sizes of the gametes differ (anisogamy). Ae One pair of gametes fused. Af-Ah Formation of zygotes. Ai Two zygotes. Aj Two auxospores discard the parent valves, new valves form later. (After Subrahmanyan 1947)

Some algae do have an alternation of haploid gametophytes and diploid sporophytes, but the life history is often complicated because of accessory asexual modes of reproduction which may help repeat the same phase (gametophyte or sporophyte) for a considerable time before the second phase develops. Such a mixed life history often depends on habitat as well as seasonal changes in environmental parameters. A typical example is that of *Ectocarpus*. Classical works of Knight (1929) and Schussning and Kothbauer (1934), on the life history of *E. siliculosis,* show the impact of habitat. Papenfuss (1935) indicated a seasonal phenomenon.

The Fucales represent a unique type of life history among the Phaeophyceae. Here the thallus is diploid and shows considerable morphological specialization and produces gametangia, antheridia and oogonia in special cavities, the conceptacles, which may be formed in special parts of the thallus called the receptacles. Meiosis precedes the formation of antherozoids in the antheridium and the ovum in the oogonium. Usually 64 antherozoids are formed in an antheridium. In the oogonium, meiosis is followed by one mitotic division of the nucleus to give rise to eight nuclei. In *Fucus* eight eggs (Fig. 14 Ga-Gc) are formed by accumulation of cytoplasm around each nucleus, but in other genera functional nuclei are reduced and one may get four (*Ascophyllum, Xipophora, Bifurcaria, Hormosira* and *Durvillea*) or two (*Pelvetia*) or even a single egg per oogonium (*Cystoseira, Sargassum, Turbinaria, Hormophysa*). In many cases the unused or supernumerary nuclei persist until fertilization but degenerate during or after fertilization.

An attempt to consider the life history of the Fucales as one of alternation of generations consisting of a dominant sporophytic phase and a highly reduced gametophytic phase as in the Laminariales, has been

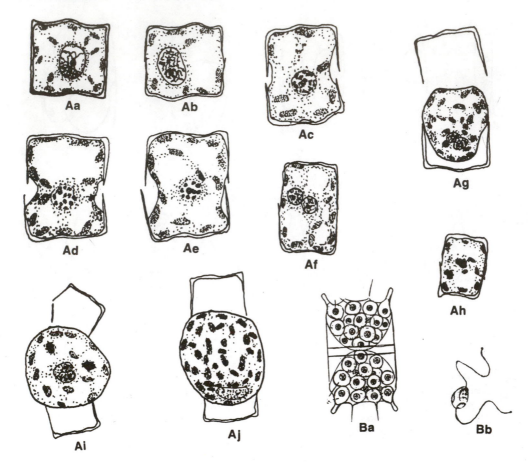

Fig. 13 A-B.   **Sexual reproduction in centric diatoms. A** *Cyclotella meneghiniana.* **Aa-Ae Successive stages in meiotic division of nucleus. Af Two of the four nuclei formed survive. Ag Pre-separation of valves. Ah, Ai Auxospore with two fused nuclei. Aj Auxospore with a single diploid nucleus. B** *Biddulphia mobiliensis.* **Ba Microspore formation. Bb Single biflagellate microspore. (A after Subrahmanyan; 1947, B after Fritsch 1935).**

made by Kylin (1938) and Svedelius (1921, 1927, 1931). The first two divisions of the nucleus in the oogonium were considered to be meiotic division leading to the formation of a tetrad of spores. The subsequent division was considered to be the haploid phase which, however, does not remain vegetative but becomes converted into gametes. This explanation is satisfactory for the antheridium only, and not for the events in the oogonium (Fritsch 1945). Therefore, the Fucales are here considered as haplobiontic diploid individuals with gametic meiosis.

In all the above types, there is only one vegetative phase. In many algae, however, two alternating vegetative phases occur, one haploid and leading to gamete formation, and the other diploid, leading to the formation of spores after meiosis, thus exhibiting an alternation of generations. This has been very clearly demonstrated in *Enteromorpha* (Ramanathan 1939). Such an alternation of a haploid gametophyte with a diploid sporophyte involves sporic meiosis. Such alternation is also reported in Dictyotales.

9.4   Sexual Reproduction and Life History in the Rhodophycota

Among the Rhodophycota, many genera of the Bangiophyceae exhibit only an asexual life cycle. Some,

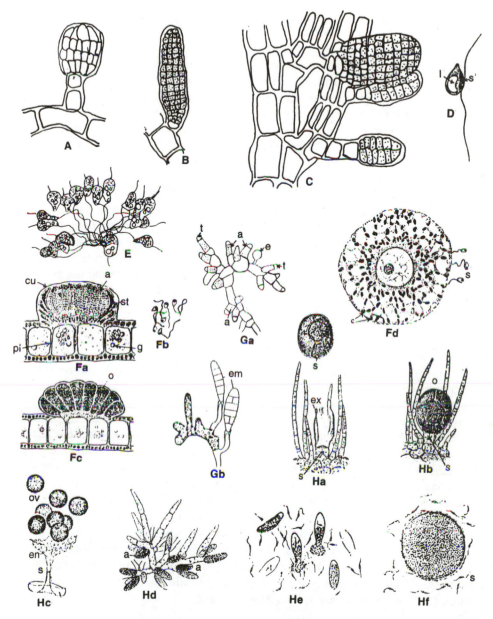

Fig. 14 A-G. Sexual reproduction in Phaeophyceae. A *Ectocarpus breviarticulatus*, plurilocular gametangia. **B** *Giffordia mitchelliae*. **C** *Sphacelaria furcigera*, plurilocular gametangia. **D** *Fucus spiralis*, antherozoid. **E** *Ectocarpus siliculosis*, Clumping of microgametes with macrogamete before fertilization. **F.** *Dictyota dichotoma,* **Fa** Cross section shows oogonia. **Fb** Antherozoids. **Fc** Antheridia. **Fd** Fertlization of an oovum. **G.** *Laminaria digitata.* **Ga** Antheridia. **Gb** Oogonial filament, two fertilized eggs develop into embryos. **H.** *Fucus vesiculosis.* **Ha** Packet of eggs, liberated by rupture of oogonial wall. **Hb** Oogonium and the associated paraphysis. **Hc** Liberation of eggs. **Hd** Antheridial branch with several mature antheridia. **He** Liberation of antheridia, some ruptured to release antheridia. **Hf** Process of fertilization; see egg surrounded by antherozoids. **a** antheridium, **cu** cuticle, **e, em** embryo, **en** endochite, **ex** exochite, **s** stalk cell, **st** sterlie cells, **o** oogonium, **ov** ova. (A–C original, D–G after Fritsch, 1935).

like *Porphyropsis* (Murray et al. 1972), exhibit more than one somatic phase, which occur in succession but without syngamy or reduction. Very few members of the Bangiophyceae have a sexual life cycle and the best studied one is that of *Porphyra*. In this genus there are two morphologically dissimilar vegetative phases, the leafy phase and the conchocelis phase which is a filamentous phase usually penetrating and growing within shells of various molluscs. The leafy phase is the gametophytic phase, while the conchocelis phase is considered as the sporophytic phase. On the leafy phase, an unmodified cell of the thallus functions as the female reproductive cell designated as carpogonium (a term used in the case of Florideophyceae). Many species are homothallic while a few are heterothallic. The spermatangium develops from another unmodified cell of the thallus. The contents of the spermatangium divides to give rise to 4–128–256 spermatia. The liberation of spermatia occurs by the gelatinization of the walls of the spermatangia. A liberated spermatium comes to rest on a carpongonial cell and develops a fertilisation tube. Although actual nuclear fusion has not been observed, it is assumed and the diploid zygote undergoes division to produce 2–32 carpospores. On liberation, each carpospore develops into the conchocelis phase. This phase gives rise to the conchospore formation. Recent studies on hybridization in *P. yezonesis* by Ohmi et al. (1986) using a red colour mutant of the species and a normal plant indicate that the leafy phase developed by the germination of conchospores from the hybrid showed segregation of red and green cells from a linear tetrad developed from the conchospore. This is an indication that meiosis occurs during the germination of the conchospores. Thus, the 'conchocelis phase' is merely an extension of the carpospores which represent an amplification of the zygote. This is a unique pattern of life history among the red algae.

In the Rhodophyceae there is considerable uniformity in the mode of sexual reproduction. The sex organs are the spermatangia and the carpogonia. The spermatangium is a small unicellular organ formed in packets inside an unmodified cell (Fig. 15Gd) of the thallus (*Porphyra*) or superficially on the surface of the thallus forming, along with similar spermatangia, a soms, or may be borne on cells of a branch system (Fig. 15A, B) known as a spermatangial branch or cluster (*Polysiphonia*). The capogonium is a single cell formed from an unmodified cell (Fig. 15Gb) of the thallus (*Porphyra*) or forming an end cell of a branch (*Acrochaetium*) which may be specialized (Fig. 15Gc) and known as the carpogonial branch (*Bactrachospermum, Polysiphonia*). The carpogonial branch may be one to many-called. The carpogonium (except in the Bangiales) is a flask-shaped cell with a basal dilated portion containing the gametic nucleus and a terminal tubular portion, the trichogyne.

The spermatangia, on liberation, come into contact with the trichogyne of the carpogonium (Fig. 15Da, Db) and at the point of contact, the wall of the spermatangium dissolves, the spermatium flows into the carpogonium. The process of fertilization is not well documented. Wille (1894) first observed the passage of the spermatial nucleus to the basal part of the carpogonium, the apposition of the nuclei of the spermatium and the carpogonium and their subsequent fusion in *Nemalion*. Such direct observations have not been made by other workers for the red algae. Quite frequently, the spermatial nucleus consists of a dense aggregate of chromosomes at the time of fusion while the carpogonial nucleus remains in a resting condition (*Polysiphonia*, Yamanouchi 1906). However, in many other algae both nuclei are in a resting condition.

The fertilized carpogonium may then show a division of its contents to give rise to a packet of carpospores as in *Porphyra* (Berthold 1881; Thuret and Bornet 1900, Krishnamurthy 1959). In the Florideophycideae, however, the fertilized carpogonium gives rise to a number of filaments which may branch variously, the gonimoblasts. Carpospores are borne on these. The gonimoblasts may be formed indirectly also from other cells of the thallus, known as auxillary cells. In such forms, the diploid nucleus or its division product from the fertilized carpogonium is first transferred to the auxillary cell by means of connecting cell (*Polysiphonia*) or filament (*Grateloupia*) and gonimoblasts are produced from this auxillary cell. The cells of the gonimoblasts carry diploid nuclei derived from the zygotic nucleus.

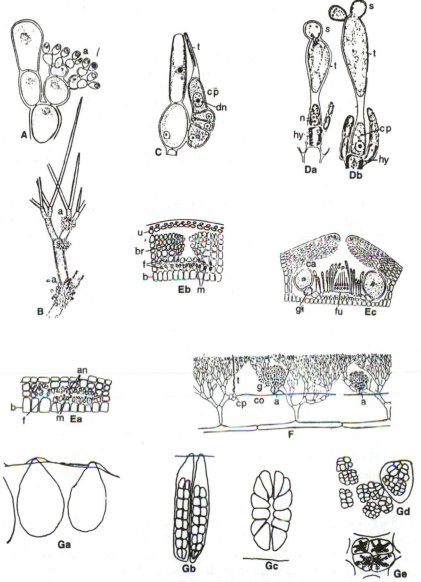

Fig. 15 A-G. Sexual reporduction in Rhodophyceae. A *Helminthocladia calvadosii*, spermatangial branch with spermatangia. B *Callithamnion corymbosum*, spermatangial clusters borne on the apical ends of cells. C *Helminthora divaricata*, carpogonial branch with a diploid nucleus in carpogonium. D *Batrachospermum moniliforme*, Da, Db Post-fertilization stages. E *Epilithon membranaceum*. Ea Eccentric vertical section of male conceptacle to show spermatangia borne on spermatangial mother cells. Eb Median section. Ec Vertical section of female conceptacle shows carpogonial branches, fusion cell and mature carposporangia. F *Platoma bairdii*, post-fertilization development shows fertilized carpogonium, connecting filament and successive auxillary cells and gonimoblasts. G *Porphyra* sp. Ga Mature carprogonia. Gb-Ge stages of durscon in carprogonium. *a* antheridia, *b* basal cells, *cp* carpogonium, *dn* diploid nucleus. *g* gonimoblast, *gi* gonimoblast initial, *Ca* carposporangium, *Co* connecting filament, *fu* fusion of cell, *hy* hypogynous cell, *t* trichogyne (after Fritsch 1935).

The position of the auxilary cell may vary. In many red algae, the auxiliary cell is either a hypogynous cell of the carpogonial branch or the cell from a sterile filament borne by the bearing cell in addition to the carpogonial branch or a cell on an accessory branch which may be spatially separated from the branch system. When the auxiliary cell is produced from the branch system arising from the supporting cell, the whole branch system is termed a procarp and the alga is described as procarpic. Other types are described as non-procarpic. Thus, *Polysiphonia* is procarpic while *Grateloupia* is non-procarpic. The gonimoblasts may be associated with sterile filaments, nutritive cells and filaments, and may be surrounded by a sterile multilayered cellular sheath, the whole structure being described as a cystocarp.

The life history in *Rhodochorton* consists of a gametophytic phase which may be homothallic or heterothallic, and the fertilized carpogonium (zygote) gives rise to two elongate cells which have been described as gonimoblasts. The latter produce filaments bearing tetrasporangia (West 1969). Meiotic division in the tetrasporangium gives rise to four haploid tetraspores. The life history shows a peculiar feature of the extension of the zygote into a tetraspore-producing diploid phase. According to Desikachary et al. (1990), this life history is closely similar to that of the Palmariaceae. On this basis, they place Rhodochortonaceae with the single genus *Rhodohchorton* in the order Palmariales.

The typical life cycle in the Rhodophyceae is diplobiontic, a haploid gametophyte alternates with a diploid tetrasporophyte as exemplified by *Polysiphonia*. In this genus the gametophyte is heterothallic, the carpogonial branch develops on one thallus and the spermatangial branch on another. The fertilized carpogonium fuses with an auxillary cell produced after fertilization, and then with other cells of the carpogonial branch system to produce a large fusion cell. This gives rise to cellular outgrowths called gonimoblasts and they produce carposporangia bearing single carpospore. The post-fertilization development culminates in carpospore production, and is of the nature of an amplification of diploid reproductive cells which are knwon as carpospores.

According to traditional interpretation, the product of development of the fertilized carpogonium is a distinct phase in the life cycle which is considered triphasic. However, recent interpetations. (Hawkes and Scagel 1986; Krishnamurthy 1988, 1996) suggest that the carposporophyte is the result of zygote amplification and need not be considered as a distinct phase in the life cycle. This is especially so, because the carposporophyte is not a free-living phase of the alga.

The carpospore develops into the tetrasporophyte, which in most Rhodophyceae, as in *Polysiphonia*, is isomorphic with the gametophyte.

In *Polysiphonia* and all procarpic members of all taxa of the Rhodophyceae, the zygote amplification is limited and takes place on the carpogonial branch system itself. However, in non-procarpic members of the Gigartinales and Cryptonemiales, auxillary cells are produced on special branches and the fertilized carpogonium establishes contact with the auxillary cell which is produced on a special branch, by means of a connecting filament which transmits to it, a division product of the diploid nucleus.

In many taxa more than one auxillary cell is produced, and either several connecting filaments issue from the zygote, or the same connecting filament, after contacting one auxillary cell, grows further and contacts another auxillary cell and this process continues. Thus, a single fertilization leads to the formation of carpospore at a number of sites. *Platoma bairdii* (Fig. 15f) figured by Kuckuck (1912) is a classic example.

In some orders of the Rhodophyceae such as Batrachospermales, Bonnemaissoniales, Nemalionales and Chaetangiales, the alteration of generations is hetermorphic, the gametophyte and the sportophyte differ markely in morphology. Heteromorphic alternation was first observed in *Asparagopsis* and *Bonnemaissonia* by J. Feldmann and G. Feldmann (1942). In both, the gametophyte is a well-developed phase, and zygote amplification is on the carpogonial branch system itself. Carpospores on germination develop into a filamentous phase which was for a long time described as a distinct genus, *Falkenbergia*. This phase produces the tetraspores. Similarly, the tetrasporic phase of *Bonnemaissonia* was known under the name *Trailliella*.

Recent studies indicate that some other members of the Nemalionales and Chaetangiales may also show such a heteromorphic life cycle.

The life history of *Palmaria* was elucidated only recently. In *P. palmata* only spermatangial and tetrasporic plants were known, and a carpogonial plant was unknown. Recently, it was observed that the carpogonia are formed on the crustose base of the thallus. After fertilization, the zygote develops directly into a foliose growth which ultimately produces tetrasporangia. Zygote amplification has not been observed, and there is direct development of the tetrasporophyte from the zygote (Van der Meer and Todd 1980).

An extremely reduced type of life history is known in *Rhodophysema* (which also belongs to the Palmariaceae). The thallus is crustose and discoid and from its cells arise erect filaments. Some of the erect filaments terminate in spermatangial mother cells which produce spermatangia and other cells differentiate into carpogonia with hair-like trichogyne.

After fertilization the zygote divides transversely to give rise to a stalk cell and a tetrasporangial cell. The stalk cell is the only representative of a diploid thallus, and the tetrasporangium gives rise to the crustose thallus after meiosis (DeCew and West 1982). This genus has a highly condensed life cycle and is haplobiontic.

Now, the question is whether the Palmariales represent a primitive, or a most advanced life cycle among the Rhodophyceae? The trend within the Palmariales suggests that this order represents a reduction series. However, the absence of a carposporophyte or any zygote amplification brings this order nearer the Bangiophyceae. Therefore, one is led to conclude that the Palmariales are perhaps a connecting link between the Bangiophyceae and the Florideophyceae (Desikachary et al. 1990).

In the Corallinales the situation is unique. The procarps, the antheridial branches, and the tetrasporic branches develop inside special cavities, the conceptacles. The conceptacle may be terminal, as in articulated genera like *Jania*, or superficial as in genera like *Amphiroa*, and in the crustose coralline algae. The base of the conceptacle is lined with procarpic or antheridial branches (Fig. 15Ea-Ec). These are overarched by a conceptacular wall which may be several layers with a terminal ostiole. The procarp consists of a supporting cell with one or two two-called carpogonial branches, and sometimes with one or two sterile cells. After fertilization the fertilized carpogonium establishes connection with its supporting cell by means of a connecting filament. Then there is a general fusion between the supporting cells of all the carpogonial branches resulting in a large irregular fusion cell. Even a single fertilization is sufficient to produce the fusion cell. The carpospores are produced in succession from the margin of the fusion cell. This type of zygote amplification is unique, and is not reported in any other order of the Rhodophyceae.

10.   Present State of Knowledge and Future Research

The foregoing brief account of the reproductive mechanisms and life cycle patterns in the algae indicate that there is considerable diversity in the (1) modes of reproduction, (2) morphology of the reproductive units, (3) function of these units, (4) formation of zygotes, (5) germination, and in the (6) evolution of the life cycle. However, the algae are still in a primitive state since no recognizable embryonic state is known in the germination of the zygote unless one considers the early stages of development of the thallus in the Fucales as embryonal stages.

Future research on reproductive mechanisms and life history of eukaryotic algae will revolve around (1) identification of new patterns of reproductive biology, (2) further investigations into the nature of sexuality and the role of pheromones, (3) the role of nutritive cells and tissues (in the Rhodophycota) in the nurture of reproductive cells such as spores, and (4) ultrastructural and biochemical changes consequent on fertilization.

Genetic studies on eukaryotic algae are now confined to a very few genera and that needs extension to other genera, especially those which are considered economically important. Genetic engineering, using the tools of modern biotechnology should also be extended to these algae.

11.   Glossary

Adaxial. Facing the axis

Agglutination. Coming together in a mass (of flagella during fusion of gametes)

Akinetes. Thick-walled resting spores

Androsporangium. A discoid cell producing androspores (in Oedogoniales)

Androspores. Spores produced in an androsporangium, germinating to produce dwarf male plants (Oedogoniales)

Antherozoids. Male reproductive cells with flagella

Aplanospores. Nonflagellated, asexual reproductive cells

Autocolony. Formation of a new colony from a cell of a parent colony by divisions of the cell and the rearrangement of the daughter cells

Autogamy. A form of sexual reproduction in which two nuclei derived by meiosis of the nucleus of a cell, function as gametic nuclei and fuse to give rise to a zygote (Bacillariophyceae)

Autospores. Spores formed from each cell of a colony which rearrange themselves to form a new colony (Chlorococcales, Chlorophyceae)

Auxiliary cell. An accessory cell from which the carposporophyte develops, formed either as part of the carpogonial branch system or on an accessory branch (Rhodophycota)

Auxospore. Zygote of a diatom which enlarges to several times the size of the parental cells

Axioneme. The central microtubular strand of a flagellum

Carpogonium. Female reproductive organ of the Rhodophycota

Carpospores. Spores produced by a carposporophyte (Rhodophycota)

Carposporophyte. A filamentous stage formed by amplification of a fertlized carpogonium (Rhodophycota)

Centric diatoms. Diatoms with valves showing radial symmetry

Chemokinesis. Activity induced by a chemical stimulant

Chemotaxis. Movement towards a chemical stimulant

Conceptacle. A cavity in which sex organs are produced (Phaeophyceae) and sometimes, even sporangia (Rhodophycota, Corallinales)

Conchospores. Spores produced, usually in series, by the Conchocelis phase of some Bangiophyceae (Rhodophycota)

Conjugation. Union of gametes or sometimes of gamete producing cells of a filament or of unicellular algae

Coronula. Crown cells surmounting the oogonium in the Charophyceae

Cysts. Reproductive units, formed by either rounding off of cell contents with formation of wall or by division of contents. In the Dasycladales (Chlorophyceae), the cysts produce gametes

Cystocarp. A compound structure consisting of a multilayered wall enclosing the carposporophyte (Rhodophycota)

Diplobiontic. A life history consisting of an alternation of two vegetative phases, one constituting the gametophyte and the other a sporophyte

Diploid. A cell or cells with a double complement of chromosomes in the nucleus

Diploidy. The diploid condition

Eukaryotae. Organisms in which the cells include membrane-bound organelles, particularly, the nucleus

Fragmentation. Breaking up into pieces (in filamentous and parenchymatous algae) leading to vegetative reproduction

Gametangium. A special cell in which gametes are produced

Gamete. Sexual reproductive cell

Gametophyte. A plant body producing gametes

Gemma. A vegetative organ of reproduction, usually a group of cells of a definite shape

Gonimoblast. Unicellular to multicellular structure produced from either a fertilized carpogonium or an auxillary cell or a fusion cell, often forming a branched filament and forming the carposporophyte (Rhodophycota)

Haplobiontic. A life history consisting of only one vegetative phase

Haploid. Cell or cells with a single complement of chromosomes in the nucleus

Haploidy. The haploid condition

Heterogamy. Fusion of dissimilar gametes

Heterokont. Possessed of flagella of different types

Heteromorphic. A life history involving two alternating vegetative phases which are dissimilar

Heterothallism. Production of compatible gametes on different thalli

Homothallism. Production of compatible gametes from the same thallus

Hypnospores. Thick-walled asexual resting spores

Hypogynous. Below the female reproductive organ

Isogamy. Fusion of similar gametes

Isomorphic. A life history of two similar alternating vegetative phases

Isthmus. Narrow median connecting region of two semicells (Desmidioideae, Zygnematales, Chlorophyceae)

Karyogamy. Fusion of nuclei

Macrandrous. With antheridia in large filaments (Oedogoniales, Chlorophyceae)

Manubrium. A cell attached to the shield cell of the antheridium in Charophyceae, which bears the head cells and ultimately the spermatogenous filaments.

Mastigonemes. Fibrillar appendages on heterokont flagella (in brown algae)

Microspore. Flagellated spores of small size in some diatoms considered to be male gametes.

Microtubules. Tubular units seen in ultrastructue of cells, involved in the structure of the flagellar apparatus, nuclear spindle and in the phycoplasts of dividing algal cells

Monospore. Asexual spore formed singly inside a cell

Nannandrium. A dwarf male plant (Oedogoniales, Chlorophyceae)

Nannandrous. A life history involving formation of a dwarf male

Oogamy. Fusion of a flagellated spermatozoid (or antherozoid or male gamete) with a much larger, non-flagellated ovum

Palmella stage. A developmental stage in the life cycle of various unicellular algae in which division products of the cell lie in an extensive mucilagenous envelope

Paraspores. Irregular masses of spores found in some Rhodophytes, considered as clusters of monospores

Parthenospores. Unfertilized aplanogametes functioning like zygotes

Pennate diatoms. Diatoms with valves of bilateral symmetry

Pheromone. A substance functioning as a sex attractant

Planogametes. Flagellated gametes

Plasmogamy. Fusion of cytoplasm of gametes

Polyhedron. Polyhedral cells formed by a zoospore and serving as a resting stage (Hydrodictyaceae, Chlorococcales, Chlorophyceae)

Polysporangium. A sporangium with several spores, in multiples of four, considered as an extended development of the tetrasporangium

Proboscis. A swelling at the base of the anterior flagellum in members of the Fucales (Phaeophyceae)

Procarp. A branch system including the carpogonial branch, sterile branches and an auxiliary cell or auxiliary mother cell, all borne on a single supporting cell

Prokaryotae. Organisms with cells not including membrane-bound organelles, especially a nucleus, but with genetic material found diffuse in the central portion of the protoplast

Propagule. A cellular structure of various shape, serving for vegetative reproduction

Protonema. A filamentous growth obtained by germination of a spore or an oospore from which the adult thallus is developed

Pseudocilia. Flagella-like structures fournd in cells of colonial algae (Tetrasporales, Chlorophyceae) but not serving for movement

Recombination. Bringing together of genes from different parents

Receptacle. Special part of the thallus bearing conceptacles containing sex organs (Phaeophyceae, Fucales)

Replication. Process of ring-like infolding of end walls of adjacent cells in a filament, leading ultimately to the fragmentation of the filament

Spermatium. Male reproductive cell of the red algae

Spermatogenous filament. Ultimate filament of discoid cells, each producing two spermatozoids developed inside the antheridium (Charophyceae)

Spermatozoid. Flagellated male gamete

Spores. Asexual reproductive units formed within sporangia, generally without flagella

Sporophyte. Vegetative thallus, ultimately producing spores

Statospores. Resting spores of diatoms, usually thick-walled with spinous outgrowths

Stephanokontan. Multiflagellated condition of reproductive unit, usually a ring of small flagella around the anterior end

Synzoospore. A large compound zoospore with small flagella all round, disposed in pairs (*Vaucheria*)

Tetrasporophyte. Thallus bearing tetrasporangia

Theca. A hard shell surrounding an organism. Also a name applied to the valve of a diatom cell

Trichogyne. The tubular apical portion of a carpogonium serving as a receptor organ for the spermatium (Rhodophycota)

Triphasic. A life history involving three vegetative phases occurring in succession (Rhodophycota)

Zoosporangium. A cell, sometimes specialized, giving rise to zoospores

Zoospores. Flagellated asexual spores

Zygote. Fusion product of gametes

References

Bal BP, Kundu BC, Sundaralingam VS, Venkataraman GS (1962) Charophyta. Indian Council Agric Res, New Delhi. (ICAR), pp i – x + 1–123.

Belajeff, W (1894) Ueber Bau und Entwicklung der spermatozoiden der Pflanzen. Flora 79: 1–48

Berthold G (1881) Die geschlechtliche Fortpflanzung der eigentlichen Phaeosporen. Mittheil Zool Stat Zu Neapel 2: 401–413

Brown RM, Johnson C, Bold HC (1968) Electron and phase-contrast microscopy of reproduction in *Chlamydomonas moewusii*. J Phycol 4: 100–120

Canter-Lund H, Lund JWG (1995) Fresh water algae—their microscopic world explored. Biopress, Bristol, pp xv + 360.

De Bary A (1871) Ueber den Befruchtungsvorgang bei den Charen. Monatsbar. Akad Wiss Berlin 187: 227–329

DeCew TC, West JA (1982) Sexual life history in *Rhodophysema* (Rhodophyceae): a reinterpretation. Phycologia 21: 67–74

Desikachary TV, Krishnamurthy V, Balakrishnan MS (1990) Rhodophyta. Madras Science Foundation, Madras, Part 1: pp 1–277, Part IIA, pp 1–279

Doraiswamy S (1940) On the morphology and cytology of *Eudorina indica* Iyengar. J Indian Bot Soc 19: 113–139

Feldmann J, Feldmann G (1942) Recherches sur les Bonnemaisoniacees et leur alternance de generation. Ann Sci Nat Bot Ser 113: 115–128

Friedmann I, Colwin AL, Colwin LH (1968) Fine structural aspects of fertilization in *Chlamydomonas reinhardii*. J Cell Sci 3: 115–128

Fritsch FE (1929) The genus *Sphaeroplea*. Ann Bot (Lond) 43: 1–26

Fritsch FE (1935) Structure and reproduction of the Algae. Cambridge Univ Press, Cambridge (UK), Vol 1: i-xvii + 1–791

Fritsch FE (1945) Structure and reproduction of the algae Vol 2. Cambridge Universty Press, Cambridge, pp 939

Gerloff J (1940) Beitrage zur Kenntnis der Variabilitat und Systematik der Gattung *Chlamydomonas*. Arch Protistenkd 94: 311–502

Groves J (1931) On the antheridium of *Chara zeylanica* Willd. J Bot London, 68: 97–98

Groves J, Bullock-Webster R (1920) The British Charophyta. 1.

Hawkes MW, Scagel RF (1986) The marine algae of British Columbia and northern Washington. Division Rhodophyta (red algae), class Rhodophyceae, order Palmariales. Can J Bot 64: 1148–1173

Hobbs MJ (1971) The fine structure of *Eudornia illinoisensis* (Kofoid) Pascher. Br Phycol J 6: 81–103

Hoffman LR (1961) Studies on the morphology, cytology and reproduction of *Oedogonium* and *Oedocladium*. Diss Abstr Univ Texas 22: 2956–2957

Iyengar MOP, Ramanathan K (1951) On the structure and reproduction of *Pleodorina spherica* Iyengar. Phytomorphology 1: 215–224

Iyengar MOP, Subrahmanyan R (1944) On reduction division and auxospore formation in *Cyclotella meneghiniana* Kuetz. J Indian Bot Soc 23: 125–152

Kniep H (1928) Die Sexualitat der niederen Pflanzen. G. Fischer, Jena, 544 pp

Knight M (1929) Studies in the Ectocarpaceae 2. The life history and cytology of *Ectocarpus siliculosis* Dilw. Trans R Soc Edinb 56: 307–332

Krishnamurthy V (1959) Cytological investigation on *Porphyra umbilicalis* (L) Kutz. var. *laciniata* (lightft) JG *Ag*. Ann Bot (Lond) 23: 147–176

Krishnamurthy V (1988) Reproduction strategy in the Rhodophyta. Seaweed Res Utiln Madras 11: 55–65

Krishnamurthy V (1996) A phylogenetic consideration of the reproduction stategies in the Rhodophyta. Nova Hedwigia 112: 189–197

Kuckuck P (1891) Beitrage zur Kenntnis der *Ectocarpus* Arten der Kieler Forde. Bot Zbl 48: 1–7, 33–41, 65–71, 97–104, 125–141

Kuckuck P (1899) Ueber den Generationswechsel von *Cutleria multifida* (Engl Bot) Grev. Wiss Meeresunters Helgoland NF 3: 95–116

Kuckuck P (1912) Ueber *Platoma bairdii* (Fari). Kuck. Wiss Meeresunters Helgoland NF 5: 187–208

Kylin H (1916) Ueber der Bau der Spermatozoiden der Fucaceen. Akad Afhandl Upps 34: 194–201

Kylin H (1918) Studien über die Entwicklungsgeschichte der Phaeophyceen. Sven Bot Tidsskr 12: 1–64

Kylin H (1938) Beziehungen zwischen generationswechsel und Phylogenie. Arch Protistenk 90: 432–447

Kylin H (1956) Die Gattungen der Rhodophyceae. G.W.K. Gleerup, (Lund) 673 pp

Lewin RA (1950) Genetics of *Chlamydomonas moewusii* Gerloff. Proc Intl Bot Congr Stockholm pp 851–860

Lewin RA (1952) Studies on the flagella of algae 1. General observations on *Chlamydomonas*. Biol Bull 103: 74–79

Lewin RA (1954) Mutants of *Chlamydomonas moewusii* with impaired motility. J Gen Microbiol 11: 858–863

Lewin RA (1957) The zygotes of *Chlamydomonas moewusii*. Can J Bot 35: 795–804

Lewin RA, Meinhart JP (1953) Studies on the flagella of algae 3. Electron micrographs of *Chlamydomonas moewusii*. Can J Bot 31: 711–717

Loiseaux S, West JA (1970) Brown algal mastigonemes: Comparative ultrastructure. Trans Amer Microsc Soc 89: 524–532

Luning K, Muller DG (1978) Chemical interaction in sexual reproduction of several Laminariales (Phaeophyceae): Release and attraction of spermatozoids. Pflanzenphysiol 84: 333–334

Maier I, Müller DG (1981) Sexual pheromones in algae. Biol Bull 170: 145–175

Mitra AK (1951) Certain new members of the Volvocales from Indian soils. Phytomorphology 1: 58–64

Moestrup O (1982) Flagella structure in algae; a review, with new observations particularly on the Chrysophyceae, Phaeophyceae (Fucophyceae), Euglenophyceae and *Reckertia*. Phycologia 21: 427–528

Moewus F (1933) Untersuchungen über die Variabilitat von Chlamydomonaden. Arch Protistenkd 80: 128–171

Müller DG (1972) Befruchtungsstoff bei Braunalgen. Ber Dtsch Bot Ges 87: 363–369

Müller DG, Jaenick L, Donike M, Akintobi T (1971) Sex attractant in a brown alga: Chemical Structure. Science, 171: 815–817

Müller DG, Gassmann G (1978) Identification of the sex attractant in the marine brown alga *Fucus vesiculosis*. Naturwisenschaften 65: 389

Müller DG, Seferiadis K (1977) Specificity of sexual chemotaxes in *Fucus serratus* and *Fucus vesiculosis* (Phaeophyceae). Z Pflanzenphysiol 84: 85–94

Murray SN, Dixon PS, Scott JL (1972) The life history of *Porphyropsis coccinea* var *dawsoni* in culture. Br Phycol J 7: 323–333

Ohmi M, Kunifuji Y, Miura A (1986) Cross experiments of the color mutants in *Porphyra yezoensis* Ueda. Jpn J Phycol 34: 101–106

Oltmanns F (1898) Die Entwicklung der Sexualorgane bei *Coleochate pulvinata*. Flora 85: 1–14

Oltmanns F (1899) Ueber die Sexualitat der Ectocarpeen. Flora 86: 86–99

Papenfuss GF (1935) Alternation of generations in *Ectocarpus siliculosis*. Bot Gaz 96: 421–446

Ramanathan KR (1939a) The morphology, cytology and alternation of generations in *Entermorpha compressa* (L) Grev var *lingulata* (J.Ag.) Hauck. Ann Bot (Lond) 3: 375–398

Ramanathan KR (1939b) On the mechanism of spore liberation in *Pithophora polymorpha* Witt. J Indian Bot Soc 18: 25–29

Ramarao K (1992) Observations on the morphology of two species of *Hypnea* in Indian waters. Seaweed Res Utiln 15: 5–10

Rosenvinge LK (1924) The marine algae of Denmark. Pt 3 Rhodophyceae (Ceramiales). Det Kgl Danske Vidensk Selsk Biol Skrift 7: 287–486

Schussnig B, Kothbauer E (1934) Der Phasenwechsel von *Ectocarpus siliculosis*. Oesterr Bot Z 83: 81–97

Searles RB (1980) The strategy of the red algal life history. Ann Nat 115: 113–120

South GR, Whittick A (1967) Introduction to phycology. Blackwell Scientific London, 278 pp

Starr, RC, O'Neil RM, Miller III CE (1980) L-glutamic acid as a mediator of morphogenesis in *Volvoxx capensis*. Proc Natl Acad Sci 77: 1025–1028

Subrahmanyan R (1947) On somatic division, reduction division, auxoxpore formation and sex differentiation on *Navicula halophila* (Grun.) Cleve. J Indian Bot Soc (MOP Iyengar Comm Vol) pp 239–266

Sundaralingam, VS (1954) The developmental morphology of *Chara zeylanica* Willd. J Indian Bot Soc 33: 227–296

Sundaralingam VS (1946) The cytology and spermatogenesis in Charazeylanica Willd. J Indian Bot Soc (MOP Iyengar Comm Vol), pp 289–303

Svedelius N (1921) Einige Bemerkungen über generationswechsel und Reduktionsteilung. Ber Dtsch Bot Ges 39: 178–187

Svedellius N (1927) Alternation of generations in relation to reduction division, Bot Gaz 83: 362–384

Svedellius N (1931) Nuclear phases and alternation in the Rhodophyceae. Beih Bot Zbl 148: 38–59

Thuret G (1854) Recherches sur la fécondation des Fucacées. Ann Sci Nat Bot ser. IV, 2: 197–214

Thuret G, Bornet E (1878) Etude phycologique. G Masson, Paris

Tsubo Y (1956) Observations on sexual reproduction in a *Chlamydomonas*. Bot Mag Tokyo 69: 1–6

Tsubo Y (1961) Chemotaxis and sexual behaviour in *Chlamydomonas*. J Protozool 8: 114–121

Tsubo Y (1957) On the mating reaction of a *Chlamydomonas* with special reference to clumping and chemotaxis. Bot Mag Tokyo 70: 327–334

Tsubo Y (1961) Chemotaxis and sexual behaviour in *Chlamydomonas*. J. Protozool 8: 114–121

Van Den Ende H (1976) Sexual interactions in plants. Academic Press, London

Van der Meer JP, Todd ER (1980) The life history of *Palmaria* in culture. A new type for the Rhodophyta. Can J Bot 58: 1250–1256

West JA (1969) The life histories of *Rhodochorton purpureum* and *R. tenue* in culture. J Phycol 5: 12–21

Wille N (1894) Über die Befruchtung bei *Nemalion multifidum* (Web. et Mohr). J Ag Ber Deutsch Bot Ges 12: 57–60

Wiese L, Shoemaker DW (1970) On sexual agglutination and mating type substances (gamones) in isogamous, heterothallic *Chlamydomonas*. 2. The effect of conconvalin A upon the mating type reaction. Biol Bull 138: 88–95

Yamanonchi S (1906) The life history of *Polysiphonia violacea*. Bot Gaz 42: 401–409

6

Reproductive Biology of Fungi

A.K. Koul and G. Sumbali

1. Introduction

Reproduction is the formation of new individuals with all the characteristics typical of the species. Most fungi reproduce asexually and sexually. Asexual reproduction is more important for propagation of the species as it is repeated several times during the life cycle and results in production of numerous exact copies of the parent. Sexual reproduction provides for meiotic recombination. Generally, it takes place only once in a life cycle. Some fungi do not go through a true sexual cycle, instead many of these practice parasexuality which provides a significant source of genetic variation. Nevertheless, sexual and parasexual cycles are not mutually exclusive. Fungi that exhibit sexuality may also accumulate additional and significant variation through parasexual reproduction.

2. Types of Life Cycles in Fungi

Raper (1966a) recognized seven types of life cycles in fungi (Fig. 1 A-G) :

1. Asexual Cycle. It is based exclusively on asexual reproduction and occurs in Deuteromycetes (Fungi Imperfecti).

2. Haploid Cycle. In this cycle, nuclear fusion is immediately followed by meiosis and the life cycle is haploid with a very brief diploid phase, as in the lower fungi and some primitive Ascomycetes.

3. Haploid Cycle with Restricted Dikaryon. This type of life cycle is characteristic of higher Ascomycetes like *Neurospora;* fusion of sexual cells results in the formation of one or more pairs of nuclei. By repeated mitotic divisions in the ascogenous hyphae, several pairs of nuclei are formed in a large number of ascal primordia, where karyogamy and meiosis take place. Therefore, this life cycle is also predominantly haploid with a difference in space and time of plasmogamy and karyogamy.

4. Haploid-Dikaryotic Cycle. This type of life cycle is common to the majority of Basidiomycetes, excluding smuts. The mono- and dikaryotic phases are completely independent with unrestricted independent growth and terminate either by dikaryotization or by formation of a fruit body.

5. Dikaryotic Cycle. This type of life cycle is common to smuts (Kniep 1926) and occasionally in yeasts (Guilliermond 1940). The meiospores immediately fuse to initiate the dikaryotic phase. The life cycle is predominantly dikaryotic.

6. Haploid-Diploid Cycle. Alternation of haploid and diploid phases occurs regularly in this life cycle, which is unusual in fungi and is restricted to certain species of *Allomyces* from the Chytridiomycetes (Emerson 1941) and in the ascomycete *Ascocybe grovesii* (Dixon 1959).

7. Diploid Cycle. This life cycle is completely diploid; the haploid phase is restricted to gametes. It

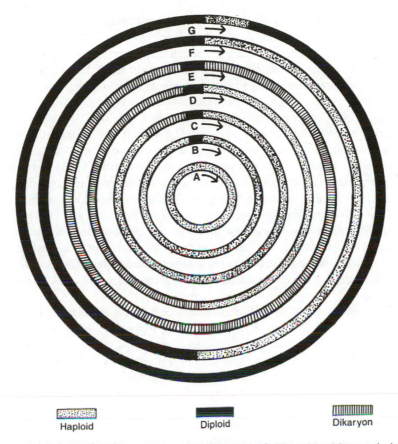

Haploid   Diploid   Dikaryon

Fig. 1 A-G.   Life cycles in fungi. A Asexual. B Haploid. C Haploid with restriction dikaryon. D Haploid-dikaryotic. E Dikaryotic. F Haploid-diploid. G Diploid (after Raper 1954).

occurs in certain yeasts (Guilliermond 1940), Myxomycetes (Alexopoulos 1962) and in majority of Oomycetes belonging to Saprolegniales and Peronosporales (Sansome 1963).

3.   Sporulation in Fungi

Spore formation is the process of differentiation of reproductive cells, and their supporting structures, from the somatic body. This ability of a fungus to switch over from the vegetative to the reproductive phase is controlled by genetic and environmental factors which create 'the internal stimulus to regulate primary and secondary pathways' (Turian 1974). The spores occupy a unique position in the fungal life cycle as they terminate reproductive and developmental cycles, and have inherent potential to launch the new generation. Compared to vegetative cells, the spore is less susceptible to adverse environmental conditions, and therefore, enables the fungus to perennate unfavourable conditions. In addition, the spore is a reproductive unit capable of getting dispersed over long distances.

A large number of fungi show seasonal sporulation, especially with respect to sexual spores. The complete set of variables that affect sporulation include:

1. Physiology of the host or nutritional status of the substratum,
2. Changes caused by aging, and
3. Physical factors such as light, temperature and water.

In his pioneering work on the factors influencing sporulation of fungi, Klebs (1899) set forth a number of principles which even today constitute an excellent framework for sporulation. Stimulation of spore formation varies from species to species, but the basic Klebsian principle that sporulation follows the limitation of growth by exhaustion of nutrients, holds true for most fungi. Fungal physiologists accept that the conditions which favour rapid mycelial growth hamper sporulation, and spore formation is induced only when the growth rate is reduced. This generalization notwithstanding, in *Gibberella zeae*, sporulation and maximal mycelial growth occur concurrently (Huang and Cappellini 1980). Basically, sporulation is a period of high rate of macromolecular synthesis and degradation, accompanied by significant changes in the molecular composition of the cells and changes in gene expression (Cole and Kendrick 1981).

Many nutritional factors control induction of fungal sporulation (Griffin 1981). Some fungi have specific carbon and nitrogen requirements, others sporulate only upon starvation/nutritional depletion or following a combination of more than one stimulus (Dahlberg 1982). Studies undertaken on the effect of starvation on sporulation of *Aspergillus nidulans* have been reviewed by Champe et al. (1981). The requirement of starvation to initiate differentiation of larger sporophores was demonstrated by an interesting study on *Schizophyllum commune* (Wessels 1965). Cleistothecium formation in *Aspergillus nidulans* follows exhaustion of glucose from the medium, at the expense of cell wall polysaccharides (Zonneveld 1972). Mullins and Warren (1975) demonstrated that the entire sexual reaction in the water mold *Achlya* occurs in the absence of exogenous nutrients.

The effects of temperature and light on sporulation have been discussed by Hawker (1966). The temperature range which supports sporulation is narrower than that which supports growth. Hawker (1966) opines that temperature may also affect the kind of spores produced, since it is likely to have differential effect on different modes of sporulation by the same fungus. Light also exhibits a profound effect on sporulation. In fact, light affects sporulation in fungi more than it affects growth (Tan 1978). Another aspect of light-induced sporulation is the phenomenon of light-entrained circadian rhythms. Endogenous rhythms of spore discharge have been described for a number of fungi, including species of *Sordaria, Daldinia* and *Pilobolus* (Ingold 1971). However, physiological mechanisms involved in the timing mechanism have yet to be understood (Lysek 1978). Carbon dioxide and humidity are two other environmental factors which influence the formation of spores and sporophores. Barnett and Lilly (1955) demonstrated that carbon dioxide is generally inhibitory to asexual reproduction in *Choanephora*, and that low humidity favours conidia formation, whereas high humidity promotes sporangia formation. Several studies implicate carbon dioxide as an inhibitor of basidiocarp development (Taber 1966). The effect of relative humidity has not been studied extensively; at present stimulatory as well as inhibitory effects are on record (Taber 1966).

4. Asexual Sporulation

In the majority of fungi, asexual reproduction takes place when a single parent forms progeny without a nuclear contribution from the second parent. Typically, nuclear alternations do not take place and, therefore, the offspring is a genetic duplicate of the parent. In contrast to the vegetative mycelium, the asexual spore (which is delimited from the thallus) is characterised by minimal metabolic turnover, low water-content, and lack of cytoplasmic movement (Gregory 1966). Towards the end of the growth cycle, when the mycelium is experiencing conditions of stress (Smith and Galbraith 1971), the specialized spore serves the triple functions of propagation, perennation and dispersal.

Fungi have evolved a variety of asexual spores and their modes of formation (Fig. 2A-F). The spores are formed by fragmentation, fission, extrusion, and cleavage (Carmichael 1971). In fragmentation the cytoplasm gets concentrated in a few cells of the mycelium; the remaining cells are exhausted. Formation by fission occurs when the cells of the hyphae break apart at double-walled septa. Extrusion involves formation of spores as extrusions from the tips or the lateral sides of the hyphae. In cleavage, the protoplasmic

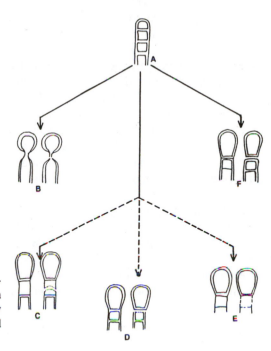

**Fig. 2 A-F.   Mechanism of conidial dehiscence. A Undifferen-
tiated hypha. B Conidia released by fracture of a
'fine' connection. C-E Conidia released by
'sacrifice' of a suporting cell. F Conidia released
by fission of a double septum.**

contents of a cell split or divide into fragments and eventually each fragment becomes surrounded by a wall.

Asexual spores arise entirely or partially from conidiogenous cells in one of two ways: thallic and blastic development (Vuillemin 1910, 1911). When the entire cell, including its wall, is delimited as the conidium initial, the development is said to be thallic. Examples of thallic sporogenesis are spore formation by fragmentation and fission. In holothallic spore ontogeny, as the very name suggests, all the wall layers of the sporogenous cell are involved in the formation of the spore wall. In enterothallic spore formation, the outer wall of the sporogenous cell does not become a part of the outer wall layer of the mature spore (Kendrick 1971). In contrast, extrusion or blastic development is essentially de novo growth of a spore initial from the fertile locus of a sporogenous cell (Cole 1975).

The conidia are discharged from the conidiogenic cells of the mycelium in the following ways:

1. In certain fungi, release of spores is caused by the breakdown of a fragile connecting cell (Fig. 2C–E). This intervening cell is thin-walled (Fig. 2C) or with an abscission ring (Fig. 2D) or is digested by enzymatic action (Fig. 2E).

2. In the second method, the spore is released without the loss of any cell, by splitting of the double septum (Fig. 2F).

3. In third method, the spore develops on a very delicate connective (Fig. 2B) and is released by its fracture (Carmichael 1971, Kendrick 1971).

4.1   Asexual Spores in Lower Fungi

The sexual reproduction in majority of Chytridiomycetes, Hypochytridiomycetes, Plasmodiophoromycetes and Oomycetes is characterised by the cleavage of cellular contents of the sporangium into numerous minute, typically uninucleate, fragments. Each fragment develops into a zoospore. In Zygomycetes, the contents of the sporangium cleave to form non-motile sporangiospores which develop a rigid spore-wall before their release from the sporangium (Hughes 1971). In some taxa of the Oomycetes and Zygomycetes, the sporangium acquires some or all the characteristics of the conidium. For example, in oomycetous genera *Pythium, Saprolegnia, Phytophthora* and *Aphanomyces,* zoosporogenesis is common in aquatic environments (Bartnicki-Garcia and Hemmes 1976). In members like *Peronospora* which are adapted to a terrestrial habitat, the sporangia are deciduous and germinate directly like the conidia. This conidia-like sporulation is achieved by delay or suppression of the zoosporangial state. Similarly, in Mucorales (Zygomycetes), typical sporangiospore formation characterises *Mucor, Rhizopus* and *Absidia.* There is also evidence of evolutionary development towards direct germination of the sporangium. This is achieved by reduction in the number of sporangiospores per sporangium. In the more highly evolved forms, the sporangia are only single spored and the sporangial wall is merged with the wall of the spore, as in *Cunninghamella elegans.* Reduction in the number of sporangiospores in these species is accompanied by an increase in the production of sporangioles, merosporangia and single-spored sporangia (Hughes 1971).

4.2   Asexual Spores in Higher Fungi

In the Ascomycetes, Basidiomycetes and Deuteromycetes, there are many types of spores and varied mechanisms for their formation. Deuteromycetes, in particular, have numerous spore-forming mechanisms, and produce an astounding variety of hyaline or coloured conidia which vary in size, shape and septation (Talbot 1971). The characteristic asexual spores in all these classes are conidia. Following Kananaskis Conference, the conidium has been defined as a 'specialized, non-motile, asexual propagule, usually caducous, not developing by cytoplasmic cleavage or free cell formation' (Kendrick 1971). The conidia arise from a conidiophore which may comprise a simple cell or a system of conidiogenic cells. Besides conidia, some higher fungi produce chlamydospores which are terminal or intercalary cells characterized by thick protective walls and do not possess any specific mechanism for liberation. These thick-walled asexual spores remain firmly attached to the mycelium. They are liberated by the mechanical fracture of a non-differentiated cell wall, enzymic lysis of the supporting hyphae by other microorganisms or by weathering. The chlamydospore enables the fungus to form a resistant structure under conditions which limit or inhibit macromolecular synthesis (Cochrane and Cochrane 1970).

4.3   Microcycle Conidiation

Some fungi show germination of spores by direct formation of conidia without the intervention of a mycelial phase, which is common to normal life cycles. It is a mode of asexual spore formation in which the normal life cycle of the fungus is bypassed. This phenomenon, called microcycle conidiation, can be induced artificially in the laboratory by manipulating various factors, especially temperature. A common feature observed in microcycle conidiation is the increase in size of the spore during a period of high temperature and it is only when the spores are removed to a lower temperature that the apical outgrowth (germination) occurs (Fig. 3A-C). Spores formed through sexual reproduction and species with unicellular thalli are not included in microcycle conidiation. It is known in *Aspergillus niger* (Anderson and Smith 1971), *Neurospora crassa* (Grange and Turian 1978), *Penicillium digitatum* (Zeidler and Margalith 1973), *P. urticae* (Sekiguchi et al. 1975), *Blastocladiella emersonii* (Lovett 1975) and species of *Allomyces* (Youtta 1976). Recently, microcycle conidiation has been reported in nature in a broad range of fungal species. It represents a part of the normal life cycle in several groups, including the Entomophthorales, Taphrinales, Clavicipitales, Uredinales, Ustilaginales, Tremellales and Exobasidiales (Hanlin 1994). The presence of a microcycle in such fungi provides a means of survival for spores that encounter unfavourable conditions (Hanlin 1994).

5.   Sexual Sporulation

Fungi are remarkable for their diversity of sexual process. Basically, sexual reproduction involves a cycle

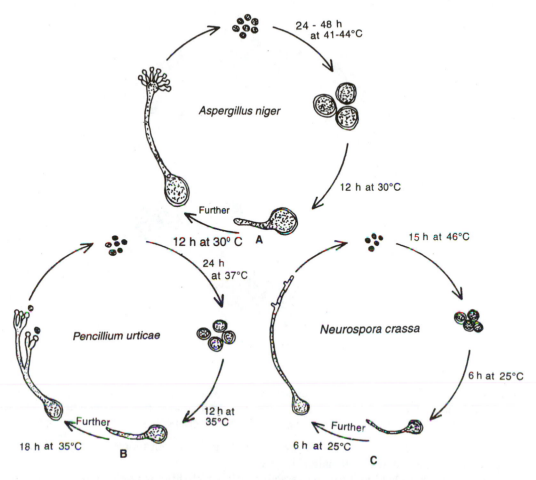

Fig. 3 A-C. Time (hour) and temperature scale for microcyclic conidiation in filamentous fungi.

of: (1) plasmogamy (accomplished by gametes or gametangia) followed by (2) karyogamy, and (3) meiosis, at specific points in the life cycle (of a species). These three events are accomplished through a variety of regulating mechanisms and range of morphological developments.

5.1   Homothallism and Heterothallism

In many fungi there is scanty or no control over the type of nuclei which fuse. Thus, self-fertilization occurs, as in *Rhizopus sexualis* (Fig. 4A-G). This situation in which each individual has the competence to elaborate sexual organs and complete the sexual cycle in isolation, was termed 'homothallism' by Blakeslee (1904). A second type of situation occurs in *Rhizopus stolonifer, Mucor mucedo, Neurospora crassa* and others. In these taxa, sexual reproduction involves two types of conjugant individuals (Fig. 5A-L) which may or may not develop morphologically distinct gametangia. Blakeslee (1904) used the term heterothallism for this situation. Blakeslee (1920) also investigated the pattern of sexuality in most of the members of Mucorales and discovered that members with a sexual stage are unambiguously divisible into heterothallic and homothallic types. Whitehouse (1949a) used the term 'heterothallism' to include all such taxa in which an intermycelial reaction is a prerequisite for sexual fusion. He distinguished two major types of heterothallism:

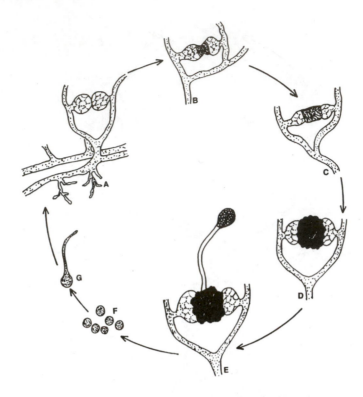

Fig. 4 A-F.   Sexual cycle of *Rhizopus sexualis* (homothallic). A Mycelium bearing progametangia. B Gametangia. C Prozygosporangium. D Zygosporangium. E Germination of zygosporangium and formation of germ-sporangium at the tip of sporangiophore. F Sporangiospores. G Germination of sporangiospore.

1. Morphological heterothallism refers to two interacting thalli with morphologically dissimilar sex organs or gametes, male and female (Fig. 6A-L).
2. Physiological heterothallism refers to interacting thalli which differ in mating type, or incompatibility factor, irrespective of the presence or absence of differentiated sex organs or gametes (Fig. 5).

Whitehouse (1949a) retained the term homothallism (in the original sense) for sexual fusion between elements of the same thallus or, in unicellular organisms between individuals of the same clone, and coined the term 'secondary homothallism' for self-fertile homokaryons.

Homothallism is predominant in the lower fungi, common in the Ascomycetes and apparently rare in the Basidiomycetes where heterothallism predominates. The majority of the fungi are homothallic and it may be that homothallism represents the primitive condition from which heterothallism evolved later (Raper 1959, 1960). Heterothallism has a selective advantage over homothallism as it increases outbreeding and resultant variability. Garrett (1963) pointed out that "heterothallism promotes outbreeding, and therefore subserves the same end as the sexual process, which it renders more efficient. Heterothallism is not the same as sex; it is a refinement superimposed upon it."

Heterothallism occurs in all the major groups of fungi: Hymenomycetes (Kniep 1920, 1922), Ustilaginales (Kniep 1919), Saprolegniales (Couch 1926), Euascomycetes (Shear and Dodge 1927), Uredinales (Craigie 1927), Blastocladiales (Harder and Sorgel 1938) and in the yeasts (Winge and Laustsen 1939a, b). Mating types have also been discovered in Myxomycetes (Fig. 7A-G) where some species such as *Physarum polycephalum* (Dee 1960, Wheals 1970) and *Didymium iridis* (Collins 1976) occur in homo- as well as

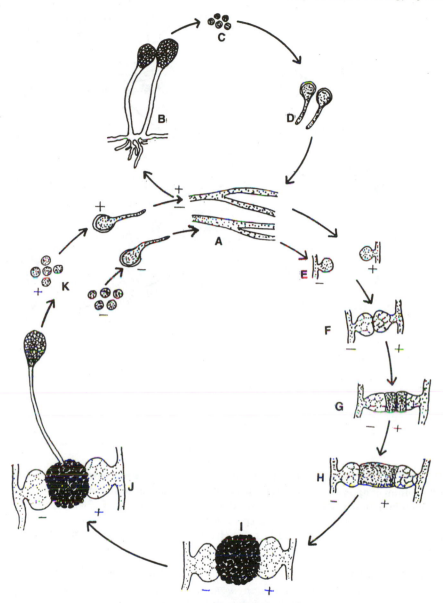

Fig. 5 A-L. Life cycle of *Rhizopus stolonifer* (heterothallic). A Somatic hyphae. B Sporangiophores with sporangia. C Sporangiospores. D Germinated sporangiospores. E Compatible zygosphores. F Progammetangia. G Gametangia. H Prozygosporangium. I Zygosporangium. J Germ-sporangium. K Two types of spores (+ and −). L Germination.

heterothallic strains. In other Myxomycetes such as *Fuligo cinerea* (Collins 1961) and *Physarum flavicomum* (Henney 1967) only homothallic or heterothallic strains have been discovered so far. In several heterothallic strains, the mating type locus has multiple alleles (Collins 1963, Dee 1966). However, the majority of Myxomycetes are homothallic (Fig. 8). In many taxa homothallism is generally assumed if monosporous cultures yield plasmodia and sporulate (Collins 1976).

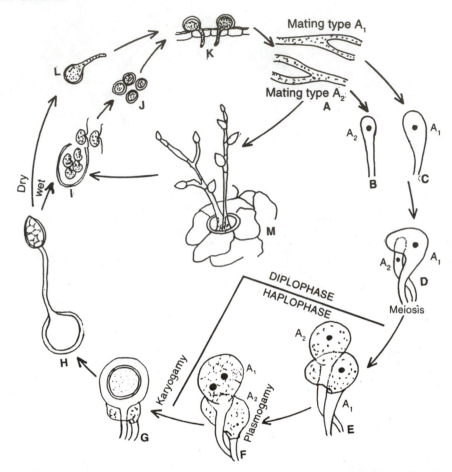

Fig. 6   Life-cycle of *Phytophthora infestans*. A Somatic hyphae of mating types A1 and A2. B, C Oogonium and antheridium of opposite mating types. D Penetration of antheridium into the Oogonium. E Development of Oogonium into a globose structure above the antheridium. F Plasmogamy. G Diploid oospore. H Germinated oospore with a germ-sporangium. I Zoospore differentiation and release. J Zoospore encystment. K Germination and host penetration. L Direct germination. M Sporangiophore bearing sporangia.

With respect to compatibility, Ascomycetes are homothallic as well as heterothallic. Among the best studied homothallic forms are *Glomerella cingulata*, *Sordaria macrospora* and *S. fimicola* (Turian 1978). In heterothallic species compatability is determined by a pair of alleles 'A$_1$A$_2$' which segregate at meiosis during ascospore formation. This is termed bipolar or unifactorial heterothallism (Fig. 9). Typical examples of Ascomycetes with bipolar mating types are *Neurospora crassa*, *N. sitophilla* (Whitehouse 1949b), *Sordaria brevicollis* (Olive and Fantini 1961), *S. heterothallis* (Fields and Maniotis 1963), *S. sclerogenia* (Fields and Grear 1966), *Ascobolus stercorarius* (Bistis and Raper 1963), *Aspergillus heterothallicus* (Raper and Fennel 1965) and species of *Chaetomium* (Seth 1967). In some bipolar heterothallic fungi an interesting mechanism operates during spore formation, whereby two nuclei of opposite mating types are incorporated in each spore. Such spores, on germination, give rise to a mycelium that contains both 'A$_1$' and 'A$_2$' nuclei and consequently behaves as if homothallic. This condition, known as secondary homothallism, is reported in the four-spored (Fig. 10) *Neurospora tetrasperma* and *Podospora anserina* (Turian 1978).

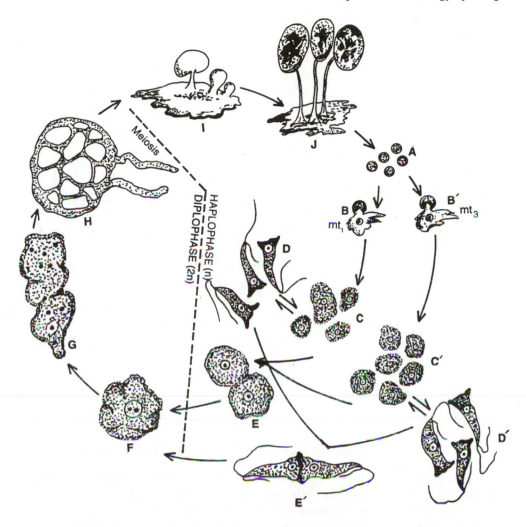

Fig. 7 A-J. Life Cycle of *Physarum polycephalum* (heterothallic strain). A Spores. B, B′ Genetically different spores germinate to release myxamoebae of mating types mt₁ and mt₃. C, C′ Formation of clone. D, D′ Swarm cells. E, E′ Fusion of myxamoebae and swarm cells. F. Zygote. G Young plasmodium. H Mature plasmodium. I Sporulation-sporangial initials. J Mature sporangia.

Basidiomycetes also include homothallic, secondary homothallic and heterothallic species. However, the majority of the species investigated are heterothallic, and from among these 25% exhibit bipolar sexual incompatibility (Alexopoulos and Mims 1979). Mating systems of Uredinales and most Ustilaginales conform to this type. Most of the other Basidiomycetes show tetrapolar (bifactoral) heterothallism (Fig. 9). In these fungi, compatibility is controlled by two pairs of factors, 'A₁A₂' and 'B₁B₂', located on different chromosomes. It is the most complicated pattern of sexuality known among fungi (Raper and Flexer 1971). Studies on *Schizophyllum commune* have revealed that each compatibility factor represents two closely-linked genes designated 'α' and 'β' (Papazian 1954, Raper et al. 1958, 1960). This two gene structure has also been demonstrated in *Pleurotus ostreatus* (Terakawa 1957, 1960), *Collybia velutipes* (Takemaru 1961) and *Coprinus lagopus* (Day 1963), and is considered to hold true for all tetrapolar forms. Secondary

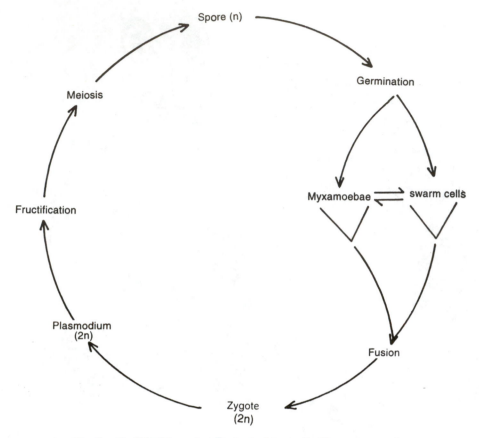

Fig. 8   Outline life-cycle of a typical homothallic myxomycete.

homothallism is known in 10–15% of species. This is exemplified by *Agaricus bisporus* which has only two basidiospores, each containing two nuclei bearing different incompatibility factors (Raper et al. 1972). The incompatibility factors are considered to be regulatory genes which control many structural genes responsible for the biosynthesis of proteins that bring about morphogenetic changes leading to dikaryon formation in higher fungi (Raper 1966b).

5.2   Dikaryon Formation

In the life cycle of most higher basidiomycetous fungi, two functionally different mycelial states can be distinguished, the mono- and dikaryon (Fig. 11). Monokaryon is the primary mycelium formed on germination of a haploid sexual spore, and contains a single nuclear type. It is capable of indefinite vegetative growth and is usually sexually sterile, but may produce abundant asexual spores. In contrast, a mycelium containing nuclei of different genotypes is termed a heterokaryon. It is the result of anastomosis of homokaryotic hyphae of different strains. The dikaryon, also called a secondary mycelium, is a specialized type of heterokaryon in which haploid nuclei of two compatible mating types associate, divide conjugately and assort one pair per cell.

Dikaryon is the predominant life-style of Basidiomycetes alone where it is capable of indefinite propagation in nature, and is the pre-requisite for sexual reproduction. The two nuclei in each cell of the dikaryon remain discrete during somatic cell divisions. For example, in the macrocyclic rusts, the aeciospores and

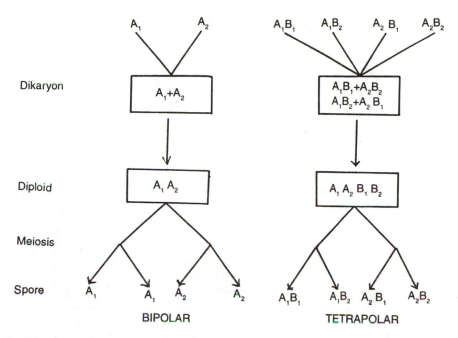

Fig. 9 **Diagrammatic representation of segregation of compatibility factors and polarity.**

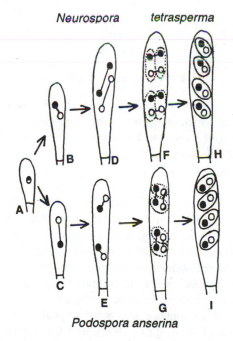

Fig. 10 **Mechanisms resulting in secondary homothallism.**

the mycelia arising from their germination, the uredospores and the mycelia arising from their germination, and the young teliospores are all dikaryotic (Fig. 12). Sometimes days, months and even years intervene

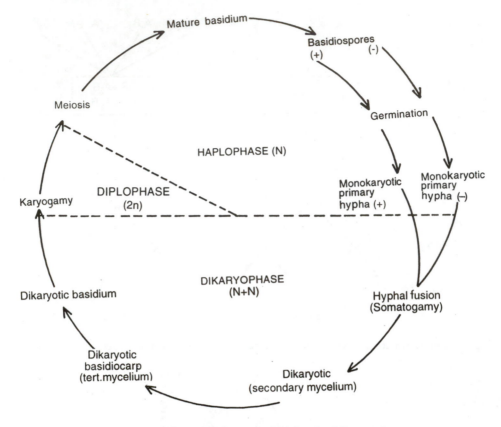

Fig. 11   Life cycle of a typical higher basidiomycete.

between plasmogamy and karyogamy. Eventually, the two nuclei fuse in special reproductive cells, the basidia, and produce diploid nuclei which undergo meiosis almost immediately, and produce sexual spores. Generally, four basidiospores develop on each basidium, and each basidiospore shares one of the four haploid nuclei produced through meiosis (Fig. 12A-I).

The events involved in formation of a dikaryon from genetically different homokaryons are sequentially illustrated in Fig. 13A-I. Dikaryotization commences with nuclear migration. During nuclear migration, the complex pore structure, termed 'dolipore' by Moore and Mc Alear (1962), gets dissolved, leaving a simple septum through which nuclei pass easily. No information is available about the factors controlling nuclear motility. Mayfield (1974) noted that nuclear migration is accompanied by flow of cytoplasmic organelles, so that the hyphae become highly vocuolated. Once the dikaryotic condition is established, regular distribution of the two nuclei to each cell is maintained by complex cell division involving formation of clamp-connections (Fig. 13). In a few species, dikaryon formation occurs in the absence of fusion between different homokaryons (Whitehouse 1949a). In some of these taxa, an (apparently) self-fertile condition results from each basidiospore which contains two non-identical nuclei. The requirement for dikaryon formation is satisfied at the time of spore germination, and the homokaryotic phase is effectively by-passed.

Attempts have been made by Raper and Esser (1961) and Wang and Raper (1969) to identify the proteins responsible for bringing about morphogenesis leading to dikaryon formation by resorting to immunological and electrophoretic techniques. Wang and Raper (1970) reported marked changes in the basic metabolism

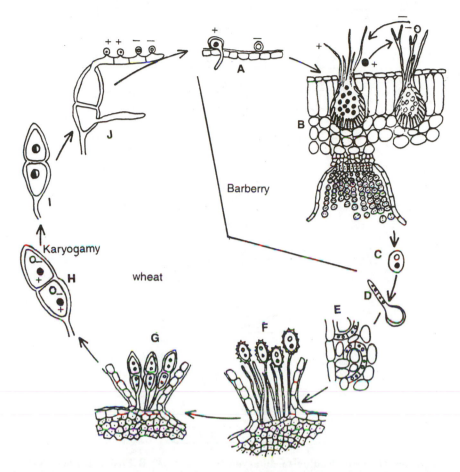

Fig. 12 A-J.   Life-cycle of macrocyclic rust, *Puccinia graminis tritici*. A Germination of basidiospores. B Pycnia showing receptive hyphae and transfer of pycniospores of opposite mating types. C Aecium bearing chains of dikaryotized aeciospores. D Germination of aeciospore. E Dikaryotic mycelium in the host. F Uredium. G Telium. H Teleutospore (binucleate). I Diploid teleutospore. J Germination of each cell of teleutospore into four-celled metabasidium, with each cell of metabasidium bearing a single basidiospore on a sterigma.

of *Schizophyllum* mycelium during this morphogenesis. Earlier, Wessels (1965) isolated the specific enzymes involved in septal dissolution and clamps cell-fusion in *Schizophyllum commune,* and considered them to be lytic (enzymes). Raper (1966b) observed that the hyphal walls become distorted, and protoplasmic extrusions are not uncommon, suggesting that the walls are weakened by the activity of lytic enzymes. The implication of R-glucanase in dikaryon morphogenesis was unravelled by the studies of Wessels and Niederpreum (1967) and Wessels (1969). It is, however, not clear whether R-glucanase is involved in clamp cell fusion also.

5.3   Hormonal Regulation of Mating

The processes by which gametes in various groups of fungi are induced to differentiate, approach each other, and fuse to form a zygote, are complex with considerable variation in form and duration. Regulation

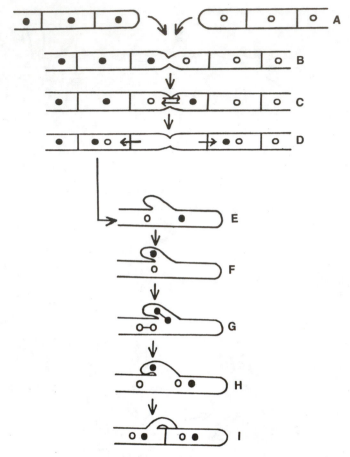

Fig. 13 A-I.   Sequence of events leading to dikaryon formation. A, B Fusion between genetically different homokaryotic hyphae. C Nuclear exchange. D Nuclear migration. E-I Clamp cell formation and apical cell division.

of these processes (in a number of fungi) is controlled by mating-type specific diffusible substances referred to as 'hormones' or 'pheromones'. Hormones are the chemical compounds which act on the same individual that produce them, while pheromones act on different individuals. O'Day (1981) favours use of the term pheromone. Bolker and Kahmann (1993) used this term for diffusible peptide mating factors. Generally, fungal sex hormones are produced in very small quantities; some constitutively and others following induction by complementary hormones. Important properties of fungal hormones are summarized in Table 1.

The simplest hormonal system known is that of *Allomyces* (Chytridiomycetes). In this water-mould the larger, motile and colourless female gametes (ca. 11 μm in diameter) produce a potent attractant for the smaller (ca. 8 μm), bright-orange male gametes. Machlis (1958) named the chemical attractant 'sirenin' and demonstrated that the male gametes are sensitive to sirenin over a very wide concentration 10^{-10} to 10^{-5} M (Machlis 1973a, b). Of the five types of swimming cells produced by the fungus during its life cycle, namely, male and female gametes, zygote and diploid and haploid zoospores, only the male gametes respond to sirenin (Carlile and Machlis 1965). The behavioural response of male gametes to sirenin has been analysed by Pommerville (1981). Sirenin secretion awaits completion of gametogenesis, and finally

Table 1   Fungal sex hormones—Structure and properties (adapted after Gooday 1983, Bolker and Kahmann 1993)

Hormone	Molecular structure	Probable precursor	Site of synthesis	Optimal yield (M)	Sensitivity of bioassay (M)
Sirenin	Sesquiterpene, $C_{15}H_{24}O_2$	Farnesyl pyrophosphate	♀ gametes *Allomyces* species	10^{-6}	10^{-10}
Antheridiol	Sterol $C_{29}H_{42}O_5$	Fucosterol	♀ cells *Achlya* species	10^{-8}	10^{-11}
Oogoniol	Sterol ester, $C_{33}H_{54}O_6$	Fucosterol	♂ cells *Achlya* species	—	10^{-6}
α_1-factor	—	—	A_1 isolates, *Phytophthora* specils	—	—
α_2-factor	—	—	A_2 isolates *Phytophthora* isolates	—	—
Trisporic acid	Apocarotenoid. $C_{18}H_{26}O_4$	Retinal	(+) /(−) cells Mucorales	10^{-6} (*M. mucedo*) 10^{-3} (*B. trispora*)	10^{-8}
α-factor	Dodeca- and tridecapeptides	Protein	α cells *Saccharomyces cerevisiae*	10^{-8}	10^{-8}
a-factor	Undecapeptide	Protein	a cells *S. cerevisae*	—	—
M-factor	Farnesylated & Carboxy-methylated nonapeptide	Protein	M cells *S. pombe*	—	—
P-factor	—	—	P cells *S. pombe*	—	—
Tremerogen A-10	Farnesyl dodecapeptide	Protein, farnesyl pyrophosphate	A cells *Tremella mesenterica*	10^{-7}	10^{-9}
Tremerogen a-13	Farnesyl tridecapeptide	Protein, farnesyl pyrophosphate	a cells *T. mesenterica*	10^{-8}	—
Rhodotorucine A	Farnesyl undecapeptide	Protein, farnesyl Pyrophosphate	A cells *Rhodosporidium toruloides*	—	—

when it is produced by female gametes it attracts and orients male gametes in a position most favourable for efficient mating. The mechanism of action of sirenin is not known, as far as identification of receptor sites and the primary biochemical reactions affected are concerned. Presumably, the receptors are located on the plasmalemma and the hormone affects the action of the flagellum in some way so as to orient the gamete so that all receptors at a given level around the circumference receive the same stimulus.

Hormonal system in the water-mould *Achlya* (Oomycetes) is more complex. In a series of experiments with heterothallic species, Raper (1951) demonstrated that sexual differentiaiton in species of this genus is regulated by complementary hormones which diffuse from some hyphae to others. He concluded that the male cells do not produce any hormone on their own. It is the female cells which produce hormone 'A' continuously that switches the male cells from vegetative growth to the production of many short antheridial branches (Fig. 14A-G). In response, the male cells release hormone 'B' which diffuses back to the female cells, switching them from the vegetative phase to production of oogonial initials. Raper (1952) estimated that as many as nine compounds are involved in regulating and maintaining this sequence of events. These include four sequentially-acting hormones and five 'modifying substances'. The female hormone 'A' was identified as 'antheridiol', and characterised as a sterol by McMorris and Barksdale (1967) and Arsenault et al. (1968). Its biological activities were assessed by Barksdale et al. (1974). The complementary hormone 'B' produced by the male cells, in response to antheridiol action, is named 'oogoniol' (McMorris 1978). It was purified from culture filtrates of *Achlya heterosexualis*, which produces it constitutively (Barksdale and Lasure 1974), and has also turned out to be a sterol (McMorris et al. 1975; McMorris 1978).

Hormonal regulation of sexual reproduction is also known of *Pythium* and *Phytophthora* of the order Peronosporales (Fig. 6). Ko (1978) observed that in heterothallic *Phytophthora parasitica*, *P. palmivora* and P. *cinnamoni* both mating types, A_1 and A_2, form oospores when they were paired with the opposite mating type of the same or different species on opposite sides of polycarbonate membranes. Successful formation of oospores across the membrane demonstrates stimulation of sexual reproduction by hormones originating in the opposite mating type and reaching the site of activity by diffusion across the polycarbonate membrane. The sex hormone produced by A_1 isolates of *Phytophthora*, designated hormone α_1, induces sexual reproduction in A_2 but not of A_1 isolates Similarly, sexual reproduction of A_1 but not A_2 isolates can be induced by hormone α_2, produced by A_2 isolates. Over the years, similar hormonal control has been demonstrated in *P. capsici* (Uchida and Aragaki 1980), *P. colocasiae* (Yu and Chang 1980), *P. megakarya* (Erselius and Shaw 1982), *P. drechsleri* (Skidmore et al. 1984), *P. infestans* (Shaw et al. 1985, Shattock et al. 1986), *P. cryptogea* (Ho and Jong 1986) and *P. megasperma* (Ho 1986). Homothallic species of *Phytophthora* are also able to induce oospore formation in A_1 and A_2 isolates, indicating their ability to produce 'α hormones' like the heterothallic species (Ko 1980). In heterothallic *Pythium splendens*, Guo and Ko (1991) observed that when different mating types were paired on opposite sides of a polycarbonate membrane, numerous oospores were produced by (+) but not (−) isolates, indicating that sexual reproduction of this fungus is the result of selfing of the (+) isolate in response to stimulation provided by the hormone produced by (−) isolate.

The first evidence about involvement of diffusible hormones in regulating sexual reproduction in Mucorales (Zygomycetes) was provided by Burgeff (1924). In fact, it was the first report implicating hormones in mating of fungi. The breakthrough came when it was demonstrated that 'trisporic acids' were involved in the induction of zygophores (Gooday 1974, Van Den Ende 1978). Trisporic acids are C_{18} terpenoids; the two major ones are trisporic acid B (a ketone) and trisporic acid C (an alcohol). These two compounds induced complete switch-over from asexual to sexual differentiation in (+) and (−) mycelia. The fact that the switch-over to sexual differentiation in Mucorales is controlled by diffusible hormones (Fig. 15) was demonstrated by zygospore formation even when the two mating types were separated physically by a membrane, or an air gap (Mesland et al. 1974, Gooday 1978). This hormone system is not species specific,

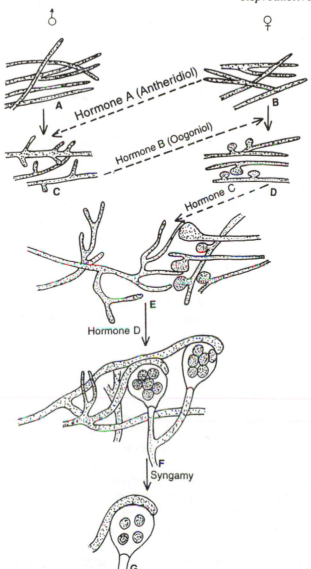

Fig. 14   **Sequence of hormonal secretions and sexual differentiation in *Achlya ambisexualis*. A Male vegetative hyphae. B Female vegetative hyphae. C Production of antheridial initials. D Production of oogonial initials. E Chemotropic growth of antheridial initials towards oogonial initials. F Differentiation of oospheres. G Maturation of oospores.**

but universal for all Mucorales, which is revealed by zygospore formation when (+) and (−) strains of different species were cultured together (Blakeslee and Cartledge 1927). The three species of the order Mucorales used extensively for research on trisporic acid include: *Blakeslea trispora* (Bu'Lock and Osagie 1973), *Mucor mucedo* (Gooday 1978) and *Phycomyces blakesleeanus* (Cerda-Olmedo 1975, Sutter 1975).

From Ascomycetes, the mechanism of pheromone production and action has been studied in *Saccharomyces* species whose mating system has been the subject of a number of recent reviews (Fields 1990, Marsh et

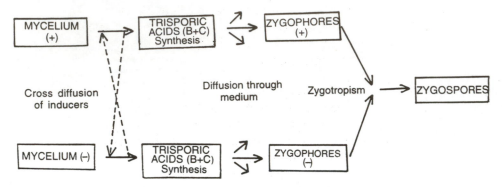

Fig. 15   **Diagrammatic representation of the sequence of events leading to sexual differentiation in heterothallic Mucorales.**

al. 1991, Kurjan 1992). Levi (1956) demonstrated, for the first time, that mating in *S. cerevisiae* is regulated by diffusible sex hormones. Haploid cells of the species are of two mating types, 'α' and 'a', which produce 'α factor' and 'a factor', respectively. When brought together, 'α' and 'a' mating type cells fuse to form the diploid zygote. In the diploid phase, most of the genes involved in pheromone production and response are 'shut off' (Herskowitz 1989). Therefore, diploid cells neither secrete nor respond to mating factor. The diffusible mating factors of *S. cerevisiae* appear to function only in the process of cell-cell recognition during the fusion of haploid cells. The 'α factor' has been characterised as a mixture of four peptides (Stotzler et al. 1976). The 'a factor' is hydrophobic, with a storng propensity to associate with very high molecular weight mannans, and is characterized as an undecapeptide (Betz et al. 1977, Betz and Duntze 1979). 'α Factor' is not species-specific; 'α factor' from *S. cerevisiae* and *S. kluyveri* is equally active on 'a cells' of both species (Mc Cullough and Herskowitz 1979).

In the fission yeast, *Saccharomyces pombe*, the cells are plus (P) and minus (M) mating types. Mating between cells of opposite types occurs only under conditions of nitrogen starvation and is mediated by pheromones that induce conjugation tube formation and cell fusion. In contrast to *S. cerevisiae*, in which mating factors function only to achieve cell fusion, in *S. pombe* the pheromones serve a continued response pathway which is necessary to proceed through meiosis (Leupold et al. 1989). The pheromone (M-factor) secreted by M-cells has been purified, sequenced and identified to be a 'farnesylated' and 'carboxymethylated nonapeptide' processed from 42- and 44-amino acid precursors encoded by two genes, *mfm* 1 and *mfm* 2 (Davey 1992). The 'P-factor' is yet to be purified. The filamentous Ascomycete, *Neurospora crassa,* also shows two different mating types, 'A' and 'a'. When this fungus enters the sexual cycle, trichogynes (female cells) and microconidia (male cells) develop. Specific recognition of opposite mating types is mediated by pheromones (Bistis 1983).

The first evidence favouring involvement of diffusible hormones in controlling syngamy in Basidiomycetes emanates from the observations of Bandoni (1965) on *Tremella mesenterica*, a heterobasidiomycetous jelly-fungus, capable of growing vegetatively as a budding yeast. Bandoni (1965) demonstrated that yeast cells of mating types A' and 'a' produce diffusible factors that cause cells of the opposite mating type to cease budding, and form conjugation tubes. Reid (1974) achieved partial purification of the hormones and suggested that they could be peptides. This has been confirmed by the characterization of two peptides, 'tremerogen A-10' (from A-type cells) and 'tremerogen a-13' (from a-type cells) (Sakagami et al. 1981a, b; Yoshida et al. 1981). Formation of conjugation tubes continues as long as tremerogens are added, but the cells revert to budding when they are removed (Tsuchiya and Fukui 1978; Flegel 1981). Other species of *Tremella* (Reid 1974; Ishibasi et al. 1984) and *Sirobasidium magnum* (Flegel 1981) also seem to follow similar hormonal regulation.

In the basidiomycetous red yeast *Rhodosporidium toruloides*, mating reaction between compatible 'A' and 'a' cells is initiated by the formation of long conjugation tubes that are directed towards the mating partner (Abe et al. 1975) The mating interaction appears asymmetric and sequential. At first, 'A' cells induce conjugation tube formation in 'a' cells, then the growing tip of the 'a' cells induces formation of conjugation tubes in 'A' cells. Eventually, the conjugation tubes fuse at their tip which marks accomplishment of sexual reproduction (Abe et al. 1975). A diffusible mating hormone, rhodotorucine A, was isolated and purified from *R. toruloides* strain 'A' (Kamiya et al. 1978; Sakurai et al. 1978), and studied for metabolic reactions (Kamiya et al. 1980). Unlike *Saccharomyces* and *Tremella* species, where both mating types constitutively produce sex-specific hormones, in *R. toruloides* only 'A' cells produce the hormone (rhodotorucine A) constitutively, and in response 'a' cells then produce their hormone, namely, rhodotorucine 'a'. This is somewhat analogous to the antheridiol-oogonial system of *Achlya* species.

The basidiomycete fungus *Ustilago maydis* also has a mating system that is controlled by two unlinked genetic loci, 'a' and 'b'. The 'a' locus with two alleles, controls fusion of haploid cells, whereas the multiallelic 'b' locus encodes two regulatory proteins that govern sexual development. Together with 'a', the 'b' locus controls the transition from yeast-like growth of haploid cells to fungal growth of the dikaryon and its maintenance (Banuett 1992). The 'a' locus is thus necessary both for cell fusion and for filament maintenance (Banuett and Herskowitz 1989). This represents the first example of a mating-type locus containing the structural genes for components that are involved directly in cell-cell communication. The presence of such genes prompted search for direct demonstration of pheromone activity (Bolker and Kahmann 1993). There is some evidence for diffusible mating factors in *Ustilago hordei* also, coupled with 'b+'-like homeodomain proteins (Bakkeren and Kronstad 1993). The most likely explanation for the mating behaviour of *U. hordei* is that the genes which trigger dikaryon-specific gene expression are linked to the genes coding for cell-signalling components.

In *Cryptococcus neoformans*, the basidiomycetous yeast, causing serious meningitis in patients who are immune-deficient, formation of dikaryon is accompanied by morphogenic transition from yeast-like to hyphal growth. Strains of the mating type *MATα* of the fungus are more virulent than those of MATa. There are indications that MATα encodes a pheromone precursor (Moore and Edman 1993). However, it is not yet clear how the pheromone precursor is processed, and it also remains to be established whether the pheromone itself affects virulence. In case it does, there will be new avenues to study the process of virulence at the molecular level.

The sexual development of cellular slime-molds (Acrasiomycetes), which produce large dormant structures, the macrocysts, is also controlled by hormones that are yet to be characterized (O'Day and Lewis 1981). *Dictyostelium discoideum*, strain NC-4, produces a volatile sex hormone which induces macrocyst formation in strain V-12 (O'Day and Lewis 1975; Lewis and O'Day 1977). Similarly, *D. purpureum* strains Dp6 and Dp7 produce a sex hormone that induces macrocyst formation in strain Dp2 (Lewis and O'Day 1976). In neither of these species has reciprocal activity ever been noticed. All the four mating types of *D. giganteum* secrete specific sex-hormones to which each of the other three strains respond by forming macrocysts (Lewis and O'Day 1979).

6.   Parasexuality

Despite the fact that Deuteromycetes (Fungi Imperfecti) apparently lack true sexual reproduction, they are amazingly abundant in all types of terrestrial and aquatic habitats, either as saprophytes or parasites. It is unlikely that such adaptation was possible in the absence of some mechanism to generate genetic variation. Moreover, the efforts to breed varieties of plants resistant to deuteromycete pathogens have often been frustrated by the appearance of new races capable of causing severe disease symptoms in 'resistant' varieties. There is ample evidence to suggest that the new pathogenic forms may be the result of genetic recombination generated by parasexuality–a process in which plasmogamy, karyogamy and haploidazation take place, but not at specified points in the thallus or the life cycle. Therefore, many such fungi which do not pass through a true sexual cycle derive benefits of sexuality through the parasexual process.

It was the discovery of heterozygous diploids in filamentous fungi which sparked a series of studies leading to the formation and elucidation of the concept of parasexual cycle. The novel process of parasexuality was first discovered in 1952 by Pontecorvo and Roper of the University of Glasgow (Scotland) in the filamentous fungus *Aspergillus nidulans*, where parasexuality coexists with the normal sexual cycle. The term 'parasexual cycle' was applied by Pontecorvo (1954) to the sequence of heterokaryosis, fusion of genetically dissimilar nuclei, followed by recombination and segregation. Pontecorvo (1958) even made tentative estimations and suggested that in *A. nidulans* recombination via the parasexual cycle is nearly .002 part of that through the sexual cycle.

The discovery of parasexuality led biologists to realise that transfer of genetic material from one individual or cell to another is not the monopoly of sexual reproduction (Pontecorvo 1958).

The essential events of the parasexual cycle are:

6.1 Heterokaryosis

An important prerequisite for recombination is the fusion of haploid nuclei which, as a first step, require inclusion of genetically different nuclei in the same cytoplasm. This is achieved through heterokaryosis. There are several ways in which a heterokaryotic mycelium is formed. The most common way is by anastomosis of somatic hyphae of different genetic constitution. Anastomosis often occurs between different strains of the same species (Boyer 1961; Garza-Chapa and Anderson 1966), but Tinline (1962) has indicated that this occurs less frequently than anastomosis within a strain. Anastomosis has been reported even between different fungal species (Hansen and Smith 1934, Ishitani and Sakaguchi 1956). Evidence is now also available to support the finding that environment influences the incidence of anastomosis (Schreiber and Green 1966; Suzuki 1967). Another method by which a homokaryotic mycelium changes into a heterokaryon is by mutation in one or more nuclei, as demonstrated in some Ascomycetes (Olive 1956). Recently, Rizwana and Powell (1995) demonstrated experimentally that UV-light enhances heterokaryon formation and parasexuality in *Cryophonectria parasitica*. Finally, heterokaryosis may be caused by inclusion of genetically dissimilar nuclei in a single spore following meiosis, as in *Podospora anserina* and *Neurospora tetrasperma* (Davis 1966).

6.2 Fusion and Formation of Diploids

Roper (1952) first produced heterozygous diploid nuclei in *A. nidulans* heterokaryon following exposure to d-camphor vapour. This chemical compound seems to increase the yield of diploids but not necessarily the frequency of nuclear fusions (Pontecorvo and Roper 1953). Using multinucleated conidia of *Aspergillus sojae* and *A. oryzae*, Ishitani et al. (1956) produced diploids by d-camphor vapour and UV treatments. The mechanism of action of these treatments on nuclear fusion is not clear but Roper (1966) noticed a selective action of these agents. Uchida et al. (1958) attempted to produce interspecific diploids between *A. sojae* and *A. oryzae*. However, there are possible barriers to interspecific parasexual recombination at heterokaryosis, nuclear fusion, or recombination. Any success along these lines would open a new approach for unravelling phylogeny and species relationship (Roper 1962).

The frequency of spontaneous somatic diploids is usually low. In *A. nidulans* it is about 1 in 10^6–10^7 conidia (Pontecorvo 1956), in *A. niger* 3.5 in 10^5 (Pontecorvo et al. 1953), in *Penicillium chrysogenum* 2.5 in 10^8 (Pontecorvo and Sermonti 1953) and in *Coprinus lagopus* 1 in 10^3–10^4 basidiospores (Casselton 1965). In the smut fungus, *Ustilago maydis*, diploids have been obtained as a consequence of failure of normal meiosis in as many as 10% of spores (Holliday 1961). Heterozygous diploids are relatively stable at mitosis and their colonies carry mainly diploid conidia of the parental type. The proof of diploidy is provided by nuclear DNA estimation. Heagy and Roper (1952) estimated DNA in known haploid and putative diploid conidia and reported a 1:2 ratio. Roper (1952) points out that the conidia of diploid strains are significantly larger than those of haploids. The ratio of their diameters is almost 1.3:1.0 and the volume is in the ratio 2:1. Pontecorvo et al. (1954) employed conidial size as a measure of ploidy and concluded that it is unambiguous. Successful detection of diploids on the basis of conidial size has also been reported by Field and Stromnaes (1966), Day and Jones (1968) and Ingram (1968). These findings notwithstanding, Ishitani et al. (1956) reported that haploid and diploid conidia of *A. sojae* do not exhibit significant size difference. The size difference has also not been observed between haploid and diploid spores of *Cochliobolus sativus* (Tinline 1962) and *Ustilago maydis* (Holliday 1961). Clutterbuck and Roper (1966) reported similar results for the hyphal tip cells of *A. nidulans*; the mean cell volume per haploid nucleus is about half that per diploid nucleus.

6.3 Mitotic Segregation

Mitotic segregation represents an extremely imporatnt step in the parasexual cycle. Pontecorvo and Roper (1953) proposed that segregants arise by mitotic crossing-over (Fig. 16A-D), a process originally discovered by Stern (1936) in *Drosophila*. Mitotic crossing-over is a rare event occurring at the four strand stage of mitosis. At any one event, crossing-over is almost invariably confined to a single exchange in one chromosome arm of the entire chromosome complement. There are numerous difficulties in estimating the absolute frequency of mitotic crossing-over. Kafer (1958, 1961) estimated that crossing-over occurs once in every 50 mitotic divisions. The only fungi for which rational estimates are available are *A. nidulans* and *A. niger*. The mitotic recombination index, estimated by the method of Pontecorvo (1958), is 2×10^{-5} for the former and 2×10^{-3} for the latter species (Pontecorvo 1958, Lhoas 1967). An increase of mitotic segregation can be achieved by treatment of diploids with a variety of physical and chemical agents (Kafer 1963; Holliday 1961). Many of these agents, particularly high energy radiations, carry substantial risk of producing

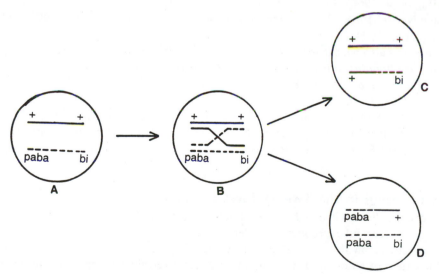

Fig. 16 A-D. **Diagrammatic representation of mitotic crossing over between a single pair of homologous chromosomes. A Diploid nucleus heterozygous for the genes paba (paraminobenzoic acid-requiring) and bi (biotin-requiring). B Crossing-over at chromosome replication (four-strand stage). C, D Genetically different nuclei formed after chromosome segregation.**

chromosomal aberrations (Tector and Kafer 1962). At present, the most effective agent, apparently free of side effects, is p-fluorophenylalanine (Morpurgo 1961; Lhoas 1961).

6.4   Haploidization

Analyses by Pontecorvo and Kafer (1958) have provided valuable data on the frequency of multiple exchanges at mitotic crossing-over, and on the co-incidence of crossing-over and non-disjunction or haploidization. Kafer (1961) carried out experiments to analyse how haploidization is caused. She observed that non-disjunction results in two daughter nuclei, one with 2n + 1 and the other with 2n−1 chromosomes, respectively. The hyperdiploid nucleus reverts to the 2n condition but the hypodiploid continues to lose chromosomes, probably one at a time, until the haploid condition is restored. In A. nidulans with n = 8, aberrants carrying appropriate marker genes have been detected with 17 (2n + 1), 15(2n − 1), 12, 11, 10, 9 and 8 chromosomes. Compared to diploid or haploid mycelia, all the hypodiploids showed reduced growth. On the basis of her analysis, Kafer (1961) has estimated that non-disjunction occurs once in every 50 mitoses.

7.   Occurrence of Parasexuality in Various Other Fungi

Claims have been made about the occurrence of parasexuality as an operative mechanism of variation in a number of filamentous fungi. Imperfect fungi in which diploids and mitotic reassortments occur include *Aspergillus niger* (Hutchinson 1958, Lhoas 1961), *A. amstelodami* (Lewis and Barron 1964), *A. rugulosus* (Coy and Tuveson 1964), *A. fumigatus* (Stromnaes and Garber 1963), *A. sojae* and *A. oryzae* (Ishitani et al. 1956), *Penicillium chrysogenum* (Pontecorvo and Sermonti 1953, 1954; Sermonti 1957), *P. italicum* (Stromnaes et al. 1964), *P. digitatum* (Stromnaes et al. 1964), *P. expansum* (Garber and Beraha 1965; Field and Stromnaes 1966), *Ascochyta imperfecta* (Sanderson and Srb 1965), *Cephalosporium mycophilum* (Tuveson and Coy 1961), *Verticillium alboatrum* (Hastie 1964), *V. dahliae* var. *longisporum* (Ingram 1968), *Fusarium oxysporum* f. sp *pisi* (Tuveson and Garber 1959), *F. oxysporum* f. sp *cubense* (Buxton 1962), *F. fujikuroi* (Ming et al. 1966), *Phymatotrichum omnivorum* (Hosford and Gries 1966), *Pyricularia oryzae* (Yamasaki and Niizeki 1965), *Pseudocercosporella herpotrichoides* (Hocart et al. 1993) and *Cladosporium fulvum* (Arnau et al. 1994).

Parasexuality has also been detected in Ascomycetes and Basidiomycetes which undergo normal sexual reproduction: *Cochliobolus sativus* (Tinline 1962), *Glomerella cingulata* (Stephan 1967), *Ustilago maydis* (Holliday 1961), *U. violacea* (Day and Jones 1968), *U. hordei* (Dinoor and Person 1969), *U. scabiosae* (Garber and Ruddat 1992), *Coprinus lagopus* and *C. radiatus* (Casselton 1965) and *Schizophyllum commune* (Middleton 1964). There is also suggestive evidence for parasexual cycle in *Puccinia graminis tritici* (Ellingboe 1961, Watson and Luig 1962). Similarly, some indications of parasexuality are also reported in coenocytic fungi, like *Phytophthora cactorum* (Buddenhagen 1958) and *Phycomyces blakesleeanus* (Park et al. 1968).

8.   Study of Fungi in the Twenty-First Century

The last few decades have witnessed a spurt of activity in the field of plant reproductive biology. This is natural, because successful reproduction underlies perpetuation of the species on one hand and their evolutionary plasticity on the other. Reproduction is being related to population structure and species diversity. The extent of auto- and xenogamy of a species determines its energy requirements and energy budgeting during the life cycle. Such phenomena as sibling rivalry, parent-offspring conflict, mate choice, etc., which are well established in the animal kingdom were unheared of in plants. The progress in reproductive biology of flowering plants during the last few decades has proved that these also operate in plants. Fungi are distinct from all other plants. Despite their heterotrophic mode of nutrition, they are ubiquitous and outnumber many higher plant groups, which reflects their evolutionary success, despite the fact that many fungi have foresaken sexuality which is the major source of variability in all other living organisms. Discovery of a parasexual cycle in some Deuteromycetes does not resolve this enigma fully.

Most of the filamentous fungi sporulate under varied spore inducing conditions; a few remain sterile. The relationship between spore inducing environmental factors and the genetic potential of the fungus for sporulation is not fully understood and, therefore, needs to be explored. Studies on microcycle conidiation can prove rewarding for unravelling the physiology and biochemistry of fungal sporulation.

Many studies have been carried out on the complex septum of Basidiomycetes, yet better understanding of its structure and biochemistry is necessary to fully appreciate the processes involved in septal dissolution, nuclear migration and clamp cell formation involved in dikaryotization. In fact, the dikaryon is a structure unique to fungi.

Problems related to sexual compatibility and hormonal regulation of syngamy need to be probed using molecular techniques. Although diffusible sex hormones have been discovered in a number of fungi, their characterization is awaited (in many cases). The existence of similar hormones in other species is yet to be expected to fill the gaps in existing knowledge.

The knowledge of reproduction in Myxomycetes is fragmentary. Discovery of heterothallic strains in some species has opened new perspectives for further research in this intriguing group of fungi.

The parasexual cycle operating in some fungi compensates for the lack of sexuality in imperfect fungi, and supplements sexuality in some other fungi with a perfect stage. Studies on the existence of parasexuality in other taxa, including coenocytic forms, are called for. The role of environmental factors in increasing or reducing the efficiency of the parasexual cycle operating in natural populations needs to be established. Should the operation of parasexuality in nature be established, it will account for the variation existing in the imperfect fungi which are hitherto known to reproduce only asexually in nature. Fuller realization of the potential of parasexuality requires research to assess the cytology and genetics of the recombination and standardization of criteria for its detection in nature. The field is fertile, though challenging.

References

Abe K, Kusaka I, Fukui S (1975) Morphological changes in the early stages of the mating process of *Rhodosporidium toruloides*. J Bacteriol 122: 710–718

Alexopoulos CJ (1962) Introductory Mycology, 2nd edn. Wiley, New York, pp 613

Alexopoulos CJ, Mims CW (1979) Introductory Mycology. Wiley Eastern, New Delhi, pp 632

Anderson JG, Smith JE (1971) The production of conidiophores and conidia by newly germinated conidia of *Aspergillus niger* (microcycle conidiation). J Gen Microbiol 69: 185–197

Arnau J, Housego AP, Oliver RP (1994) The use of RAPD markers in the genetic analysis of the plant pathogenic fungus *Cladosporium fulvum*. Curr Genet 25: 438–444.

Arsenault GP, Biemann K, Barksdale AW, McMorris TC (1968) The structure of antheridiol, a sex hormone in *Achlya bisexualis*. J Am Chem Soc 90: 5635–5636

Bakkeren G, Kronstad JW (1993) Conservation of the b-mating type gene complex among bipolar and tetrapolar smut fungi. Plant Cell 5: 123–136.

Bandoni RJ (1965) Secondary control of conjugation in *Tremella mesenterica*. Can J Bot 43: 627–630

Banuett F (1992) *Ustilago maydis*, a blight that is a delight. Trends Genet 8: 174–180.

Banuett F, Herskowitz I (1989) Different 'a' alleles of *Ustilago maydis* are necessary for maintenance of filamentous growth but not for meiosis. Proc Natl Acad Sci USA 86: 5878–5882

Barkasdale AW, Lasure LL (1974) Production of hormone B by *Achlya heterosexualis*. Appl Microbiol 28: 544–546.

Barksdale AW, McMorris TC, Seshadri R, Aranachalam T, Edwards JA, Sundeen J, Green JM (1974) Responses of *Achlya ambisexualis* E87 to the hormone antheridiol and certain other steroids. J Gen Microbiol 82: 295–299

Barnett HL, Lilly VG (1955) The effects of humidity, temperature and carbon dioxide on sporulation of *Choanephora cucurbitarum*. Mycologia 47: 26–29

Bartnicki-Garcia S, Hemmes DE (1976) Some aspects of the form and function of Oomycete spores. *In*: Weber DJ, Hess WM (eds) The fungal spore: form and function. Wiley, New York, pp 593–641.

Betz R, Duntze W (1979) Purification and partial chararacterization of 'a'-factor, a mating hormone produced by mating type-'a' cells from *Saccharomyces cerevisiae*. Eur J Biochem 95: 469–475

Betz R, Mackay VL, Duntze W (1977) 'a-Factor' from *Saccharomyces cerevisiae* : partial characterization of a mating hormone produced by cells of mating type-'a'. J Bacteriol 132: 462–472.

Bistis GN (1983) Evidence for diffusible, mating-type-specific trichogyne attractants in *Neurospora crassa*. Exp Mycol 7: 292–295

Bistis GN, Raper JR (1963) Heterothallism and sexuality in *Ascobolus stercorarius*. Am J Bot 50: 880–891

Blakeslee AF (1904) Sexual reproduction in the Mucorineae. Proc Natl Acad Sci USA 40: 205–319

Blakeslee AF (1920) Sexuality in mucors. Science 51: 375–382

Blakeslee AF, Cartledge JL (1927) Sexual dimorphism in Mucorales. 2. Interspecific reactions. Bot Gaz 84: 51–57

Bolker M, Kahmann R (1993) Sexual pheromones and mating responses in fungi. Plant Cell 5: 1461–1469

Boyer MG (1961) Variability and hyphal anastomosis in host-specific forms of *Marssonina populi* (Lib.) Magn. Can J Bot 39: 1409–1427

Buddenhagen IW (1958) Induced mutations and variability in *Phytophthora cactorum*. Am J Bot 45: 355–365

Bu'Lock JD, Osagie AU (1973) Prenols and ubiquinones in single strain and mated cultures of *Blakeslea trispora*. J Gen Microbiol 76: 77–83

Burgeff H (1924) Untersuchungen Über Sexualitat and Parasitismus bei Mucorineen. 1. Bot Abh 4: 5–155

Buxton EW (1962) Parasexual recombination in the banana-wilt *Fusarium*. Trans Br Mycol Soc 45: 274–279

Carlile MJ, Machlis L (1965) A comparative study of the chemotaxis of the motile phases of *Allomyces*. Am J Bot 52: 484–486

Carmichael JW (1971) Blastospores, aleuriospores, chlamydospores. *In* : Kendrick B (ed) Taxonomy of fungi imperfecti, University of Toronto Press, Toronto, pp 50–70

Casselton LA (1965) The production and behaviour of diploids of *Coprinus lagopus*. Genet Res 6: 190–208

Cerda-Olmedo E (1975) The genetics of *Phycomyces blakesleeanus*. Genet Res 25: 285–296

Champe SP, Kurtz MB, Yager LN, Butnick NJ, Axelrod DE (1981) Spore formation in *Aspergillus nidulans*. Competence and other developmental processes. *In* : Turian G, Hohl HR (eds) The fungal spore: morphogenetic controls. Academic Press, London, pp 255–276

Clutterbuck AJ, Roper JA (1966) A direct determination of nuclear distribution in heterokaryons of *Aspergillus nidulans*. Genet Res 7: 185–194

Cochrane VW, Cochrane JC (1970) Chlamydospore development in the absence of protein synthesis in *Fusarium solani*. Dev Biol 23: 345–354

Cole GT (1975) The thallic model of conidiogenesis in the Fungi Imperfecti. Can J Bot 53: 2983–3001

Cole GT, Kendrick B (1981) Biology of conidial fungi, vol 2, Academic Press, New York, pp 660

Collins OR (1961) Heterothallism and homothalism in two Myxomycetes. Am J Bot 48: 674–683

Collins OR (1963) Multiple alleles at the incompatibility locus in the myxomycete *Didymium iridis*. Am J Bot 50: 477–480

Collins OR (1976) Heterothallism and homothallism. A study of 27 isolates of *Didymium iridis*, a true slime mold. Am J Bot 63: 138–143

Couch JN (1926) Heterothallism in *Dictyuchus*, a genus of the water moulds. Ann Bot (Lond) 40: 848–881

Coy DO, Tuveson RW (1964) The effects of supplementation and plating densities on apparently aberrant meiotic and mitotic segregation in *Aspergillus rugulosus*. Genetics 50: 847–853

Craigie JH (1927) Discovery of the function of pycnia of the rust fungi. Nature 120: 765–767

Dahlberg KR (1982) Physiology and biochemistry of fungal sporulation. Annu Rev Phytopathol 20: 281–301

Davey J (1992) Mating pheromones of the fission yeast *Schizosaccharomyces pombe* : Purification and structural characterization of M-factor and isolation and analysis of two genes encoding the pheromone. EMBO J 11: 951–960

Davis RH (1966) Mechanisms of inheritance. 2. Heterokaryosis. In: Ainsworth GC, Sussman AS (eds) The fungi, an advanced treatise. Academic Press, New York, pp 567–588.

Day AW, Jones JK (1968) The production and characteristics of diploids in *Ustilago violacea*. Genet Res 11: 63–81

Day PR (1963) The structure of the A-mating type factor in *Coprinus lagopus:* wild alleles. Genet Res 4: 323–325

Dee J (1960) A mating type system in an acellular slime mold. Nature 185: 780–781

Dee J (1966) Multiple alleles and other factors affecting plasmodial formation in the true slime mould *Physarum polycephalum*. Schw. J Protozool 13: 610–616

Dinoor A, Person C (1969) Genetic complementation in *Ustilago hordei*. Can J Bot 47: 9–14

Dixon PA (1959) Life history and cytology of *Ascocybe grovesii*. Ann Bot Lond 23: 509–520

Ellingboe AH (1961) Somatic recombination in *Puccinia graminis* var. *tritici*. Phytopathology 51: 13–15

Emerson R (1941) An experimental study of the life cycle and taxonomy of *Allomyces*. Lloydia 4: 77–144

Erselius LJ, Shaw DS (1982) Protein and enzyme differences between *Phytophthora palmivora* and *P. megakarya*: evidence for self-fertilization in pairings of the two species. Trans Br Mycol Soc 78: 227–238

Field A, Stromnaes O (1966) The parasexual cycle and linkage groups in *Penicillium expansum*. Hereditas 54: 389–403

Fields S (1990) Pheromone response in yeast. Trends Biochem Sci 15: 270–273

Fields WG, Grear JW (1966) A new heterothallic species of *Sordaria* from Ceylon. Mycologia 58: 524–528

Fields WG, Maniotis J (1963) Some cultural and genetic aspects of a new heterothallic *Sordaria*. Am J Bot 50: 80–85.

Flegel TW (1981) The pheromonal control of mating in yeasts and its phylogenetic implication: a review. Can J Microbiol 27: 373–389

Garber ED, Beraha L (1965) Genetics of phytopathogenic fungi. 16. The parasexual cycle in *Penicillium expansum*. Genetics 52: 487–492

Garber ED, Ruddat M (1992) The parasexual cycle in *Ustilago scabiosae* (Ustilaginales). Int J Plant Sci 153: 98–101

Garrett SD (1963) Soil fungi and soil fertility. 2nd edn. Pergamon Press, Oxford (UK), pp 150

Garza-Chapa R, Anderson NA (1966) Behaviour of single basidiospore isolates and heterokaryons of *Rhizoctonia solani* from flax. Phytopathology 56: 1260–1268

Gooday GW (1974) Fungal sex hormones. Annu Rev Biochem 43: 35–49

Gooday GW (1978) Functions of trisporic acid. Philos Trans R Soc Lond B 284: 509–520

Gooday GW (1983) Hormones and sexuality in fungi. In: Bennett JW, Ciegler A (eds) Secondary metabolism and differentiation in fungi. Marcel Dekker, New York, pp 239–266

Grange F, Turian G (1978) Differential deoxyribonucleic acid synthesis during microcycle conidiation in *Neurospora crassa*. Arch Microbiol 119: 257–261

Gregory PH (1966) The fungus spore: what it is and what it does. In: Madelin MF (ed) The fungus spore. Butterworths, London, pp 1–13

Griffin DH (1981) Fungal physiology. Wiley, New York, pp 383

Guilliermond A (1940) Sexuality, developmental cycle and phylogeny of yeasts. Bot Rev 6: 1–24

Guo LY, Ko WH (1991) Hormonal regulation of sexual reproduction and mating type change in heterothallic *Pythium splendens*. Mycol Res 95: 452–456

Hanlin RT (1994) Microcycle conidiation—a review. Mycoscience 35: 113–123

Hansen HN, Smith RE (1934) Interspecific anastomosis and the origin of new types in imperfect fungi. Phytopathology 24: 1144

Harder R, Sorgell G (1938) Über einen neuen planoisogamen Phycomyceten mit Generationswechsel und seine phylogenetische Bedeutung. Biologie 3: 119–127

Hastie AC (1964) The parasexual cycle in *Verticillium albo-atrum*. Genet Res 5: 305–315

Hawker LE (1966) Environmental influences on reproduction. *In*: Ainsworth GC, Sussman AS (eds) The fungi: an advanced treatise. Academic Press, New York, pp 435–469

Heagy FC, Roper JA (1952) Deoxyribose nucleic acid content of haploid and diploid *Aspergillus* conidia. Nature 170:713

Henney MR (1967) The mating type system of the myxomycete *Physarum flavicomum*. Mycologia 59: 637–652

Herskowitz I (1989) A regulatory hierarchy for cell specialization in yeast. Nature 342: 749–757

Ho HH (1986) Notes on the heterothallic hehaviour of *Phytophthora megasperma* from alfalfa. Mycologia 78: 306–309

Ho HH, Jong SC (1986) A comparison between *Phytophthora cryptogea* and *P. drechsleri*. Mycotaxon 27: 289–319

Hocart MJ, Lucas JA, Peberdy JF (1993) Characterization of the parasexual cycle in the eyespot fungus, *Pseudocerocosporella herpotrichoides*. Mycol Res 97: 967

Holliday R (1961) Induced mitotic crossing-over in *Ustilago maydis*. Genet Res 2: 204–231

Hosford RM, Gries GA (1966) The nuclei and parasexuality in *Phymatotrichum omnivorum*. Am J Bot 53: 570–579

Huang BF, Cappellini RA (1980) Sporulation of *Gibberella zeae*. 6. Sporulation and maximum mycelial growth occur simultaneously. Mycologia 72: 1231–1235

Hughes SJ (1971) Phycomycetes, Basidiomycetes and Ascomycetes as Fungi Imperfecti. *In*: Kendrick B (ed) Taxonomy of fungi imperfecti. University Toronto Press, Toronto Buffalo, pp 7–36

Hutchinson JM (1958) A first five-marker linkage group identified by mitotic analysis in the asexual *Aspergillus niger*. Microb Genet Bull 15: 17

Ingold CT (1971) Fungal spores: their liberation and dispersal, Clarendon Press, Oxford (UK), pp 302

Ingram R (1968) *Verticillium dahliae* var. *longisporum*, a stable diploid. Trans Br Mycol Soc 51: 339–341

Ishibashi Y, Sakagami Y, Isogai A, Suzuki A (1984) Structures of tremerogens A-9291-1 and A-9291-VIII: peptidal sex hormones of *Tremella brasiliensis*. Biochem 23: 1399–1404

Ishitani C, Sakaguchi K (1956) Hereditary variation and genetic recombination in Koji-molds (*Aspergillus oryzae* and *A. sojae*). 5. Heterocaryosis. J Gen Appl Microbiol (Tokyo) 2: 345–400

Ishitani C, Ikeda Y, Sakaguchi K (1956) Hereditary variation and genetic recombination in Koji-molds (*Aspergillus oryzae* and *A. sojae*). 6. Genetic recombination in heterozygous diploids. J Gen Microbiol (Tokyo) 2: 410–430

Kafer E (1958) An 8-chromosome map of *Aspergillus nidulans*. Adv Genet 9: 105–145

Kafer E (1961) The process of spontaneous recombination in vegetative nuclei of *Aspergillus nidulans*. Genetics 46: 1581–1609

Kafer E (1963) Radiation effects and mitotic recombination in diploids of *Aspergillus nidulans*. Genetics 48: 27–45

Kamiya Y, Sakurai A, Takahashi N (1980) Metabolites of mating pheromone, rhodotorucine A, by 'a' cells of *Rhodosporidium toruloides*. Biochem Biophys Res Commun 94: 855–860

Kamiya Y, Sakurai A, Tamura S, Takahashi N, Abe K, Tsuchiya E, Fukui S (1978) Isolation of rhodotorucine A, a peptidyl factor inducing the mating tube formation in *Rhodosporidum toruloides*. Agric Biol Chem 42: 1239–1243

Kendrick B (1971) Arthroconidia and meristem arthroconidia, *In*: Kendrick B (ed) Taxonomy of fungi imperfecti. University Toronto Press, Toronto, Buffalo, pp 160–175

Klebs G (1899) Zur Physiologie der Fortpflanzung einiger Pilze. Jahrb Wiss Bot 33: 71–156

Kniep H (1919) Untersuchungen über den Antherenbrand (*Ustilago violacea* Press.). Z Bot 11: 257–284

Kniep H (1920) Uber morphologische und physiologische Geschlechts-differenzierung. (Untersuchungen an Basidiomyzeten). Verh Phys Med Ges Wurzburg 46: 1–18

Kniep H (1922) über Geschlechtsbestimmung und Reduktionsteilung. Verh Phys Med Ges Wurzburg 47: 1–28

Kniep H (1926) über Artkreuzungen bei Brandpilzen. Z Pilzkd 5: 217–247

Ko WH (1978) Heterothallic *Phytophthora*: evidence for hormonal regulation of sexual reproduction. J Gen Microbiol 107: 15–18

Ko WH (1980) Hormonal regulation of sexual reproduction in *Phytophthora*. J Gen Microbiol 116: 459–463

Kurjan J (1992) Pheromone response in yeast. Annu Rev Biochem 61: 1097–1129

Leupold U, Nielsen O, Egel R (1989) Pheromone-induced meiosis in P-specific mutants of fission yeast. Curr Genet 15: 403–405.

Levi JD (1956) Mating reaction in yeast. Nature 177: 753–754

Lewis KE, O'Day DH (1976) Sexual hormone in the cellular slime mold *Dictyostelium purpureum*. Can J Microbiol 22: 1269–1273

Lewis KE, O'Day DH (1977) Sex hormone of *Dictyostelium discoideum* is volatile. Nature 268: 730–731

Lewis KE, O'Day DH (1979) Evidence for a hierarchical mating system operating via phermones in *Dictyostelium giganteum*. J Bacteriol 138: 251–253

Lewis LA, Barron GL (1964) The pattern of the parasexual cycle in *Aspergillus amstelodami*. Genet Res 5: 162–163

Lhoas P (1961) Mitotic haploidization by treatment of *Aspergillus niger* diploids with para-fluoro-phenylalanine. Nature 190: 744

Lhoas P (1967) Genetic analysis by means of the parasexual cycle in *Aspergillus niger*. Genet Res 10: 45–51

Lovett JS (1975) Growth and differentiation of the water mold *Blastocladiella emersonii*: cytodifferentiation and the role of ribonucleic acid and protein synthesis. Bacteriol Rev 39: 345–404

Lysek G (1978) Circadian rhythms. *In*: Smith JE, Berry DR (eds) The filamentous fungi, vol 3. Edward Arnold, London, pp 376–388

Machlis L (1958) A study of sirenin, the chemotactic sexual hormone from the water mold *Allomyces*. Physiol Plant 11: 845–854

Machlis L (1973a) Factors affecting the stability and accuracy of the bioassay for the sperm attractant sirenin. Plant Physiol 52: 524–526

Machlis L (1973b) The chemotactic activity of various sirenins and analogues and the uptake of sirenin by the sperm of *Allomyces*. Plant Physiol 52: 527–530

Marsh L, Neiman AM, Herskowitz I (1991) Signal transduction during pheromone response in yeast. Annu Rev Cell Biol 7: 699-728

Mayfield JE (1974) Septal involvement in nuclear migration in *Schizophyllum commune*. Arch Microbiol 95: 115–124

Mc Cullough J, Herskowitz I (1979) Mating pheromones of *Saccharomyces kluyveri* and *S. cerevisiae*. J Bacteriol 138: 146–154

Mc Morris TC (1978) Antheridiol and the oogoniols, steroid hormones which control sexual reproduction in *Achlya*. Philos Trans R Soc Lond B 248: 459–470

McMorris TC, Barksdale AW (1967) Isolation of a sex hormone from the water mould *Achlya bisexualis*. Nature 215: 320–321

McMorris TC, Seshadri R, Weiche GR, Arsenault GP, Barksdale AW (1975) Structures of oogoniol-1, -2 and -3 steroidal sex hormones of the water mold, *Achlya*. J Am Chem Soc 97: 2544–2545

Mesland DAM, Huisman JG, Van Den Ende H (1974) Volatile sex hormones in *Mucor mucedo*. J Gen Microbiol 80: 111–117

Middleton RB (1964) Sexual and somatic recombination in common-AB heterokaryons of *Schizophyllum commune*. Genetics 60: 209–210

Ming YN, Lin PC, Yu TF (1966) Heterokaryosis in *Fusarium fujikuroi* (Saw.) Wr. Sci Sin (Peking) 15: 371-378

Moore RT, McAlear JH (1962) Fine structure of Mycota. 7. Observations on septa of Ascomycetes and Basidiomycetes. Am J Bot 49: 86–94.

Moore TDE, Edman JC (1993) The α-mating type locus of *Cryptococcus neoformans* contains a peptide pheromone gene. Mol Cell Biol 13: 1962–1970

Morpurgo G (1961) Segregation induced by p-fluoro-phenylalanine. *Aspergillus* News L 2: 10

Mullins JT, Warren CO (1975) Nutrition and sexual reproduction in the water mold *Achlya*. Am J Bot 62: 770–774

O'Day DH (1981) Modes of cellular communication and sexual interactions in eukaryotic microbes. *In* : O'Day DH, Horgen PA (eds) Sexual interactions in eukaryotic microbes. Academic Press, London, pp 3–17

O'Day DH, Lewis KE (1975) Diffusible mating type factors induce macrocyst development in *Dictyostelium discoideum*. Nature 254: 431–432

O'Day DH, Lewis KE (1981) Pheromonal interactions during mating in *Dictyostelium*. *In* : O'Day DH, Horgen PA (eds) Sexual interactions in eukaryotic microbes. Academic Press, New York, pp 199–221

Olive LS, (1956) Genetics of *Sordaria fimicola*. 1. Ascospore colour mutants. Am J Bot 43: 97–106

Olive LS, Fantini AA (1961) A new heterothallic species of *Sordaria*. Am J Bot 48: 124–128

Papazian HP (1954) Exchange of incompatibility factors between the nuclei of a dikaryon. Science 119: 619–693

Park S, Kenehan P, Goodgal S (1968) Heterocaryons and recombination in *Phycomyces blakesleeanus*. Genetics 60: 209–210

Pommerville J (1981) The role of sexual hormones in *Allomyces*. *In* : O'Day DH, Horgen PA (eds) Sexual interactions in eukaryotic microbes. Academic Press, London, pp 53–92.

Pontecorvo G (1954) Mitotic recombination in the genetic systems of filamentous fungi. Caryologia (Suppl) 6: 192–200

Pontecorvo G (1956) The parasexual cycle in fungi. Ann Rev Microbiol 10: 393–400

Pontecorvo G (1958) Trends in genetic analysis. Columbia University Press, New York, pp 145

Pontecorvo G, Kafer E (1958) Genetic analysis by means of mitotic recombination. Adv Genet 9: 71–104

Pontecorvo G, Roper JA (1952) Genetic analysis without production by means of polyploidy in *Aspergillus nidulans*. J Gen Microbiol 6: 7 (Abst.)

Pontecorvo G, Roper JA (1953) Diploids and mitotic recombination. Adv Genet 5: 218–233

Pontecorvo G, Sermonti G (1953) Recombination without sexual reproduction in *Penicillium chrysogenum*. Nature 172: 126

Pontecorvo G, Sermonti G (1954) Parasexual recombination in *Penicillium chrysogenum*. J Gen Microbiol 11: 94–104

Pontecorvo G, Roper JA, Forbes E (1953) Genetic recombination without sexual reproduction in *Aspergillus niger*. J Gen Microbiol 8: 198–210

Pontecorvo G, Tarr-Gloor E, Forbes E (1954) Analysis of mitotic recombination in *Aspergillus nidulans*. J Genet 52: 226–237

Raper CA, Raper JR, Miller RE (1972) Genetic analysis of the life cycle of *Agaricus bisporus*. Mycologia 64: 1088–1117

Raper JR (1951) Sexual hormones in *Achlya*. Am Sci 39: 110–120

Raper JR (1952) Chemical regulation of sexual processes in the Thallophytes. Bot Rev 18: 447–543

Raper JR (1954) Life cycles, sexuality and sexual mechanisms in the fungi. *In*: Wenrich DH, Lewis IF, Raper JR (eds) Sex in microorganisms. American Association for the Advancement of Science, Washington DC, pp 42–81

Raper JR (1959) Sexual versatility and evolutionary processes in fungi. Mycologia 51: 107–124

Raper JR (1960) The control of sex in fungi. Am J Bot 47:794-808

Raper JR (1966a) Life cycles, basic patterns of sexuality, and sexual mechanisms. In: Ainsworth GC, Sussman AS (eds) The fungi: an advanced treatise, vol 2. Academic Press, New York, pp 473–511

Raper JR (1966b) Genetics of sexuality in higher fungi. Ronald Press, New York, pp 283

Raper JR, Esser K (1961) Antigenic differences due to the incompatibility factors in *Schizophyllum commune*. Z Vererbungsl 92: 439–444

Raper JR, Flexer AS (1971) Mating systems and evolution of the Basidiomycetes. *In*: Peterson RH (ed) Evolution in the higher Basidiomycetes. University of Tennessee Press, Nashville, pp 149–176

Raper JR, Baxter MG, Ellingboe AH (1960) The genetic structure of the incompatibility factors of *Schizophyllum commune*: The 'A' factor. Proc Natl Acad Sci USA 46: 833–842

Raper JR, Baxter MG, Middleton RB (1958) The genetic structure of the incompatibility loci in *Schizophyllum*. Proc Natl Acad Sci USA 53: 889–900

Raper KB, Fennel DI (1965) The genus *Aspergillus*, Williams and Wilkins, Baltimore, pp 686

Reid ID (1974) Properties of conjugation hormones (erogens) from the Basidiomycete *Tremella mesenterica*. Can J Bot 52: 521–524

Rizwana R, Powell WA (1995) Ultraviolet light-induced heterokaryon formation and parasexuality in *Cryphonectria parasitica*. Exp Mycol 19: 48–60

Roper JA (1952) Production of heterozygous diploids in filamentous fungi. Experientia 8: 14–15

Roper JA (1962) Genetics and microbial classification. Symp Soc Gen Microbiol 12: 270–288

Roper JA (1966) Mechanisms of inheritance. 3. The Parasexual cycle. *In* : Ainsworth GC, Sussman AS (eds) The fungi: an advanced treatise. Academic Press, New York, pp 589–617

Sakagami Y, Yoshida M, Isogai A, Suzuki A (1981a) Structure of tremerogen 'a-13', a peptidal sex hormone of *Tremella mesenterica*. Agric Biol Chem 45: 1045–1047

Sakagami Y, Yoshida M, Isogai A, Suzuki A (1981b) Peptidal sex hormones inducing conjugation tube formation in compatible mating-type cells of *Tremella mesenterica*. Science 212: 1525–1527

Sakurai A, Tamura S, Takahashi N, Abe K, Tsuchiya E, Fukui S, Kitada C, Fujino M (1978) Structure of rhodotorucine 'A', a novel lipopetide, inducing mating tube formation in *Rhodosporidium toruloides*. Biochem Biophys Res Commun 83: 1077–1083

Sanderson KE, Srb AM (1965) Heterokaryosis and parasexuality in the fungus *Ascochyta imperfecta*. Am J Bot 52: 72–81

Sansome ER (1963) Meiosis in *Pythium debaryanum* Hesse and its significance in the life history of the Biflagellatae. Trans Br Mycol Soc 46: 63–72

Schreiber LR, Green RJ Jr (1966) Anastomosis in *Verticillium alboatrum* in soil. Phytopathology 56: 1110–1111

Sekiguchi J, Gaucher GM, Costerton JW (1975) Microcycle conidiation in *Penicillium urticae*: an ultrastructural investigation of spherical spore growth. Can J Microbiol 21: 2048–2058

Sermonti G (1957) Analysis of vegetative segregation and recombination in *Penicillium chrysogenum*. Genetics 42: 433–443

Seth HK (1967) Studies on the genus *Chaetomium*. 1. heterothallism. Mycologia 59: 580–584

Shattock RC, Tooley PW, Fry WE (1986) Genetics of *Phytophthora infestans:* determination of recombination, segregation and selfing by isozyme analysis. Phytopathology 76: 410–413

Shaw DS, Fyfe AM, Hibberd PG, Abdel-Stattar MA (1985) Occurrence of the rare A_2 mating type of *Phytophthora infestans* on imported Egyptian potatoes and the production of sexual progeny with A_1 mating types from the UK. Plant Pathol 34: 552–556

Shear CL, Dodge BO (1927) Life histories and heterothallism of the red-bread mold fungi of the Monilia group. J Agric Res 34: 1019–1042

Skidmore DI, Shattock RC, Shaw DS (1984) Oospores in cultures of *Phytophthora infestans* resulting from selfing induced by the presence of *P. drechsleri* isolated from blighted potato foliage. Plant Pathol 33: 173–183

Smith JE, Galbraith JC (1971) Biochemical and physiological aspects of differentiation in the fungi. Adv Microb Physiol 5: 45–134

Stephan BR (1967) Untersuchungen Uber die Variabilitat bei *Colletotrichum gloeosporioides* Penzig in Verbindung mit Heterokaryose. 3. Versuche zum Nachweis der Heterokaryose. J Bakteriol Parasitenkd 121: 73–83

Stern C (1936) Somatic crossing over and segregation in *Drosophila melanogaster*. Genetics 21: 625–730

Stotzler D, Klitz HH, Duntze W (1976) Primary structure of α-factor peptides from *Saccharomyces cerevisiae,* Eur J Biochem 69: 397–400

Stromnaes O, Garber ED (1963) Heterocaryosis and the parasexual cycle in *Aspergillus fumigatus*. Genetics 48: 653–662

Stromnaes O, Garber ED, Beraha L (1964) Genetics of phytopathogenic fungi. 9. Heterocaryosis and the parasexual cycle in *Penicillium italicum* and *P. digitatum*. Can J Bot 42: 423–427

Sutter RP (1975) Mutations affecting sexual development in *Phycomyces blakesleeanus* Proc Natl Acad Sci USA 72: 127–130

Suzuki H (1967) Studies on biologic specialization in *Pyricularia oryzae* Cav. Institute of Plant Pathology, Tokyo, pp 235

Taber WA (1966) Morphogenesis in Basidiomycetes. *In* : Ainsworth GC, Sussman AS (eds) The fungi: an advanced treatise, vol 2. Academic Press, New York, pp 387–412

Takemaru T (1961) Genetic studies on fungi. 10. The mating system in Hymenomycetes and its general mechanism. Biol J Okayma Univ 7: 133–211

Talbot PHB (1971) Principles of fungal taxonomy. Macmillan, London, pp 274

Tan KK (1978) Light-induced fungal development. *In* : Smith JE, Berry DR (ed) The filamentous fungi, vol 3. Edward Arnold, London, pp 334–357

Tector MA, Kafer E (1962) Radiation-induced chromosomal aberrations and lethals in *Aspergillus nidulans*. Science 136: 1056–1057

Terakawa H (1957) The nuclear behaviour and the morphogenesis in *Pleurotus ostreatus*. Sci Pap Coll Gen Educ, University of Tokyo 7: 61–68

Terakawa H (1960) The incompatability factors in *Pleurotus ostreatus*. Sci Pap Coll Gen Educ, University of Tokyo 10: 65–71

Tinline RE (1962) *Cochliobolus sativus*. 5. Heterokaryosis and parasexuality. Can J Bot 40: 425–437

Tsuchiya E, Fukui S (1978) Biological activities of sex hormone produced by *Tremella mesenterica*. Agric Biol Chem 42: 1089-1091

Turian G (1974) Sporogenesis in fungi. Annu Rev Phytopathol 12: 129–137

Turian G (1978) Sexual morphogenesis in the Ascomycetes. *In*: Smith JE, Berry DR (eds) The filamentous fungi, vol 3. Edward Arnold, London, pp 315–333

Tuveson RW, Coy DO (1961) Heterokaryosis and somatic recombination in *Cephalosporium mycophilum*. Mycologia 53: 244–253

Tuveson RW, Garber ED (1959) Genetics of phytopathogenic fungi. 2. The parasexual cycle in *Fusarium oxysporum* f. *pisi*. Bot Gaz 112: 74–80

Uchida JT, Aragaki M (1980) Chemical stimulation of oospore formation in *Phytophthora capsici*. Mycologia 72: 1103–1108

Uchida K, Ishitani C, Ikeda Y, Sakaguchi K(1958) An attempt to produce interspecific hybrids between *Aspergillus oryzae* and *A. sojae*. J Gen Appl Microbiol (Tokyo) 4: 31–38

Van Den Ende H (1978) Sexual morphogenesis in the Phycomycetes, *In*: Smith JE, Berry DR (eds) the filamentous fungi, vol 3. Edward Arnold, London, pp 257–274

Vuillemin P (1910) Les conidiophores. Bull Soc Sci (Nancy, France) 11: 129–172

Vuillemin P (1911) Les Aleurispores. Bull Soc Sci Nancy 12: 151–175

Wang CS, Raper JR (1969) Protein specificity and sexual morphogenesis in *Schizophyllum commune*. J Bacteriol 99: 291–297

Wang CS, Raper JR (1970) Isozyme patterns are sexual morphogenesis in *Schizophyllum*. Proc Natl Acad Sci USA 66: 882–889

Watson IA, Luig NH (1962) Asexual intercrosses between somatic recombinants of *Puccinia graminis*. Proc Linn Soc NSW 87: 99–104

Wessels JGH (1965) Morphogenesis and biochemical processes in *Schizophyllum commune* Fr. Wentia 13: 1–113

Wessels JGH (1969) Biochemistry of sexual morphogenesis in *Schizophyllum commune:* effect of mutations affecting the incompatibility system on cell wall metabolism. J Bacteriol 98: 697–704

Wessels JGH, Niederpruem DJ (1967) Role of a cell wall glucan degrading enzyme in mating of *Schizophyllum commune*. J Bacteriol 94: 1594–1602

Wheals AE (1970) A homothallic strain of the Myxomycete *Physarum polycephalum*. Genetics 66: 623–633

Whitehouse HLK (1949a) Heterothallism and sex in the fungi. Biol Rev Camb Philos Rev Camb Philos Soc 24: 411–447

Whitehouse HLK (1949b) Multiple allelomorph heterothallism in the fungi. New Phytol 48: 212–244

Winge O, Laustsen O (1939a) On 14 new yeast types, produced by hybridization. CR Trav Lab Carlsberg Ser Physiol 22: 337–355

Winge O, Laustsen O (1939b) *Saccharomycodes ludwigii,* a balanced heterozygote. CR Trav Lab Carlsberg Ser Physiol 22: 357–370

Yamasaki Y, Niizeki H (1965) Studies on variation of the rice blast fungus, *Pyricularia oryzae* Cav. 1. Karyological and genetical studies on variation. Bull Natl Inst Agric Sci (Japan) 13: 231–273

Yoshida M, Sakagami Y, Isogai A, Suzuki A (1981) Isolation of tremerogen 'a-13', a peptidal sex hormone of *Tremella mesenterica*. Agric Biol Chem 45: 1043–1044

Youatt J (1976) Sporangium formation in *Allomyces* throughout the growth cycle. Trans Br Mycol Soc 67: 159–161

Yu JY, Chang HS (1980) Chemical regulation of sexual reproduction in *Phytophthora colocasiae*. Bot Bull Acad Sin 21: 155–158

Zeidler G, Margalith P (1973) Modification of the sporulation cycle in *Penicillium digitatum*. Can J Microbiol 19: 481–483

Zonneveld BJM (1972) Morphogenesis in *Aspergillus nidulans*. The significance of α-1, 3-glucan of the cell wall and α-1, 3-glucanase for cleistothecium development. Biochem Biophys A 273: 174–187

7

Reproductive Biology of Lichens

K.V. Krishnamurthy and D.K. Upreti

1. Introduction

Lichens are among the most widely distributed eukaryotic organisms in the world (Galun 1988) and already about 14000 lichen species are known. Since the symbiotic association of a mycobiont and a photobiont is a pre-requisite to the existence of a lichen, for its dispersal either both partners must be distributed simultaneously or certain adaptations must ensure contact and relichenisation after separate dissemination of mycobionts and photobionts (Jahns 1988). Although the symbiotic association provides the lichen the ability to grow on a wide variety of substrata in diverse climate conditions, the dispersal unit is decided primarily by the type of reproduction, sexual or non-sexual. Since in the lichens the mycobiont is the 'exhabitant' and the photobiont the 'inhabitant', and since it is the exhabitant which normally retains the capacity to reproduce sexually in mutualistic symbiosis (Law and Lewis 1983), sexual reproduction of lichens involves only the mycobiont which produces a fruiting body that contains spores. The fungal spores are liberated and under suitable conditions and substrata they germinate and acquire compatible algae and develop new vegetative thalli. In nonsexual reproduction both mycobiont and photobiont multiply and remain associated in producing the vegetative propagules such as soredia, isidia and hormocysts.

When lichenised, for the photobionts, even the common modes of propagation encountered in their free-living state may be absent or restricted to a great extent (Tschermak-Woess 1988). In lichenised *Gloeocapsa*, bipartition of cells may occur, but neither akinetes nor nanocytes have been recorded, probably due to the absence of conditions favouring active vegetative multiplication. Lichenised *Nostoc*, Rivulariaceae, *Stigonema* and *Scytonema* do not produce hormogonia; neither are there reports of akinetes in the lichenised *Stigonema* and *Scytonema* (Cyanobacteria). Species of *Heterococcus* (Xanthophyceae) and *Cocomyxa*, *Pseudotrebouxia*, *Dictyochloropsis*, *Myrmecia* and *Gloeocystis* (Chlorophyceae) reproduce in the lichenised state only through autospore formation but never through zoospores. Similarly, *Trebouxia* produces only aplanospores. Although there are reports of zoosporangia and/or gametangia in the lichenised *Trentepohlia* and *Coenogonium*, they are either probably wrong or need confirmation (Tschermak-Woess 1988).

2. Vegetative Reproduction

The vegetative reproductive organs of lichens are classified, based on their structure, as isidia, soralia and hormocystangia, respectively, producing vegetative propagules known as isidia, soredia and hormocysts. Both the organs and the propagules they produce are lichenised structures containing the photo- and myco-bionts in a symbiotic state. These propagules can produce a new vegetative thallus, after dispersal, under suitable conditions.

Fragmentation. Fragmentation is one of the commonest methods of vegetative reproduction in lichens.

The lichen thalli become dry and brittle before crumbling into smaller bits. After dispersal, new thalli develop from these bits by regeneration either at the same locality or at a new site. Any portion of the lichen thallus containing both the symbionts can potentially act as a source of regeneration of a new thallus. Vegetative reproduction through fragmentation is also very common before the death of the original thallus which has already produced fruit-bodies, by the progressive death of older and older parts in the thallus. One of the very interesting aspects about such regeneration from bits of original thallus pertains to the criticality of the size of the bits; extremely small bits obviously cannot regenerate a new thallus.

Soredia. Soredia are diaspores that are more common in lichens. They are produced in specialised areas or organs of the thallus called 'soralia' which are open excrescences of the thallus. Soralia are common in foliose and fruiticose lichens, less frequent in crustose lichens and are not formed by gelatinous lichens (Jahns 1988). According to their position on thallus, soralia may be laminal or marginal (= border soralia, Fig. 1A, B); usually formed on the surface of the thallus but occasionally in cracks that develop on the thallus (=fissure soralia). According to their size they may be 'farinose' or 'granular'. On the basis of their morphology, the soralia may be maculiform (elongated or rounded flat depressions on the thallus (Fig. 1C), spherical (Fig. 1 D), capitate (Fig. 1 F), subcapitate, cuff-shaped (Fig. 1 G), reniform, labriform (Fig. 1E), or punctiform. Soredia are invariably formed from the medullary region along with the algal layer of the thallus (Fig. 1H). In the primitive crustose lichens like *Lepraria,* the thallus is made up of a continuous layer of diffuse soredia. In some lichens soredia may develop from the tips of isidia.

Soredia are rounded balls with a few algal cells enveloped by fungal hyphae. Depending on the species they range between 25–100 μm in diameter. Often, several soredia adhere together to form a globular mass. Soredia are released by breaks or pores or cracks in the soralia and are dispersed by wind, water and/or animals.

The formation of soralia and the development of soredia in them depends on the micro-climatic conditions that exist near the lichen. Higher humidity by day at moist sites may favour production of soredia. Shade, increased moisture and other environmental factors tend to stimulate algal growth and soredial formation may increase. There is no direct evidence that the production of soredia has a genetic basis, but the nature and structure of soralia and soredia, specific to specific lichens, indicates the involvement of probable genetic control.

Soredia are often distributed in bulk so that when they fall on any new substratum they are often deposited in clusters. This mode has special advantages. Full thallus regeneration is hardly possible from the germination product of a single soredium. This is overcome by the simultaneous germination of several closely placed soredia and subsequent fusion of a number of germinated soredial products to form a viable lichen thallus (Jahns 1982). This methodology is adopted by many soredial taxa (in India too).

Isidia. Isidia are finger-like outgrowths of the upper cortex (Fig. 2A). They have well-developed fungal tissues enclosing algal cells and correspond in their anatomy with the thallus structure (Fig. 2E). Isidia are of widespread occurrence in crustose, foliose and fruiticose lichens; gelatinous lichens form only isidia but no soredia (Jahns 1988). Isidia may be marginal or laminal. In rare instances they may develop inside soralia (e.g. *Bryoria* and *Usnea*). Isidia are simple (Fig. 2A), branched (Fig. 2B) or coralloid (Fig. 2D), solid (occasionally hollow in some lichens), corticated, 0.01 to 0.3 mm in diameter and 0.5 to 3.0 mm in height. They may be spherical (Fig. 2C), globular, cylindrical (Fig. 2A), lobulate, terete or sometimes squamulose.

Isidia break off in one piece; therefore each isidium is a single diaspore. The breaking occurs only after the degeneration of subtending cortical tissue of the thallus. Under such condition isidia break off from the main thallus through slashing rain water drops or by the action of wind. Isidia cannot be easily dispersed especially by wind due to their relatively larger weight. Consequently reproduction of lichens by isidia is

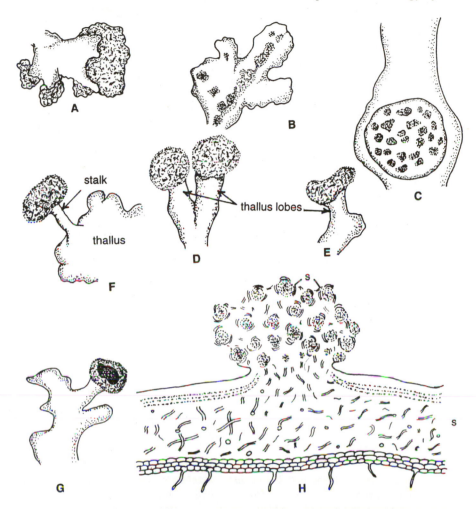

Fig. 1 A-H.   Soralia and Soredia. A *Physcia tribacoides,* marginal soralia. B *Parmelia* sp., laminal soralia. C *Ramalina* sp., maculiform soralia. D *Physcia* sp., spherical soralia. E *Hypogymnia* sp., labriform soralia. F *Parmelia* sp., capitate soralia. G *Menegazzia terebrata,* cuff-shaped soralia. H Transection of thallus of *Lobaria* sp. shows soralium with exposed soredia(s) (diagrammatic).

not as efficient as that by soredia. It is also important to mention that propagated isidia cannot directly develop into a new thallus but must follow a more complicated way of differentiation (Jahns 1988).

In many foliose and gelatinous lichens the isidia fail to get disseminated and remain attached to the thallus. Such isidia not only increase the surface area of the thallus, thus contributing to an increased assimilative capacity of the thallus, but also are important for regeneration of new thallus lobes, especially in old, dying thalli.

The causes for the formation of isidia are not known but apparently genetically determined. Awasthi (1984) reported that in Indian and Napalese lichens there are a few sorediate and isidiate forms amongst the crustose genera, but the foliose and fruiticose genera have a higher percentage of sorediate and isidiate taxa. The foliose genera of the tropical and sub-tropical regions have a higher percentage of sorediate and isidiate taxa while fruiticose genera of temperate region have a higher percentage of sorediate-isidiate taxa according to another estimate (see Bowler and Rundel 1975).

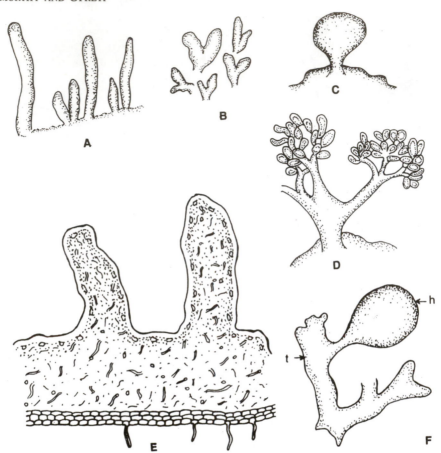

Fig. 2 A-F Isidia.   A *Heterodermia isidiophora,* cylindrical isidia. B *Caloplaca* sp., branched isidia. C *Parmelia* sp., spherical isidium D *Umbilicaria pustulata* corolloid isidia. E Transection of thallus of *Heterodermia isidiophora* through two isidia cut longitudinally (diagrammatic). F *Hormocystangium* of *Lempholemma* sp. (adapted after Jahns 1973).

Hormocysts. Hormocysts are so far known only in *Lempholemma,* where the photobiont is a cyanobiont (Jahns 1973). Hormocysts consist of short trichomes or individual cells often with a gelatinous sheath, which are in turn covered by fungal hyphae. These develop in chains in specialised hormocystangia which are swollen structures on the margins of the thalli (Fig. 2F). The hormocysts are released by the dehiscence or decay of hormocystangia. Further details on hormocysts are not known. Probably they behave like soredia in establishing a new thallus.

Significance of Vegetative Reproduction. The vegetative reproductive methods enumerated above provide the lichens with an exceptional regenerative capacity that enables them not only to colonise new sites but also to maintain them in the originally colonised sites; in many taxa even after several cycles of sexual reproduction. Consequently vegetative reproduction enables the lichen to overcome the influence of competitors, if any. Soredia, and in some cases isidia also, provide lichens with the potential to colonise newer areas after dispersal.

It was generally presumed that dissemination of lichens through a vegetative diaspore is much more favourable and efficient than propagation by ascospores, basidiospores or conidiospores. Establishment of the former (vegetative propagules) is straight-forward as the two symbionts remain closely attached, while that of the latter is limited by the availability of specific algal partners in locations where these spores are deposited after dispersal. In other words, in such instances relichenisation processes must start afresh every time. It should also be mentioned that some photobionts are scarce in an unlichenized state and, therefore, relichenisation is a very difficult process, even if it is facilitated by certain adaptive mechanisms (Bubrick and Galun 1986; Bubrick et al. 1985; Ott 1987; Ott et al. 1993). In fact, the formation of a new lichen thallus by the ascospore-derived mycelia was reported in one out of several million cases (Scott 1971). There is , therefore, an enormous wastage of resources allocated for the production of ascospores, basidiospores and conidiospores as well as for the organisation of their respective productive structures (ascomata, basidiomata and conidiomata, respectively). This observation indicates that vegetative propagation and propagules not only save wastage of excessive biomass allocation but they also aid in a more rapid colonisation of newer sites by the lichens. This also explains the already known fact that the lichen species that produce soredia are widely distributed while esorediate taxa have a very limited geographical area of distribution (Poelt 1970). In South India it has also been observed by us that sorediate taxa have a greater distributional range (unpubl. observ.). The speed of recolonization is also an important factor in competition.

As already stated, isidia, because of their greater mass, corticated nature and possession of fewer algal cells in proportion to the fungal mass, do not often get dispersed. Even if they do so, they invariably fail to germinate and produce a new lichen thallus. However, in terms of overcoming unfavourable environmental periods/conditions, isidia are better than soredia. Also, as already stated, isidia help in producing newer lichen lobes in old and decaying thalli, helping in the successful retention of the species at the same site.

3. Sexual Reproduction

Lichenisation reaches its zenith in the Ascomycotina (Hawksworth 1988a) and, therefore, the majority of lichens (98% of all lichens) are ascomycetes. They produce fruiting bodies (ascocarps or ascomata) like apothecia or perithecia which are more or less similar in organisation to the corresponding structures of the non-lichenised ascomycetous fungi. A small group of lichens, known as 'lichen imperfecti', however, do not produce any sexual diaspores. The common genera of this group are *Lepraria, Leprocaulon* and *Lichenothrix.*

In the above two groups of lichens special structures called 'conidiomata' are also produced as part of the reproductive cycle. These structures form conidia[1]. The role of conidia in lichens has been the subject of much debate and uncertainty for many years (Hawksworth 1988b). Although they function as spermatia (male cells) in many lichens, and as diaspores in yet others, when they form new thalli on coming into contact with a suitable photobiont, these two roles should not be assumed to be mutually exclusive. However, in the lichenised members of Deuteromycotina (with no sexual stage), the conidia can be assumed to act only as diaspores. Some foliicolous lichens produce macro- and micro-conidia either in the same conidiomata or on different conidiomata. In such instances, the micro-conidia certainly act as spermatia and the macro-conidia as diaspores. The third group, basidiolichens, comprising about 20 species in about 10 genera, have basidiocarps that produce haploid basidiospores as 'sexual' diaspores.

The conidiomata may be globose or flask-shaped ('pycnidia'), cupuliform ('acervulus'), cushion-like ('camyiidia') or erect ('synnemata' or 'hyphophores'). Rarely, conidia arise directly from filaments of sporogenous cells. The pycnidial conidiomata, which are most common, may be immersed in the thallus

---

[1]The term 'conidia', according to Hawksworth (1988b), should be preferred over other obsolete terms such as 'microspores' and 'spermatia'.

or raised above its level. Five basic types of pycnidial conidiomata are known (Hawksworth 1988b): Lecanactis type, Roccella type, Umbilicaria type, Lobaria type, and Xanthoria type (Fig. 3A-E).

3.1   Sexual Reproduction in Ascolichens

The basics of sexual reproduction in ascolichens and ascomycetous fungi are similar (Fig. 4), although the different steps leading to the formation of ascomata and ascospores in lichens are rather poorly-known when compared with the fungi. The "sexual cycle" in ascolichens is trigenetic, as in the Florideae of red algae (Chadefaud 1960). It consists of alternation involving (1) male and 'female' elements produced by the thallus, (2) sporophyte I (or prosporophyte) with cells containing 'male' and 'female' nuclei, and (3) sporophyte II (or ascosporophyte) containing dikaryotic cells and asci. In the latter, the two nuclei fuse,

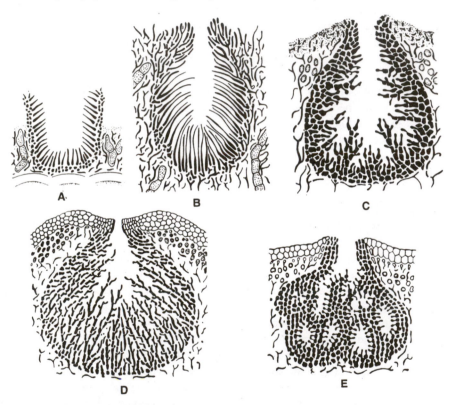

Fig 3 A-E.   Five types of pycnoconidiomata (vertical sections) in lichen-forming fungi. A Lecanactis type common in *Lecanactis, Arthonia* and *Byssoloma* is intermediate between an acervulum and pycnidium, and is more open and disc-like unlike the remaining types of pycnidia which are flask-shaped, rare among lichenised fungi. B Roccella type, common in the family Roccellaceae, the conidiophores are very sparsely-branched. C Umbilicaria type, occurs in *Acrosycyphus, Cetraria, Hypogymnia* and *Parmelia,* the conidiophores either look like a monochasium and bear terminal conidiogenous cells as in *Cetraria,* some species of *Parmelia* are sparsely-branched conidiophores bear intercallary conidiogenous cells as in *Acrosyphus, Hypogymnia, Parmelia* and *Umbilicaria* (cf. Figs 4 and 5 A-H). D Lobaria type, the conidiophores are branched extensively, and conidiogenous cells are produced intercalarily. E Xanthoria type, the conidiomata (appear) comprise several chambers with floors bordered by isodiametric cells which bear conidiogenous cells in *Dermatocarpon* and *Endocarpon.* The thalli have an ostiole through which conidia are released (after Vobis 1980)

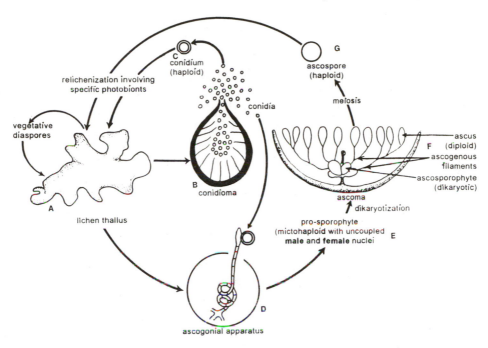

Fig. 4. A-G.   A Basic scheme of reproduction in ascolichens, the lichen thallus is haploid (both photobiont and mycobiont) and produces vegetative diaspores like soredia, isidia and hormocysts. They contain both the symbionts. These diaspores germinate and establish a new thallus. The trigenetic 'sexual cycle' in ascolichens consists of an alternation involving (i) 'male' (B, C) and 'female' elements D produced by the thallus, (ii) sporophyte I (or pro-sporophyte) with cells containing 'male' and 'female' nuclei (stage E in 4) and (iii) sporophyte II (or ascosporophyte) containing dikaryotic cells and asci; the latter give rise to haploid ascospores.

undergo meiosis, and give rise to haploid ascospores. The ascospores produce new thalli after relichenization. The male elements are presumed to be produced as conidia (= spermatia) in specialized conidiomata of the vegetative thallus. The female elements are produced in the form of ascogones in specialized ascogonial apparatus. From the available information, the details of sexual reproduction of ascolichens are summarised below:

Gametophytic Reproductive Apparatus. The gametophyte consists of the thallus which produces two types of reproductive structures.

Conidiomata, Conidiophores, Conidiogenous Cells and Conidia. These are multi-hyphal conidium-bearing structures (Fig. 4B) of various types. The conidiomata consists of non-specialised hyphae similar to the vegetative hyphae, and bear specialised hyphae called conidiophores which support the conidiogenous cells. The shape and branching of the conidiophores and conidiogenous cells vary considerably, depending upon the lichen taxa. Until now eight basic types have been recognised (Fig. 5A-H), although it is possible to expect more types when many more lichens are intensively studied. Conidia (Fig. 4C) of different shape and size which may either be colored or colourless are produced by the conidiogenous cells by a process called conidiogenesis. Conidia are always exogenously produced. The conidia may be sub-globose, bacilliform, elipsoid, falcate, sigmoid, thread-like, etc. When ready, the conidia are discharged and dispersed by water, wind and/or animals.

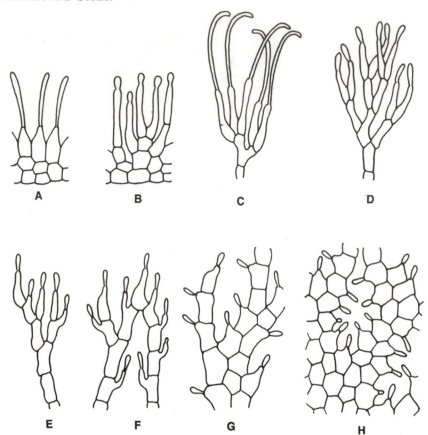

Fig. 5 A-H.   Conidiophores and the disposition of conidiogenous cells in lichen-forming fungi. The types represented in A-D can be called 'exobasidial' because conidiogenous cells are terminal with apically-produced conidia. E, F Represent types which are intermediate between endo- and exobasidial types with both terminal and intercalary conidiogenous cells. G, H Represent 'endobasidial' types where the conidiogenous cells are intercalary and bear lateral conidia. A is represented by *Arthonia;* B by *Byssoloma, Lecanactis, Peltigera* and *Thermutis;* C by *Dirina* and *Roccella;* D by *Cladonia* and *Ramalina;* E by *Alectoria. Cetraria* and some species of *Parmelia;* F by *Acroscyphus, Hypogymnia,* some species of *Parmelia* and *Umbilicaria;* G by *Lobaria, Mephoma* and *Anaptychia;* H by *Xanthoria, Dermatacarpon* and *Teloschistes* (after Vobis 1980).

Ascogonial Apparatus. The ascogonial apparatus (also called ascogonial filaments) represents the female reproductive system of lichens. It has nearly the same basic structure as in Collema type, in almost all species of ascolichens (Bellemére and Letrouit-Galinou 1988). The ascogonial apparatus is a filamentous multicellular structure with all the cells cylindrical, isodiametric and uninucleate. It comprises mainly of three parts (but not all the parts are always present in all lichens; Fig. 6): (1) a poorly-differentiated foot of sterile cells, (2) the ascogone (previously called carpogonium), which is the most essential part, composed of several fertile ascogonial cells with very heavy cytoplasm and (3) the trichogyne which is a straight, slender, erect, sterile and multicellular structure often protruding beyond the thallus; the trichogyne tip may be simple or branched. The ascogonial cells simulate the role of 'female' fertile elements, while the trichogyne receives the male elements (Fig. 4D).

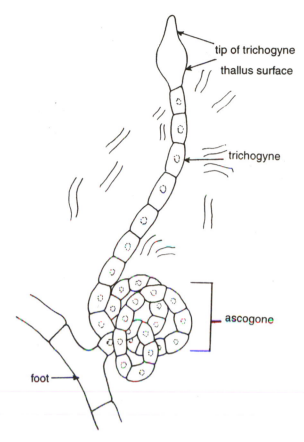

tip of trichogyne

thallus surface

trichogyne

ascogone

foot

Fig. 6   **Typical ascogonial apparatus of the Collema type (diagrammatic). It consists of poorly-differentiated foot, a much-coiled ascogone composed of a number of fertile ascogonial cells that function as female elements and a multicellular erect filamentous trichogyne protruding beyond the thallus surface; it receives the male elements. All the cells of the ascogonial apparatus are uninucleate.**

One other type of ascogonial apparatus called the 'Peltigera type' is also reported. The ascogonial cells are elipsoidal and multinucleate.

In both Collema and Peltigera types, the ascogonial filaments may be simple or, more commonly, they are profusely branched and all the branches together organise the complicated ascogonial apparatus. The branches may remain free from one another (arbuscular type), or the branches become intertwined into a glomerule (glomerular type). Rarely, the ascogonial filaments give rise to paraplectenchymatous or pseudoparenchymatous compact structures as in Lobariaceae and Parmeliaceae. In the glomerular type only the central elements in the glomerule are fertile.

The multicellular nature of trichogynes and ascogonial filaments (especially the fertile region), is a very significant distinction between the ascogonial apparatus of ascolichens and ascomycetous fungi. In addition, the foot is generally poorly differentiated in ascolichens than in the non-lichenised fungi. The highly coiled structure of the fertile filaments, the absence of auxillary cells, and the multicellular trichogyne distinguish the ascogonial system of lichens from the somewhat similar structure in the Rhodophyceae.

There is no report in ascolichens of either heterothallism, so common in ascomycetous fungi, or of a dioecious condition [see, however, the reports of dioecism in *Tylophoron crassiusculum* and *Lecidea verruca* (Poelt 1980; Hawksworth 1988b)]

Fertilization. There is much uncertainty about the actual occurrence of fertilization of the ascogone of ascolichens. Many investigators consider that conidia (microconidia in some taxa) released from conidiomata serve as the male cells (in which case they are called spermatia and the conidiomata producing them spermogonia) (Fig. 4D). In many cases the attachment of the conidia to the protruding trichogyne has been observed. There is also other circumstantial evidence for this contention (see literature in Letrouit-Galinou 1973). SEM studies have indicated the entry of conidial contents into the trichogynes of *Cladonia furcata* (Honegger 1984). This process of union of conidia with the trichogyne of the ascogonial apparatus has been termed 'conidial trichogamy'.

F. Moreau and Mme F. Moreau (1928) initiated the opposite opinion that there is no fertilization and that 'apogamy' is the rule in ascolichens, especially when conidia are not formed. Even if present, they are presumed to act as asexual spores, or as non-functional spermatia. The chief evidence for this opinion is that the initial hyphae of the ascosporophyte derived from the female element (ascogonial cells) are all uninucleate and not dikaryotic.

Since this aspect of fertilization has been least studied in lichens, one cannot exclude the possibility of either of the phenomena mentioned above occurring in lichens. Until then it is better to follow one of the two viewpoints: (1) If the preliminary hyphae of the ascosporphyte are dikaryotic from the outset, one could assume the fertilization of the ascogone either by the conidium (conidial trichogamy) or by a hyphal cell from the female gametophyte (somatogamy, e.g. species of *Lecidea*), or by the transformation of some nuclei of the ascogone into male nuclei and some others into female nuclei as in the case of multinucleate ascogonial cells of probably *Peltigera* (autogamy). (2) If the preliminary hyphae of the ascosporophyte are uninucleate from the outset and subsequently become dikaryotic through fusion between cells of the ascosporophyte (perittogamy), then one can assume the occurrence of apogamy. Since there is a gradual degeneration of sexuality with all the above phenomena reported in the nonlichenised ascomycetes, it is likely that the same phenomena occur even in lichenised ascomycetous fungi.

Ascomata. The stage recognisable immediately after the maturation of the ascogonial apparatus is the multinucleate mictohaploid condition of one or more of the fertile ascogonial cells. Mictohaploid refers to a condition in which separate haploid male and female nuclei, which have not yet conjugated, have been observed. Such a condition results by either fertilization, apogamy or peritogamy of the uninucleate ascogonial cells of the Collema type. Mictohaploidy also results in ascogonial cells which are multinucleate as in the Peltigera type, again by fertilization, apogamy or perittogamy. This multinucleate mictohaploid ascogonial cell is reported to represent the sporophyte I stage or the pro-sporophyte of ascolichens (Chadefaud 1953, 1960; Letrouit-Galinou 1973; Bellemère and Letrouit-Galinou 1988; E in Fig. 4). In species of *Peltigera,* the pro-sporophyte is slightly more elaborate. The multinucleate mictohaploid ascogonial cell gives rise to a number of vasicular tubular extensions which again are multinucleate and mictohaploid (Fig. 7A). Either in the more elaborate condition, or in the reduced condition, the pro-sporophyte is known to develop so far in only about 15 discolichens. The pro-sporophyte stage is intercalated between uninucleate ascogonial cell stage and the dikaryotic ascogenous filaments which are the forerunners in the production of asci in the ascomata.

The sporophyte II stage, or the ascosporophyte, develops very quickly either from the vesicles or directly from the multinucleate mictohaploid ascogonial cell of the pro-sporophyte. A number of ascogenous filaments with dikaryons (one male and one female nuclei from the multinucleate mictohaploid pro-sporophytic structure) are formed (Fig. 7A). The dikaryons multiply by conjugated mitosis. These filaments generally consist of hyphae with clamp connections and are composed of successive dangeardia (= croziers) with dikaryotic cells. Some of the dangeardia situated at the tip of hyphae are ascogenous dangeardia (Fig. 7 F-H); each of their dikaryotic cells gives rise to an ascus (Fig. 7I). Repeated production of asci (Fig. 7I), along with the associated sterile structure, results in the formation of ascomata (previously known as ascocarps), the fruiting bodies of ascolichens.

Ascomata arise invariably in the same thallus that have condiomata, but ontogenetically a little later than conidiomata. They invariably occur on the upper surface of the thallus except in some species of *Rhizocarpon.*

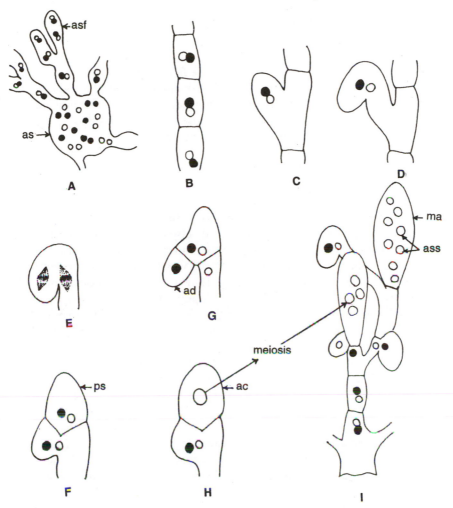

Fig. 7 A-I.   Events in the ascosporophyte that result in the production of mature asci (diagrammatic). (Based on unpubl. obs. in a species of *Collema*). A Multinucleate, mictohaploid prosporophyte (*as*) produces a number of ascogenous filaments (*asf*) which branch repeatedly. The ascogenous filaments have several dikaryotic nuclei, which undergo repeated conjugated divisions as the filaments further extend and branch. B Transverse septation at the apical region of ascogenous filaments into dikaryotic cells. C-H Stages in the formation of ascogenous dangerdia. C, D The more terminally located dikaryotic cells of the ascogenous filaments produce inverted V-shaped lateral extensions called croziers or ascogenous dangeardia (*ad*). E The dikaryons move into the ascogenous dangerdium and undergo one conjugate mitotic division. F. The cross wall between the two pairs of daughter nuclei is formed in such a way that the apical part of the dangerdium (crozier hook) forms a cell with a dikaryon. This cell forms the Proascus (ps). G Basal cell of the dangerdium (crozier hook) with one nucleus fuses (=clamp connection) with the remaining part of the cell to form another dikaryotic cell. This latter cell may repeat the crozier formation several times and result in several proasci. H The two nuclei of the proascus fuse and result in a diploid nucleus; simultaneously the cell enlarges into an ascus (*ac*). I Development of ascus and formation of ascospores. The diploid nucleus of the ascus (*ac* in H) undergoes meiosis and result in a variable number of haploid ascospores (in this case 8 ascospores) (*ass*) in the mature ascus (*ma*).

Ascomata may be scattered over the thallus, concentrated in the centre of the thallus or marginally located as in *Peltigera*. In some genera such as *Cladonia, Stereocaulon, Beomyces* and others, ascomata are formed on special outgrowths of the thallus called podetia. The typical components of an ascomata are (Fig. 8):

(a) Hymenium. This consists of asci and interascal sterile filaments of diverse ontogenies. The filaments may be uni- or multi-cellular, simple or branched, and free or anastomosed. Depending on ontogeny, these filaments may be paraphyses which are free filaments produced from the hymenium and disposed parallel to the asci, pseudoparaphyses which are free or anastomosed filaments produced from the epihymenial part, and paraphysoids which develop from the ascosporophyte region.

(b) Hypothecium or sub-hymenium. This lies below the hymenium and consists of ascosporophytic apparatus mentioned earlier. In addition, it may contain sterile filaments.

(c) Excitulum or exciple. This envelops the hymenium and hypothecium and is devoid of ascosporophytic apparatus. In some groups of lichens this may be absent.

The ascomata are principally either perithecia or apothecia. Pertithecia appear as black dot-like structures on the thallus and are typically flask-shaped. They range between 0.2 to 3.0 mm in diameter, immersed in the thallus and open by a pore or ostiole from the hymenium at the upper surface of the thallus. The perithecial wall is made up of thallus tissues of 3–5 layers which may be carbonised in some taxa.

Apothecia are open, disc- or cup-shaped structures. They are brightly coloured or dark and are either prominent on the thallus surface or immersed. The size of apothecial disc varies between 0.5 and 3.0 mm in crustose lichens, while in some foliose and fruiticose lichens it may vary up to 20 mm. Apothecia may be laminal or marginal or found on the lower surface of the thallus as in *Nephroma*. Stipitate apothecia occur in Cladoniaceae. There are a few noteworthy modifications in apothecial structure in certain lichens. In Caliciales, the apothecial disc disintegrates at maturity and the liberated spores and paraphysoid mass form a mazaedium. In Graphidaceae, the apothecial disc elongates and forms lirellae. Generally the lichenised fungi do not basically differ from non-lichenised ascomycotina as far as ontogeny and organisation of ascocarps are concerned. Apothecia that contain only the fungal tissue without the photobiont are called lecidine or biatorine apothecia and these are exactly like fungal apothecia; apothecia of lichens which have a thalline rim (with photobiont component) around the apothecial disc are called lecanorine apothecia. The lecanorine type with thalline rim is the most common type in lichens but is poorly represented among non-lichenised ascomycotina. Similarly, the eudiscopodian type of apothecium which is present in ascomycotina is absent in lichens.

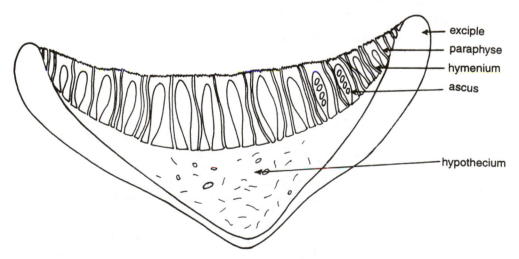

Fig. 8   Vertical section of an ascoma with various constituent parts (diagrammatic).

(d) The Ascus. It is also called the **sporocyst**. In lichens, the asci are formed from dikaryotic pro-ascal cells. A single ascus is formed from an ascogenous terminal dangeardium, or many are often formed because the clamp of the dangeardium produces several secondary ascogenous dangeardia and so on. Karyogamy of the two dikaryons takes place at the beginning of ascus development, followed by meiosis and one or more mitotic divisions. Some of the nuclei may degenerate so that the mature ascus has 1, 2, 3, 4, 6, 8 or many haploid ascospores.

The asci are usually small (approximatly 50 μm long), as in many Arthoniaceae, but in some taxa such as *Pertusaria* the asci attain a length of up to 300 μm. Asci are generally clavate, but in others sub-globose (Arthoniaceae), flask-shaped (Thelocarpaceae), or cylindrical (Caliciales). The asci have a two-layered wall (bitunicate wall) as in many Ascomycotina. The outer wall is called exoascus or exotunica and inner endoascus or endotunica. The outer wall is very thick and often inextensible while the inner wall is extensible. Recent electron microscopic studies indicate that both the walls are multilayered.

The apical apparatus of the ascus tip, where dehiscence invariably happens, is very varied and is of great taxonomic value. It refers to the types of thickening and differentiation of the ascus wall at the tip of the ascus, involving either exoascus, endoascus or both. In some cases the cytoplasmic contents also are characteristic in the apical apparatus region. The different types of apical apparatus in the asci of lichens (Bellemère and Letrouit-Galinou 1988) are represented in Fig 9A-F.

Teloschistes Type (Fig. 9A). In the Teloschistes type, the ascus tip has a thick cap in the exoascus and a subapical thickening in the endoascus. The later is highly positive when tested for total insoluble polysaccharides. Examples: *Teloschistes* and *Caloplaca*.

Catillaria Type (Fig. 9B). In the Catillaria type, the exoascus has no apical cap. The endoascus is extremely thick at the apical region and the thickening material is homogeneous and positive to PATAg staining (which indicates the presence of total insoluble polysaccharides). The ocular chamber mentioned in Lecanora type is absent in this type. Example: *Catillaria*

Lecanora Type (Fig. 9C). In the Lecanora type, the exoascus has no apical cap. The endoascus is similar to the Catillaria type, except for the presence of an ocular chamber (oc) surrounded by a prominent axial body which is PATAg negative. In the slight variations of the Lecanora type, the axial body may be either smaller or larger than the one shown in Fig. 9C. Example: *Lecidiella, Lecanora, Candelariella* and *Physcia*.

Psora Type (Fig. 9D). In the Psora type, the exoascus has no apical cap. The ocular chamber is reduced. Around the axial body a special and prominent polysaccharide thickening develops. The degree and appeerence of this polysaccharide thickening result in variations of this type, as seen in *Lecidea* and *Collema*.

Opegrapha Type (Fig. 9E). In the Opegrapha type, the exoascus has no apical cap. There is no axial body but the ocular chamber is variable in size; there is also the absence of the special polysaccharide thickening seen in Psora type. The endoascus is very thick at the ascus apex and the degree of thickening is almost the same throughout the length of the ascus. It is always finely stratified and subdivided into two sublayers. A variation of this type is seen in *Rhizocarpon,* where there is a very slight development of the special polysaccharide thickenings at the extreme apex of the ascus. A variation of this is seen in *Peltigera* which lacks sublayering of endoascus but there is an insoluble polysaccharide positive 'pendant' present at the apex of the cytoplasmic beak.

Phlyctis Type (Fig. 9F). In the Phlyctis type, both exoascus and endoascus are uniformly thin throughout the length of the ascus.

The apical apparatus plays an important role in the dehiscence of asci and release of ascospores. Several

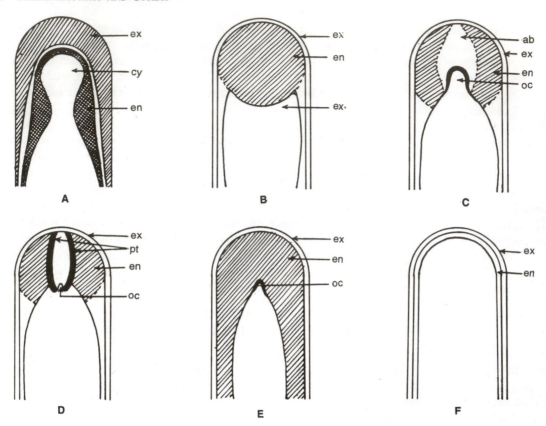

Fig. 9 A-F. Different types of apical apparatus of asci met within lichenised Ascomycotina. A Teloschistes type. B Catillaria type. C Lecanora type. D Psora type. E Opegrapha type. F Phlyctis type (adapted after Bellemère and Letrouit-Galinou 1988).

types of ascus dehiscence are known in lichens (Eriksson 1981). Some occur only in lichens (e.g. Teloschistes type and Rostrum type), some are common to both lichenised and non-lichenised ascomycotina (Fissitunicate type) and some others are known in ascomycotina but have not been observed in lichens (e.g. Operculate type). The types recognised in lichens are:

Jack-In-The-Box or Fissitunicate Dehiscence (Fig. 10B). The exoascus ruptures at the tip and endoascus after some elongation reaches beyond the hymenium. The spores are then ejected by the evagination of endoascus This type of dehiscence, with or without slight modifications, is noticed in members of Arthoniales, Dothideales, Gomphillales, Lecanidiales, Opegraphales, Peltigerales and Pyrenulales (Hafellner 1988).

Bivalvate or Bilabiate Dehiscence (Fig. 10D). The exoascus splits longitudinally into two valves. The endoascus does not come out. This type is noticed in some Helotiales and Pertusariales (Hafellner 1988).

Rostrate Dehiscence (Fig. 10C). This is very common in lichens but has not been reported so far in any ascomycotina. The exoascus and the external layers of the endoascus split only at the tip. Then the inner layer lengthens into a rostrum until it reaches the hymenial surface and the apical dome bursts out. This type is seen in Lecanorales.

Poricidal Dehiscence (Fig. 10E). It is also called the eversion type. In this type, there is almost no

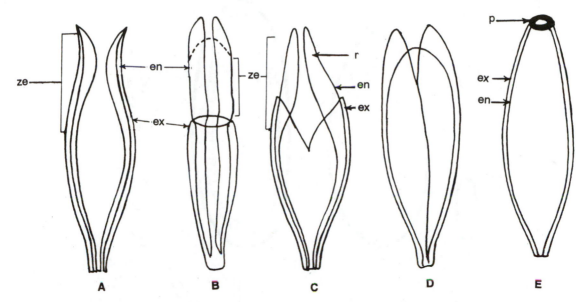

Fig. 10 A-E. Types of ascus dehiscence recorded in Ascolichens (Based on Eriksson 1981). A Teloschistis type. B Jack-in-the-box type. Dotted line indicates the position of the endoascal tip before evagination and its subsequent split. C. Rostrate type. D Bivalvate type. E Poricidal type. *en* endoascus, *ex* exoascus, *p* pore, *r* rostrum, *v* valves, *ze* zone of elongation in the ascus.

elongation of the ascus wall. Dehiscence is by a more or less well-defined apical pore and a slight eversion of the apical thickening. This type is known in Graphidales, *Gyalectus,* some Caliciales and some Helotiales.

Teloschistes Type (Fig. 10A). This is restricted only to lichens and has not been observed so far in Ascomycotina. In this type all the wall layers lengthen equally and reach the hymenial surface where the wall ruptures. This type is seen in Teloschistales and some species of *Pertusaria*.

Mazaedium Type. In this type there is no active dehiscence of asci. There is a rupture, disarticulation or gellification of the wall resulting in a Mazaedium formation. This type is noticed in some Caliciales.

Ascospores. The ascospores vary in shape and size. The size ranges from barely exceeding 1 μm as in *Acarospora* to about 500 μm long in the folicolous lichen, *Bacidia marginalis*. The ascospores may be variously-shaped and some of the more common forms are shown in Fig 11A-V. They may be single-celled or more than one-celled, due to septation. The septate spores are called muricate. Spore content has rarely been studied, although the ascospores may also be coloured in addition to being colourless. The pigmentation may be uniform throughout the spore or in distinct bands and is supposed to be due to melanines. Since lipids are frequent in ascospores, their quantity, quality and distribution pattern within the spore may yield very valuable taxonomic data. The other spore feature that has been neglected till now is the presence or absence of germ pores, perhaps because of their extremely small size.

4. Sexual Reproduction in Basidiolichens

As already stated, there are about 20 species of basidiolichens. The families of Basidiomycotina whose members are involved in lichenization are: Tricholomataceae, Clavariaceae, Dictyonemataceae and Atheliaceae (Oberwinkler 1970). In view of their rarity, basidiolichens have not been studied in any detail. It has been assumed that the development of fruit bodies or basidiocarps are along the same developmental pathway

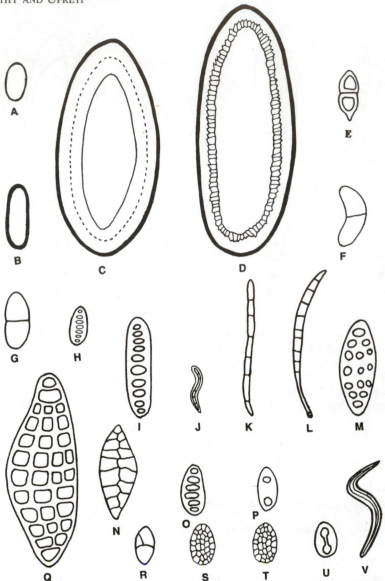

Fig. 11 A-V. Types of ascospores in Ascolichens (J and V modified from Bellemère and Letrouit-Galinou 1988 rest after Hariharan 1991). A. *Usnea picta*, single -celled, thin-walled and almost-globular ascospore. B *Lecanora allophana*, thick-walled, elliptical, single-celled ascospore. C. *Pertusaria macouna*, very large, single-celled ascospore with thick stratified wall. D *Pertusaria xanthostoma*, large, single-celled and thick-walled spore with dentations. E *Anaptychia ciliaris f. nigrescens*, bicelled thick-walled spore. F *Ramalina subpusilla*, bicelled, thin-walled clavata spore. G *Ramalina celastri*, bicelled and thin-walled spore. H *Graphis persicina*, multilocular spore. I *Glyphis cicatricosa*, multicellular, thick-walled spore, with locules in a linear row. J *Scolociosporum umbrinum*, vibriform, multicellular spore. K *Haematomma puniceum*, filamentous, multicellular spore. L *Bacidia fusconigrescens*, filamentous, multicellular and slightly-clavate spore. M *Anthracothecium ochrotropum*, multicellular spore. N *Leptogium moluceanum*, multicellular muriform spore. O *Letrouitia transgressa*, submuriform spore. P *Pyxine nilgiriensis*, two-celled thick-walled spore. Q *Graphina fissofurcata*, large muriform spore. R *Phaeographina sp.*, muriform spore with 3 chambers. S *Diploschistes actinostomus*, muriform spore. T *Leptotrema microglaenoides*, muriform spore. U *Caloplaca cerina*, polaribilocular spore. V *Sarrameana paradoxa*, 'S'-shaped single-celled spore.

as in Basidiomycetes: haploid basidiospores → haploid primary mycelium → secondary dikaryotic mycelium through intercellular fusions (perittogamies) with lateral clamp connections → aggregation of dikaryotic mycelia into carpophores → basidia production in the fertile layer of the carpophore → fusion of dikaryones followed by meiosis and mitosis → basidiospores (Fig. 12). There is no gametophytic phase as in ascolichens, and the primary and secondary mycelia are homologous to the primary and secondary sporophytes of perittogamous ascolichens. The photobiont may or may not be involved in the organisation of basidiocarp. The basidiocarps may be of the agaricoid, clavaroid or aphyllophoroid types and the basidiospores globose or cylindrical. The spores are invariably hyaline, rarely yellowish or greenish-blue.

5. Dispersal of Diaspores

The very wide distribution of lichens suggests that they have a very efficient means of dispersal of diaspores (Pyatt 1973). The frequency of a lichen in any area, therefore, is an indication to the degree of success of dispersal of its vegetative diaspores (thallus fragments, isidia, soredia, or hormocysts) or asco- or basidiospores and, of course, to the successful subsequent establishment of these diaspores. In the case of asco- or basidiospores, as already stated, the availability of suitable photobionts and favourable conditions for relichenisation are also very critical (Armstrong 1988). It also depends to some extent on the degree of viability and the capacity of the propagules to remain in a dormant state. Generally, soredia lose their viability quickly (i.e. less dormant) while ascospores (and basidiospores ?) are able to survive for a long time in a dormant and viable state. There is insufficient evidence to generalise about the most frequent propagules dispersed, the method of dispersal, the distances the propagules travel in the field, or the viability period of the propagules. Lichen propagules are reported to be dispersed by wind, water, or animals (Bailey 1976). There are no known instances of active dispersal of vegetative propagules, although it is known for ascospores (see Pyatt 1973). It has also been confirmed that there does not appear to be any

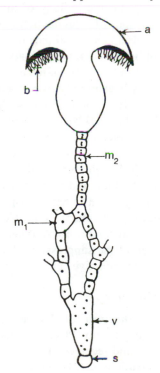

Fig. 12 **Basic reproductive cycle (diagrammatic) of a basidiolichen (modified from Chadefaud 1960). b basidia; m_1 primary mycelium, m_2 secondary mycelium with dikaryons, v multinucleate vesicle, s basidiospore.**

rhythm of dispersal of vegetative propagules, although this has been noticed in the case of ascospore dispersal (see Pyatt 1973). The only control that the thallus can exert is with regard to the time of production of propagules (Pyatt 1973).

Wind is assumed to be the most common agent for dispersing the lichen propagules. Tapper (1976) reported that the soredia of *Evernia prunastri* and *Ramalina farinacea* were dispersed by wind up to 30 and 20m, respectively. Water, especially rain water, is a common agent for the dispersal of the lichen propagules, but (perhaps) to a very short distance only.

Animals have long been reported to disperse lichen propagules (see literature citations in Pyatt 1973; Seaward 1988). In fact, several vertebrate and invertebrate animals have been shown to be permanently, temporarily, or casually associated with lichens (Seaward 1988). Of all these, the arthropods (belonging to Insecta and Acarina) are the most important, many of them are constantly associated with the lichens in the so-called lichen ecosystem or microniche (Kocheril 1994. Krishnamurthy et al. 1993). Many insects and mites carry lichen propagules that get attached to various parts of their body, for short or long distances. In certain moths, belonging to Psychidae, as well as in some *Neuroptera,* lichen thalli are organised on the cuticle of the larvae and during ecdysis the cuticle along with the lichen thalli with the propagules contained in them are cast off. They are subsequently carried by wind to new sites (Slocum and Lawrey 1976; Krishnamurthy et al. 1993, see also Seaward 1988). Adequate experimental studies so far have not been conducted to assess the viability of propagules so carried, and their successful establishment in a new life.

6.   Concluding Remarks: Studies in the 21st Century

Investigations carried out on lichens up to now have largely been centerd around the collection and documentation of lichen taxa throughout the world. Very little attention has been paid so far to study the reproductive biology of this interesting group of plants. Although vegetative, non-sexual and "sexual" methods of reproduction have been recorded in different taxa of lichens, no systematic attempt has been made so far to estimate the relative roles of these three modes of reproduction in the efficiency of survival of lichens and their colonisation of fresh sites. In certain parts of the world, including India, which are very rich in lichen flora, none of these modes of reproduction has been studied in detail in any lichen member. This is especially true of sexual reproduction. Details of ascocarp development have so far been studied in only about a dozen ascolichens out of about 16 000 lichen taxa known through out the world. Out of the about 20 basidiolichens reported so far, in none of them are the details of basidiocarp development available. The extent of degeneration of sexuality in the lichenised fungi also needs to be investigated thoroughly. These aspects should be concentrated for investigation in the coming years.

Although some studies have been made on the synthesis of lichens by bringing together the specific photobionts and mycobionts, they have not substantially added any information to the lichenization process that goes on in nature, especially after ascospores and basidiospores formation, after dispersal, and after falling on suitable sites. Concerted effort should, therefore, be focussed on studying lichenization under natural conditions. Another area that needs to be investigated in the coming years concerns the factors that lead to the suppression of sexual reproduction in the lichenised algae, and the roles of such suppression in the effective functioning of the lichen thallus and its very survival.

Glossary

Apothecium (Pl. Apothecia). An ascoma in which the hymenium is exposed at maturity.
Ascogonium (Pl. Ascogonia). The cell/cells of an ascolichen that function (s) as the female gametangium and receives(s) the male nucleus/nuclei; it subsequently gives rise to the ascogenous hyphae.
Ascolichen. A lichen in which the mycobiant is an ascomycetous fungus.

Ascoma (Pl. Ascomata). Multihyphal structure of Ascolichens in which asci are produced and aggregated; previously known by the term Ascocarp.

Ascosporc. A haploid diasporc developed inside an ascus (scc Ascus) of an Ascolichcn.

Ascosporophyte. The second phase (Sporophytes II) of the reproductive cycle of an ascolichen developed from the prosporophytes (Sporophyte I). It is formed by ascogenous filaments with dikaryons.

Ascus (Pl. Asci). The 'sexually' reproducing structure of the mycobiants of ascolichens; ascospores are produced inside ascus.

Basidiolichen. A lichen in which the mycobiont is a basidiomycetous fungus.

Basidioma (Pl. Basidiomata). Multihyphal structure of Basidiolichens in which basidia are produced and aggregated; prevously known by the term Basidiocarp.

Basidiospore. A haploid diaspore developed outside a basidium of a Basidiolichen.

Basidium (Pl. Basidia). A cell in which karyogamy and meiosis takes before the exogenous produciton of the basidiospores.

Conidioma (Pl. Conidiomata). Multihyphal structures of lichenised fungi in which conidia are produced and aggregated.

Conidium. A spore produced from condiomata of lichenised fungi serving as a diaspore, a spermatium or both and varies greatly in shape and size.

Crustose. A type of lichen where the thallus is flattened to the substrate and adhering strongly to it by its entire lower surface. It is very difficult to obtain the thallus free from the substrate without some amount of damage to the thallus.

Foliose. A type of lichen where the leaf-like thallus is loosely attached to the substratum so that it can be easily separated from the latter.

Fruticose. A type of lichen where the thallus is not leaf-like and is loosely attached to its substratum only through a small portion of its thallus.

Gelatinous. A type of lichen where the thallus can absorb water profusely and attain a gelatinous nature.

Hormocyst. A vegetative diaspore developed inside a hormocystangium

Hormocystangium (Pl. Hormocystangia). Vesicular structures arising from the thallus of species of *Lempholemma;* several hormocysts, each containing a few cyanobionts surrounded by heavy gelatinous fungal sheaths, are produced in these structures for the purpose of vegetative propagation.

Isidium (Pl. Isidia). Small outgrowth of the lichen thallus, containing the photobiont, that invariably servs for vegetative propagation; the isidia are solid and corticated.

Lichen. A symbiotic organism formed by the successful assemblage of a fungus (Mycobiont) and an alga or cynobacterium (Photobiont). The thus-formed lichen has very little morphological resemblance to either one of its component organisms.

Lichenisation. A process whereby a mycobiont accesses and incorporates its specific photobiont to from a composite lichen thallus.

Mycobiont. The fungal partner of a lichen; it is the 'exhabitant' in a lichen thallus.

Perithecium (Pl. Perithecia). An ascoma which is closed; the ascospores are released through a pore or slit.

Photobiont. The photosynthesising algal or cynobacterial partner of a lichen; it is the, inhabitant' in a lichen thallus.

Prosporophytes. The first phase (Sporophyte I) of the reproductive cycle of an ascolichen; it consists of the fertilised ascogonial cell from which multinucleate and mictohaploid vesicles or tubes arise. In many ascolichens this phase is reduced to varying extent or even appearing almost absent.

Soralium (Pl. Soralia). Open excrescence of the lichen thallus containing many soredia which are small globulose bodies consisting of a few photobiont cells enmeshed by fungal hyphae; the soredia serve as diaspores in vegetative propagation.

Soredium (Pl. Soredia). Vegetative diaspore developed inside a Soralium.

Spermatium (Pl. Spermatia). A conidium assuming the sexual reproductive function.

Spermogonium (Pl. Spermogonia). Conidioma that is presumed to have a sexual role i.e. the conidia produced inside this act as spermatia.

Trichogamy. The type of 'sexual' union in which the male nucleus/nuclei enter(s) into the ascogonium through the trichogyne.

Trichogyne. The thin filamentous cell/cells that is/are located at the tip of an ascogonium meant to receive and pass the contents of the spermatium to the ascogonium during sexual reproduction.

References

Armstrong RA (1988) Substrate colonization, growth, and competition, *In:* Galun M (ed) Handbook of lichenology vol. 2. CRC Press, Boca Raton, Florida, pp 3–16

Awasthi DD (1984) Reproduction in lichens. Phytomorphology 33: 26–30

Bailey RH (1976) Ecological aspects of dispersal and establishment in lichens. *In:* Brown DH, Hawksworth DL (eds) Lichenology: progress and problems. Academic Press, New York pp 215–247

Bellemère A, Letrouit-Galinou MA (1988) Asci, ascospores, and ascomata, *In:* Galun M (ed) Handbook of lichenology, vol I. CRC Press, Boca Raton, Florida pp 161–179

Bowler PA, Rundel PW (1975) Reproductive strategies in lichens. Bot J Linn Soc 70: 325–340

Bubrick P, Frensdorff H, Galun M (1985) Selectivity in the lichen symbiosis. *In:* Brown DH (ed) Lichen physiology and cell biology. Plenum Press, New York pp. 319–334

Bubrick P, Galun M (1986) Spore to spore resynthesis in *Xanthoria parietina*. Lichenologist 18: 47–49

Chadefaud M (1953) Le cycle et les sporophytes des Ascomycetes. Bull Soc Mycol Fr 69: 199–219

Chadefaud M (1960) Vegetaux non Vasculaires (Cryptogamie). *In:* Chadefaud M, Emberger L (eds) Traites de Botanique Systemetique. Masson, Paris pp 429–686.

Eriksson OE (1981) The families of bitunicate ascomycetes. Opera Bot 60: 1–209

Galun M (ed) (1988) Handbook of lichenology vol 1. CRC Press, Boca Raton, Florida, 297 pp.

Hafellner J (1988) Principles of classification and main taxonomic groups, *In:* Galun M (ed) Handbook of lichenology vol 3. CRC Press, Boca Raton, Florida pp 41–52

Hariharan GN (1991) Lichens of Shervaroy hills of South India—Ph. D. Thesis, Bharathidasan Univ., Tiruchirapalli, India

Hawksworth DL (1988a) The fungal partner. *In*: Galun M (ed) Handbook of lichenology, vol 1. CRC Press, Boca Raton, Florida pp 35–38.

Hawksworth DL (1988b) Conidiomata, conidiogenesis, and conidia *In:* Galun M (ed) Handbook of lichenolgy vol 1. CRC Press, Boca Raton, Florida pp 181–193

Honegger R (1984) Scanning electron microscopy of the contact site of conidia and trichogynes in *Cladonia furcata*. Lichenogist 16: 11–19

Jahns HM (1973) Anatomy, morphology, and development. *In:* Ahmadjian V, Hale ME (eds) The lichens. Academic Press, New York pp 3–58

Jahns HM (1982) The Cyclic development of mosses and the lichen *Bacomyces rufus* in an ecosystem - Lichenologist 14: 261–268

Jahns HM (1988) The lichen thallus, *In:* Galun M (ed) Handbook of lichenology, vol 1. CRC Press, Boca Raton, Florida pp 95–143

Kocheril JT (1994) Mites associated with lichens. Ph.D. Thesis, Bharathidasan Univ., Tiruchirapalli, India.

Krishnamurthy KV, Hariharan GN, Kocheril JT (1993) Mutualism between *Metisa* sp. (Lepiodoptera: Psychidae) and *Lepraria* sp (*Lichens imperfecti*). Phytophaga 5: 97–99

Krishnamurthy KV, Kocheril JT, Mohanasundaram M (1999) Lichen-mite association. *In:* Mukerji KG (ed) Studies in cryptogamic botany, Vol VII: Lichens. (in press)

Law R, Lewis DH (1983) Biotic environments and the maintenance of sex: some evidence from mutualistic symbioses. Biol J Linn Soc 20: 249–276

Letrouit-Galinou MA (1973) Sexual reproduction. *In:* Ahmadjian V, Hale ME (eds) The lichens. Academic Press, New York pp 59–90

Moreau F, Moreau Mme F (1928) Les phenomenes cytologiques de la reproduction chez les Champignons des Lichens. Botaniste 20: 1–67

Oberwinkler F (1970) Die Gattungen der Basiodiolichen. Vortr Ges Bot 4: 139–169

Ott, S (1987) Reproductive strategies in lichens. *In:* Preveling E (ed) Progress and problems in lichenology in the eighties. Bibl Lichenol 25: 81–93. Cramer Berlin

Ott S, Treiber K, Jahns HM (1993) The development of regenerative thallus structures in lichens. Bot J Linn Soc 113: 61–76

Poelt J (1970) Das Konzept der Artenpaare bei den Flechten. Ber Dtsch Bot Ges 4: 187–198

Poelt J (1980) Eine diozische Flechte. Plant Syst Evol 135: 81–87

Pyatt FB (1973) Lichen propagules. *In:* Ahmadjian V, Hale ME (eds) The lichens. Academic press, New York pp 117–145

Scott GD (1971) Plant symbiosis 2nd edn. Arnold, London

Seaward MRD (1988) Contribution of lichens to ecosystems, *In:* Galun M (ed) Handbook of lichenology, vol 2. CRC Press, Boca Raton, Florida pp 107–129

Slocum RD, Lawrey JD (1976) Viability of the epizoic lichen flora carried and dispersed by green lacewing (*Nodita pavida*) larvae. Can J Bot 54: 1827–1830

Tapper R (1976) Dispersal and changes in the local distribution of *Evernia prunastri* and *Ramalina farinaca*. New Phytol 77: 725–734

Tschermak-Woess E (1988) The algal partner. *In:* Galun M (ed) Handbook of lichenology, vol 1. CRC Press, Boca Raton, Florida pp 39–92

Vobis G (1980) Bau und Entwicklung der Flechten-Pycnidien und ihrer Conidien. Bibl. Lich. Vaduz 14: 1–141

8

Reproductive Biology of Bryophytes

Virendra Nath and A.K. Asthana

1. Introduction

The bryophytes include three parallel groups; Musci, Hepaticae and Anthocerotae. Some workers (Crandall Stotler 1980) have considered these groups even to be independent phyla. However, plants of all of these three groups possess a basically similar pattern of life cycle, bearing the photosynthetically autonomous, usually perennial, haploid gametophyte as the dominant phase of the life cycle, which gives rise to an unbranched independent/partially-dependent diploid sporophyte carrying a single capsule. Despite wide-ranging modalities in reproduction, there are some basic constraints in the life cycle of bryophytes (Longton and Schuster 1983) due to the necessity of water for fertilization, production of a terrestrial gametophyte, short durability of the sporophyte and one-time production of spores. Some valuable contributions regarding the life cycles of bryophytes have been made by Gayat (1897), Longton and Greene (1967, 1969), Longton (1969, 1980, 1988), Udar (1976), Zander (1979), Pujos (1992), Hadderson and Longton (1995).

In Musci and Anthocerotae the sporophyte is more or less independent as it manufactures a considerable precentage of its own food by chlorophyllous cells, while in hepatics, the sporophyte is surrounded by gametophyte-borne protective devices from a very early stage of development (calyptra , perianth, etc.), and is entirely dependent on gametophyte for its nourishment.

There is a considerable reduction in sexual reproduction among dioecious taxa due to spatial separation of male and female gametophytes and limited range of fertilization. This is the chief reason behind lack of genetic diversity in bryophytes in general and in hepatics in particular. This is the reason why they are unsuccessful in forming dominant vegetation; usually a genetically diverse plant is most adaptable to climatic changes. To overcome the limitations for sexual reproduction, several taxa develop asexual devices (gemmaae, propagules, etc.) for maintenance of populations.

The reproductive biology of bryophytes has not received considerable attention in the past and this is why there is not much data available for a comparative study of Musci, Hepaticae and Anthocerotae. So, there is a fruitful avenue for future research on this aspect.

It will not be out of place to mention the contributions made by several workers on asexual reproduction (Heald 1898, Burl 1939, Allsopp and Mitra 1958, Udar and Singh 1958; Berthier et al. 1976).

Udar and Gupta (1977, 1982), Shaw (1990), Duckett and Ligrow (1992), Imura (1994), Kimmer and Young (1995), sexual reproduction (Leitgeb 1869, Bryan 1917, Brown 1919, Harvey Gibbson and Miller-Brown 1927, Saxton 1931, Muller 1948, Gemmell 1950, Allorge 1955, Udar and Chandra 1964, Longton and Greene 1967, 1969, 1979; Vitt, 1968; Allard. 1975, Reynolds 1980, Crundwell 1981, Longton 1988, Mehra and Kumar 1989 and Une 1985), on spore production, dispersal and their viability (Blackslee 1906; Chalaud 1932; Ingold 1939; Bergeron 1944, Udar 1957a, 1958a, Udar and Chandra 1964, Udar and

Srivastava 1968, 1970, 1977, Udar and Kumar 1972, Srivastava and Udar 1975; Boatman and Lark 1971, Chopra and Rawat 1977, Inoue 1961, Hoffman 1970, Udar and Singh 1978, Udar *et al.*, 1982, Gupta and Udar 1986, Mogenson 1978, 1981, Bortholomew 1986, Mehra 1987, Miles and Longton 1990, Nath and Asthana 1993; Bortholomew-Began and Stotler 1994; Bortholomew–Began 1996), hybridization in nature and by experimental methods (Allen 1935, Bedford 1938, Proskauer 1951, Anderson and Bryan 1956, Khanna 1960, Lewis 1961; Udar and Chandra 1964, a,b Longton 1976; Engel 1968, Newton 1968, 1971, 1972, Anderson and Lemmon 1972, Ando 1972 Udar *et al.*, 1974, Ashton and Cove 1977, Krzakowa and Szweykowski 1979, Cummins and Wyatt 1981 Chopra and Kumra 1988, Cameron and Wyatt 1990, Chopra and Gupta 1991, Gupta *et al.*, 1991), and on evolutionary trends in reproductive biology (Kühn 1870, Lindberg 1882, Campbell 1896, 1898, 1913, 1915, 1920; Evans 1897, Lotsy 1909; Dixon 1924; Gemingham and Robertson 1950; Fulford 1951, Mehra and Handoo 1953, Schuster 1953, 1959–1960, 1966, Srivastava and Udar 1957a, Parihar 1959, Clarke and Greene 1970, Inoue and Schuster 1971, Iwatsuki 1972, Crum 1972, Longton 1976, 1982, Miller 1979, Udar 1976, 1976 a; Anderson 1980, Richardson 1981, Udar *et al.* 1982, 1983; Longton and Miles 1982, Udar and Srivastava 1984, Chopra and Kumra 1988, Miles and Longton 1992, a.

In the present chapter, important aspects of reproductive biology of bryophytes, i.e., asexual reproduction, sexual reproduction, sporophyte production, frequency and failure, hybridization in nature, by experimental techniques, evolutionary trends in reproductive biology and studies on reproductive biology in the 21st century are discussed and critically reviewed.

2.   Asexual Reproduction

Asexual reproduction in bryophytes is of foremost importance since the need of water for fertilization and the effective range of sexual reproduction has considerably restricted this phenomenon to some populations. In such instances the relevance and necessity of development of asexual reproductive devices for population maintenance, potential and dispersal has become clearly evident. Thus, a wide range of asexual devices have evolved which are derived from modification of either leaves or stems, among the leaf-derived modifications, gemmae are the most important means of asexual reproduction in Hepaticae. These develop at leaf or thallus margins by differentiation of cells, which later on become meristematic. In the Hepaticae themselves, several types of gemmae are known, e.g. discoid gemmae in *Marchantia polymorpha* (Fig. 1G) which develop inside the gemma cup on the dorsal surface of thallus, marginal discoid gemmae in *Radula nilgiriensis* (Fig. 1 L, M), endogenous gemmae in *Riccardia* sp. and *Blasia* sp. *pusilla* (Fig. 1C, E), stellate gemmae in *Blasia pusilla* (Fig. 1D). In *Blasia* gemmae are produced in a special flask-like structure called a gemma flask (Fig. 1J) on the dorsal surface of the thallus. In addition to this, there are several other taxa which exhibit luxuriant production of leaf gemmae, e.g. *Lophocolea minor* (Fig. 1K) and *Cololejeunea formosona* (Fig. 1H). Another leaf-derived device of asexual reproduction is caducous leaves in *Plagiochila* (Fig. 1W). and a modified caducous leaf in *Rectolejeunea* (Fig. 1U). Among stem-derived modifications which give rise to a new plant are caducous brood branches, e.g. in *Drepanolejeunea* (Fig. 1O). In *Lejeunea cardoti* fragmented stems (Fig. 1P) are capable of developing a new plant. In *Pellia neesiana* and in some other species of this genus, a conspicuous regeneration tendency has been observed which gives rise to the regeneration of young thalli from the apices of thalloid plants (Fig. 1A). However, in *Metzgeria furcata*, a process of regeneration takes place on the lateral margin of the thallus (Fig. 1B). In Anthocerotae, some small caducous spongy bodies are known to develop on the lateral and sometimes on the apical margin of the thallus in *Anthoceros angustus*, *Folioceros kashyapii* and *Folioceros appendiculatus*, which sometimes gives rise to young thallus regeneration. Tubers overcome unfavourable conditions, later on giving rise to new plants. In the taxa of Marchantiales, Metzgeriales and class Anthocerotae, a tuber-forming tendency is frequently seen. In *Fossombronia*, tubers develop on the stem on the ventral surface

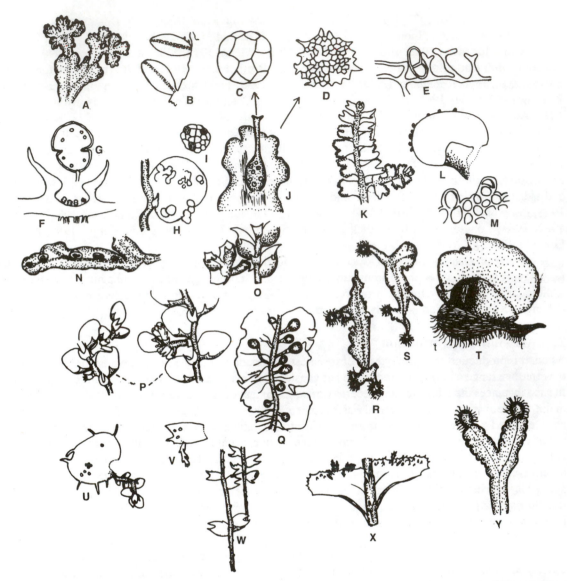

Fig. 1 A-Y. Asexual reproduction in liverworts (Hepaticae) and hornworts (Anthocerotae). A *Pellia neesiana*, regeneration of young thalli from apices, B *Metzgeria furcata*, regenerants on lateral margin of thallus. C *Blasia* sp. endogenous gemmae. D *Blasia pusilla*, stellate gemmae. E *Riccardia* sp., endogenous gemmae. F *Marchantia polymorpha*, cross section of thallus through gemmae-cup. G *M. polymorpha*, gemmae. H *Cololejeunea formosana*, a leaf with gemmae. I *C. formosana*, gemmae with three mamillose cells J *Blasia* sp., thallus with gemmae flask. K *Lophocolea minor*, gemmiparous plant. L *Radula* sp., leaf lobe with marginal gemmae. M *R. nilgiriensis*, marginal discoid gemmae. N *Marchantia polymorpha* thallus with gemma cups. O *Drepanolejeunea* sp., caducous brood branch. P *Lejeunea cardoti*, fragmenting stems. Q *Fossombronia* sp., tuber development. R, S *Phaeoceros himalayensis*, tuber development. T *Sewardiella tuberifera*, tubers. U *Rectolejeunea* sp., modified caducous leaf. V *Plagiochila* sp., regeneration from deciduous leaf. W *Plagiochila* sp., caducous leaves. X *Plagiochila* sp., propagules from postical face. Y *Riccia billardieri*, tubers in thallus (A-E after Schuster 1966, F, G after Udar, 1976, H-I after Udar *et al.* 1987 J after Udar, 1976 K-Q, U-X after Schuster 1966, R, S after Asthana and Srivastava 1991, T after Pande et al. 1955, Y after Udar 1976).

(Fig. 1Q), however, in *Riccia* and *Phaeoceros*, tubers develop on the thallus itself (Fig. 1 R, S, Y). In *Sewardiella tuberifera*, after complete vegetative growth of the plant, the apex of the plant gets thickened to form tubers (Fig. 1T) and if branching takes place, each branch bears tubers. This tuber formation phenomenon in *Sewardiella tuberifera* is annual and year after year successive tubers are formed and their position can be recognized by up and down curves followed by swellings.

In mosses, the protonema usually bears several buds that give rise to many gametophores, except in *Sphagnum* where the thalloid protonema produces only one bud. Branching also plays an important role in the development of moss colonies by the development of shoots, after breaking the contact between them. In acrocarpous mosses, two or three innovations usually develop at each shoot apex or base so that branches in a colony increase more and more. In a study by Longton and Miles (1982), the development of rhizoids at the base of young branches were found to break their connection with the parent shoot within 2 years, thus forming an independent plant. In many pleurocarpous mosses numerous short branches of determinate growth are produced frequently as compared to the long branches of indeterminate growth. These long branches give two different independent plants e.g. *Pleurozia schreberi* (Longton and Schuster 1983). There are several other methods by which mosses asexually reproduce and the ability of detached fragments of moss gametophytes to form a new plant by regeneration has also been noticed e.g. regeneration by detached leaves in *Physcomitrium* (Meyer 1942), in *Atrichum undulatum* (Gemmell 1950) and *Pleurozium schreberi* (Longton and Greene 1979). A detached shoot apex is also capable of regeneration into a new plant in *Bryum argenteum*. Several species of *Campylopus* regenerate from deciduous shoot apices and several fragile or caducous leaves. However, species of *Bryum* produce gemmae on rhizoids and other plant parts. Some types of asexual propagules like bulbils are also known in *Bryum* and *Pohlia* (Longton and Schuster 1983). Asexual propagules are commonly found in epiphytic mosses (Watson 1971). It is clearly evident that asexual reproduction contributes much to the development and maintenance of populations in the vast majority of mosses.

3.   Sexual Reproduction

Sexual reproduction in bryophytes is rather irregular as it depends upon the availability of water for fertilization. Differentiation between monoecious and dioecious states in bryophytes was first recognised by Hedwig (1782). The recent terminology presented in Fig. 2 originated from Lindberg (1882), and is based on the different position of male and female sex organs. Schuster (1966) also presented the same terminology. In unisexual (dioecious) taxa, male and female sex organs develop on two different plants. However, in bisexual plants two categories have been created, i.e. synoecious and monoecious. In the former category, male and female inflorescences usually develop together at the apex, and are subdivided into three different subcategories based on different positions of the sex organs, e.g. paroecious, in which the female inflorescence is at the apex and the male organ develops just beneath it on a separate branch; autoecious, in which male and female inflorescences develop on separate branches arising from the apex of the plant; heteroecious, in which the female inflorescence develops at the apex and male inflorescence develops on branches just beneath it, male and female inflorescences may, however, also be present separately at the apex of different lateral branches.

As in other plants, vegetative growth and sexual reproduction in bryophytes is governed by external factors like light (light duration: long-day, short-day, neutral; light intensity, light quality), temperature (temperature-photoperiod interaction) and pH, etc. All these factors interact with each other and light and temperature are the main factors for the initiation of sex organs. These factors either affect the synthesis of growth hormones in the system, or remove growth inhibitors. Photoperiodic control of gametangial induction was first reported by Wann (1925) in *Marchantia polymorpha*. On the basis of experimental studies combined with studies in nature, plants can be grouped for photoperiod-based gametangial initiation:

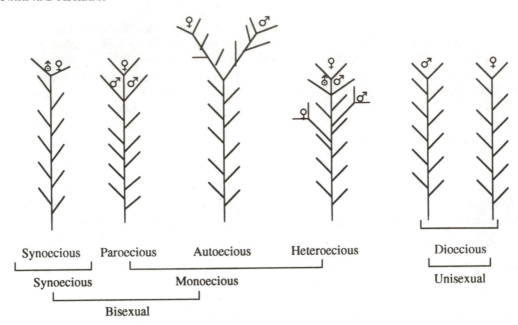

Fig. 2   Types of inflorescence in Jungermanniales, semi-diagrammatic (after Schuster 1966).

long-day plants, e.g. *Marchantia polymorpha, Conocephalum conicum, Pellia epiphylla, Riccardia multifida,* etc.; short-day plants, e.g. *Anthoceros punctatus, Phaeoceros laevis, Riccia glauca, Sphagnum plumulosum,* etc., day-neutral plants, e.g. *Cryptothallus mirabilis, Riccia crystallina, R. gangetica, R. frostii, Bryum coronatum, Funaria hygrometrica, Physcomitrium pyriformae,* etc. In addition, temperature also plays an important role in gametangial induction, as *Conocephalum conicum* and *Marchantia polymorpha* become fertile under long days at 21°C but not at 10°C. However, *Preissia quadrata* and *Pellia epiphylla* produced more gametangia at a higher temperature. Archegoniophore elongation in *Reboulia hemispherica* usually occurs in late spring, exhibiting a long-day response but it may be delayed by low temperature. In *Funaria hygrometrica,* gametangial induction takes place at a relatively low temperature with little photoperiodic response beyond a minimum light requirement. Similarly, low temperature requirements have also been demonstrated in *Philonotis turneriana, Physcomitrium pyriformae* and *Physcomitrella patens* (Nakosteen and Hughes 1978, Kumra and Chopra 1983).

Studies on the monoecius *Funaria hygrometrica* (Brown 1919) indicated that male sex organs develop before the female ones. Sex organs in bryophytes develop from superficial cells. the differentiation to the mature organ takes a considerable time after formation of its apical meristem.

As far as antheridial development is concerned, it is exogenous in Hepaticae and endogenous in Anthocerotae. In the liverworts, sex organs can be naked on the axis as in *Metzgeria pubescens* (Fig. 3E), *Fossombronia* and in some cases of Calobryales, but in most other liverworts they are protected in the axils of leaves or by specialized devices (Schuster 1966). In some cases antheridia are protected inside the androecial chamber, formed by the projected thallus tissue, e.g. *Pellia epiphylla* (Fig. 3D) and in *Anthoceros* sp. (Fig. 3G). In *Calobryum indicum* antheridia are apical in position and protected by a crown of leaves (Fig. 3A). In Jungermanniales (specially leafy plant species) antheridia borne on a small distinct condensed male branch which are protected by male bracts and bracteoles (Fig. 3B, C). In Marchantiales there is a tendency to group antheridia and archegonia into receptacles. Antheridia in most groups of Marchantideae are closely aggregated into defined and usually sessile receptacles. However, in the family Marchantiaceae, these receptacles are stalked as in *Marchantia polymorpha* (Fig. 3F), *Dumortiera,* etc.

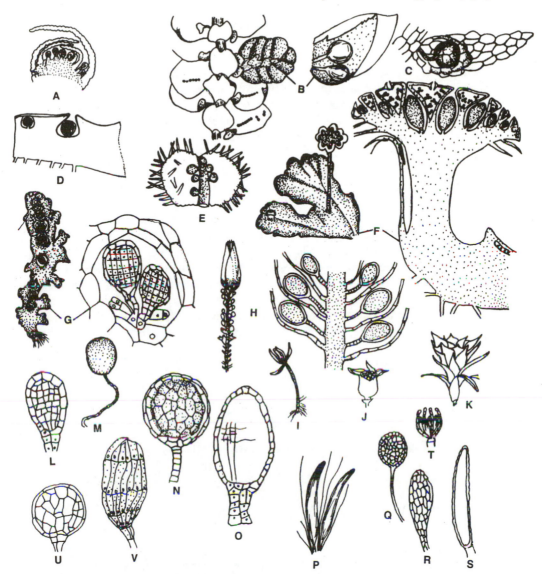

Fig. 3 A-V. Androecia and antheridia in bryophytes. A *Calobryum indicum*, VLS through antheridia. B *Frullania tamarisci*, male branch and male bract showing two antheridia. C *Aphanolejeunea cornutissima* bract with solitary antheridium. D *Pellia epiphylla*, VTS thallus through antheridia. E *Metzgeria pubescens*, ventral branch shows antheridia on costa. F *Marchantia polymorpha*, antheridiophore cross section shows position of antheridia. G *Anthoceros* sp., thallus and antheridia in androecial chamber. H *Sphagnum* sp., antheridial branch shows position of antheridia. I *Funaria hygrometrica* male, branch. J *Mnium* sp. antheridial head. K *Pogonatum* antheridial head. L-V Antheridia. L *Calobryum indicum*. M *Lejeunea* sp. N *Lophozia* sp. O *Marchantia* sp. P *Pogonatum* sp. Q *Sphagnum* sp.; R *Funaria hygrometrica*; S *Polytrichum* sp.; T *Andreaea* sp. U *Anthoceros* sp. V *Phaeoceros* sp. (A after Udar and Chandra 1965, B after Cavers, C after Schuster 1966, D, F, after Parihar 1959, E after Schuster 1966, G after Asthana and Srivastava 1991. H after parihar 1959, I after Udar 1976, J, K after Udar 1976, L after Udar and Chandra 1965, M after Müller 1948, N after Schuster 1966, O after Parihar 1959, P-T after Udar 1976, U, V after Asthana and Srivastava 1991).

Female sex organs (archegonia) in Hepaticae are exogenous in Calobryales, Jungermanniales, Metzgeriales and Marchantiales, however, they are endogenous in Anthocerotales. In Calobryales archegonia develop on the flattened or dome-shaped apex of female plant (acrogynous), and are usually 25 in number (in *C. indicum*) each with a short stalk and twisted rows of neck cells (Fig. 5L) usually with four rows of cells, in Jungermanniales and Metzgeriales there are five rows of neck canal cells, six rows in Marchantiales. However, in Anthocerotales the number is not fixed (Fig. 5R). In Musci, usually three to six archegonia form at the apex of the plant in most of the species of *Pogonatum* and *Polytrichum* (Fig. 5I) which stops the apical growth of the plant. Archegonia are interspersed with paraphysis and each mature archegonium has a long neck with six rows of neck cells. In *Funaria*, the apical cell of the female shoot is ultimately used up in the formation of the archegonium, thus it is acrogynous in nature (Fig. 5T). A mature archegonium contains six to many neck canal cells and dilated venter with venter canal cell and egg. In Andreales the apical cell itself forms the archegonium and, like Bryales, sex organs are intespersed with multicellular paraphysis. In *Sphagnum* the archegonium exhibits some interesting features by having an intermediate ontogenetical pattern resembling Hepaticae and Musci (Fig. 5U). the primary neck canal develops into an eight- to nine-celled row of neck canal cells.

Archegonia can be protected by modified leaves (perianth), hairs, stem structures or by an internal (sunken) position. The protective devices develop fully only after fertilization and sometimes complete new structures are formed to protect the young sporophytes (perigynia). In Hepaticae the perianth is of foremost importance from a taxonomic point of view. It develops by partial or total fusion of the two or three innermost leaves around the archegonial cluster, immediately after fertilization. The types of perianth

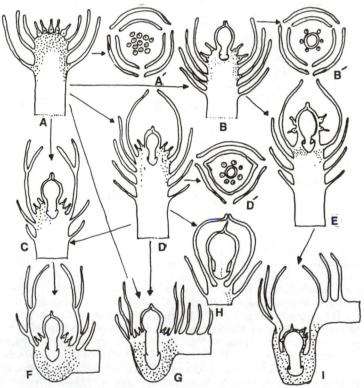

Fig. 4 A-I.   Some of the leaf and stem-derived gynoecial protective devices in Jungermanniales.
A-I Longitudinal, and A′B′D′ Transverse section (after Schuster 1966).

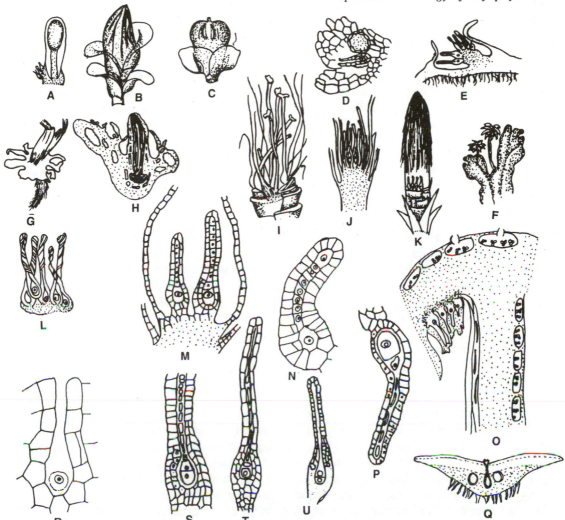

Fig. 5 A-Q. Specialized structures and devices for protection of archegonia, and structure of archegonia in bryophytes. A *Haplomitrium hookeri*, shoot calyptra. B *Solenostoma* sp., plicate perianth. C *Rectolejeunea* sp., perianth. D *Riccardia multifida*, synoeca on reduced lateral branch. E *Pellia endivaefolia*, section through involucre. F *Marchantia polymorpha*, archegoniophore. G *Anthoceros erectus*, involucre. H A section of thallus through length of young sporophyte. I *Polytrichum* sp., apex of female gametophore after dissecting perichaetial leaves. J *Funaria* sp., vertical section at the apex of female gametophore. K Apex of female shoot dissected to show archegonia and a young sporophyte enclosed by hairy calyptra. L-P Structure of archegonia. L *Calobryum indicum*. M *Frullania* sp. N *Pellia epiphylla*. O *Marchantia polymorpha* vertical section of archegoniophore. P Archegonia. Q *Riccia billardieri*, cross section of thallus showing archegonium in centre. R *Anthoceros* sp., archegonium. S *Pogonatum*, archegonium. T *Funaria* sp., archegonium. U *Sphagnum* sp., archegonium (A-D after Schuster 1966, E after Udar 1976, F after Kashyap 1929. G after Asthana and Srivastava 1991, H after Bharadwaj 1950. J after Udar 1976. K after Udar 1976. L after Udar and Chandra 1965, M after Udar, 1976, N after Udar, 1976 O, P after Parihar 1959. Q after Udar 1957 a, R after Parihar 1959, U after Udar 1976).

and their hypothetical phylogenetic relationships in Jungermanniales is illustrated in Fig. 6. Apart from this there are several other protective devices for the protection of the young sporophyte which are derived from axial tissue or at least involve the participation of stem tissue. The terms stem perigynium has been used for these. In the order Calobryales, the shoot calyptra develops to protect the sporophyte in *Haplomitrium hookeri* (Fig. 5A) and *Calobryum indicum* (Fig. 7A). A shoot calyptra is formed with the involvement of stem tissue in addition to bracts and bracteoles. In *Trichocolea* the sporophyte is protected with a coelocaule which is formed by deep penetration of the foot into the axis. In some genera like *Geocalyx* (Fig. 7G), *Jackiella* (Fig. 7Fa) and *Calypoeia*, there is the formation of specialized sporophyte protective devices, i.e. a marsupium at the tip of the main axis or ventral branches by shallow to deep penetration into the axis. In the order Metzgeriales, *Riccardia multifida* exhibits thallus margin outgrowths for the protection of syneica (Fig. 5D). In *Marchantia polymorpha* after fertilization calyptra as well as a perigynium develop to protect the young sporophyte (fig. 7K). However, in *Riccia*, archegonia and sporophytes are embedded in the thallus (Fig. 5Q). In the order Anthocerotales, soon after fertilization, thallus tissue is projected to form the involucre (Fig. 5G, H) to protect the developing sporophyte.

In the mosses sex organs are mainly protected by their location in leaf axils (Fig. 3H) or by trichomes (paraphyses Fig. 3P, T). Perichaetial leaves envelop synoicous groups of sex organs, while perigonial bracts sorround male reproductive organs. The bracts forming perichaetia and perigonia often have some useful taxonomic differentiation. Now here in mosses does the structure of the sorrounding bracts become as complex as in the Hepaticae and Anthocerotae (Goebel 1930). Paraphyses are located among the sex organs, and are mainly uniseriate cell rows, the top cell is frequently swollen (dome-shaped) and contains chlorophyll in many cases.

4. Fertilization

The male sex organ, the antheridium, encloses numerous spermatocytes within the antheridial wall. These spermatocytes differentiate into biciliate spermatozoids while the female sex organ, the archegonium, encloses in its archegonial wall an axial row of cells: the neck canal cells, the ventral canal cell and an egg. Fertilization essentially occurs in the presence of a supply of water. The archegonial apex opens and the canal cells disorganize to form a mucilagenous substance through which the spermatozoids contained in water reach the egg. Numerous spermatozoids reach the egg but usually only one fertilizes it. The fertilized egg (zygote) marks the onset of sporophyte generation. It divides to form a multicellular embryo (developing sporophyte) within the archegonial venter which provides a protective covering known as the calyptra (Udar 1976).

5. Sporophyte Production: Frequency and Failure

The zygote undergoes an initial transverse division into an upper (epibasal) and a lower (hypobasal) cell in Marchantiales and Anthocerotales. However, in Calobryales, early divisions that follow the initial transverse division are irregular in sequence, although the smaller hypobasal cells supposedly develop into only the haustorium, while the epibasal portion gives rise to the foot, seta and capsule.

In Jungermanniales and Metzgeriales there is considerable variation from group to group in the sequence of succeeding divisions (Schuster 1966). A mature sporophyte may be merely a spherical capsule embedded in the gametophyte or may be differentiated into a capsule and a foot or into capsule, seta and foot of various shapes, anchoring the sporophyte to the gametophytic tissue.

As far as the frequency of sporophyte production has been observed (Longton and Schuster 1983), there are strong tendencies for an increase in bisexuality, as bisexual taxa in most of the cases more frequently produce sporophytes. In Marchantiales there is a progressive evolution of bisexual gametophytes which have the capacity for inbreeding, e.g. Ricciaceae in which 85% of taxa are bisexual. However, less advanced

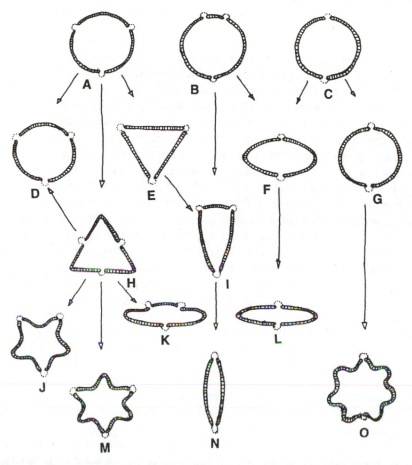

Fig. 6 A-O. Possible modes of evolution of perianth types in the Jungermanniales (schematic, transverse sections of fused leaves.) A Blepharostoma-Ptilidium type. B Jungermannia type. C Pleurozia type. D Ptilidium type. E Lophocolea type. F Diplophyllum type. G Pleurozia type I. H Mastigolejeunea-Lepidozia type. I Leptoscyphus type. J Lejeunea type. K Rectolejeunea type. L Scapania-Radula type. M Brachiolejeunea type. N Plagiochila type. O Pleurozia type II (after Schuster 1966).

taxa are unisexual, e.g. the taxa of Monocleales, Sphaerocarpales and members of Lunulariaceae (Marchantiales). Similarly, Jungermanniales, Herbertaceae, Lepidoziaceae and Lejeuneaceae, etc. are either unisexual or rarely produce bisexual species. Thus, in these groups spore production is less frequent. Among Lejeuneaceae, the subfamily Lejeuneoideae seems to be moderately advanced as it possesses 55–70% bisexual species (Longton and Schuster 1983). In addition, there is frequent production of sporophytes in advanced Cololejeuneoideae, which are highly bisexual and their gametangia ($\female$ and $\female$) develop very close to each other (within microns). Among Anthocerotales it is evident that *Phaeoceros laevis* ssp. *carolinianus* (Michx.) Prosk is a monoecious taxon, is the most widely distributed on the globe compared with dioecious taxa of the group.

A study by Gemmell (1950) regarding frequency of sporophyte production in mosses has revealed that 87% of moss species in which sporophyte production is rare are dioecious, however, 83% of monoecious moss species are known to frequently produce sporophytes.

6. Failure of Sporophyte Production

There are several reasons for the failure of sporophyte production in Hepaticae, e.g. (1) proliferation of unisexual plants from bisexual ones which in turn establishes unisexual centres. In this case asexual disseminules (gemmae, cladia, etc.) give rise to unisexual gametophytic clones, regardless of whether they are dispersed from bisexual or unisexual taxa, in such instances sporophyte formation seems to be impossible; (2) genotype depletion of unisexual taxa which were once bisexual. In such cases only sterile or only female or only male gametophytes are reported in a specified geographical region; (3) inability to produce male or female or both types of gametangia, such inability to produce gametangia in the *Plagiochila* is linked with frequent reproduction by caducous leaves (Longton and Schuster 1983).

In mosses, failure of sporophyte production in dioecious taxa may be due to incompatibility between different biotypes or through spatial separation of male and female gametangia (Gemmell 1950). Longton (1976) strongly supported the second reason for the failure in sporophyte production. Sporophyte production is seldom seen in some populations of moss species due to the occurrence of male and female gametangia in different colonies and in extreme cases there is complete absence of either antheridia or archegonia throughout the range of species, e.g. in some species of *Barbula*, *Bryum*, *Dicranum*, *Tortella*, etc.

The incidence of sporophyte production in mosses is similar to that of Hepaticae, as the percentage of sporophyte production in bisexual taxa is quite high as compared to ünisexual taxa.

The entire process of sexual reproduction, starting from the development of sex organs, fertilization, sporophyte development, sporogenesis to spore release and thereafter spore germination, protonema growth and bud formation are all affected by physical factors of environment which interact with biological processes in an individual to regulate the seasonal time table of various events and to ensure the success of the reproductive process (Longton 1988). In an intensive study made by Longton (1988), several temperate moss species reflected wide ranging phenological patterns. His study revealed that gametangia are first seen from spring through summer to autumn and winter. Male and female sex organs may appear at the same time, but usually juvenile antheridia appear several months before archegonia (*Polytrichum alpestre*) even in monoecius taxa, suggesting that different factors may stimulate induction of the two sexes. Fertilization may begin within 1 to 2 months of the appearance of archegonia, or it may be sometimes long delayed. The usual time for fertilization ranges from spring to autumn, probably to overcome the problems of sperm mobility. The period required for sporophyte development commonly ranges from 6 months to 18 months with a small resting phase and spore liberation also may occur at any time from spring through summer and autumn to winter, while tropical mosses of west Africa exhibit adaptation to a seasonal alternation of wet and dry weather. In nearly all the species studied, young gametangia were reported from March to May. At the beginning of rainy season, fertilization may occur from May to July, in some cases until September. Mature sporophytes had normally appeared by the end of the rains in October. Spores were liberated between November and March. In this way, the entire reproductive cycle was completed within 12 months.

Compared with mosses, there have been few detailed studies on reproductive phenology in Hepatics, but it is clear that there are different seasonal patterns of development in liverworts also. In most of the hepatics, development of gametangia is completed by May or June. Fertilization occurs during the rainy season (from June-August) probably due to frequent motility of sperms in rain water. Young-mature sporophytes are usually seen in October after the rains. The period of spore dispersal usually ranges from October to March.

7. Sporophyte Development

A fully-developed sporophyte is well-differentiated into foot, seta and capsule. The foot is embedded in the gametophytic tissue and derives nourishment for the developing sporophyte. The seta is the stalk of the capsule which remains small within the calyptra but with maturity it rapidly elongates, piercing the calyptra to raise the capsule. The seta elongates rapidly in liverworts, e.g. *Pellia epiphylla* (5 *mm/h*). In Anthocerotae the seta is highly reduced and represented by a constriction only, where as in Musci it is well developed except in *Sphagnum*, and *Andraea* where it is absent. The capsule has a wall enclosing the archesporium which develops into spores and elaters in Hepaticae and Anthocerotae. However, in Musci it gives rise to spores only, since elaters are absent in the group.

8.   Spore Production, Dispersal and their Viability

The process of sporogenesis starts from the stage when the zygote differentiates into the amphithecium and endothecium. In Hepaticae, the amphithecium gives rise to the capsule wall, while the endothecium forms an archesporium which develops into spores and elaters. In Anthocerotae the amphithecium differentiates into the outer and inner amphithecium. The outer amphithecium forms a capsule wall and the inner amphithecium forms an archesporium which in turn develops into spores and elaters, while the endothecium forms the columella. In Musci, sporogenesis is quite different as the amphithecium forms a capsule wall, however, the endothecium gets differentiated into outer and inner layers and the outer endothecium gives rise to the archesporium which forms spores only, since elaters are totally absent. The inner endothecium forms the columella.

An estimate of spore output per capsule has been carried out in several taxa with the help of haemacytometer counts (Ingold 1959; Longton 1962, 1976, Mogenson 1978, Longton and Miles 1982). In mosses, the number of spores per capsule is much greater as compared to Hepatics and Anthocerotae, as the size of spores is smaller in Mosses. The data indicates that the mean spore content of a single moss capsule ranges from 50000–600000. The lowest output is in *Archidium alternifolium*, i.e. 16 spores per capsule. However, *Dawsonia exhibits* the highest spore output, i.e. 50–80 million spores per capsule. But *Dawsonia* has not succeeded in extending its range from Australasia to south America. In the Hepaticae, due to the larger size of spores, the output is comparatively very low since, in this group of plants, elaters are also produced and usually the spore: elater ratio is 4:1. But in Jungermanniales, due to several mitotic divisions before meiosis, there are larger numbers of spores than elaters (36–72: 1). In Marchantiales, the spore output per capsule may be very low, e.g. in *Riccia crystallina* (246), *Corsinia* sp. (120), *Reboulia hemispherica* (2500), *Conocephalum conicum* (5300). In contrast, Jungermannialian taxa, like *Lophocolea* sp., *Diplophyllum* sp. and *Scapania undulatum* produce 23900, 400000 and 1000000 spores per capsule, respectively. In the case of members of the class Anthocerotae, spore production is continuous due to the presence of a meristematic zone between the capsule and bulbous foot, while it is a single production in Hepaticae and Musci.

The release of spores from the capsule takes place through various modes of dehiscence of the capsule. According to the pattern of dehiscence, the bryophyte taxa can be grouped as follows:

1.   *Cleistocarpous taxa*. In these taxa dehiscence of the capsule does not take place and spores are releasesd by the decay of the capsule wall mainly in mosses, e.g. *Phascum* sp., *Ephemarum* sp., some members of Sphaerocarpales, Marchantiales (*Riccia* sp. Fig. 7L) and taxa of *Notothylas*.

2.   *Schistocarpous taxa*. Dehiscence takes place either through single slit or four-valved remain intact at the apex and base (Fig. 7R-a, R-b). In *Haplomitrium hookeri*, initially the capsule splits along a single longitudinal slit (Fig. 7C-a) and a completely dehisced capsule exhibits a double spoon-head with fixed elaters at the apex (Fig. 7C-b, - C-c). The four-valved-dehiscence takes place in most of the Hepaticae. In Jungermanniales, a distinct four-valved dehiscence takes place (Fig. 7D-b, E, G, H), whereas in Metzgeriales there is a tendency to form the elaterophores, either basal or apical (Fig. 7I-b, J-b). The capsule splits into six to eight valves in *Marchantia* (Fig. 7K) and in anthocerotes it dehisces into two valves from the apex towards the base, exposing the columella and spores (Fig. 7M, N).

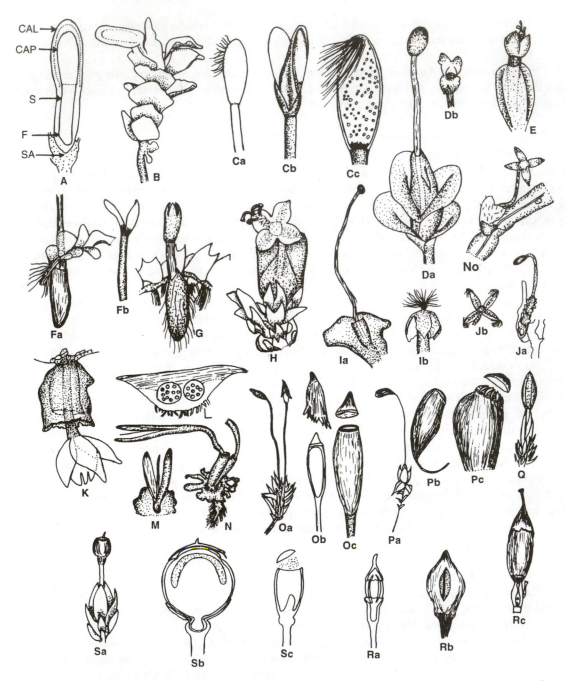

Fig. 7 A-S.   Sporophyte and its dehiscence pattern in bryophytes. A *Calobryum indicum*, VLS through young sporophyte. *CAL* calyptra, *CAP* capsule, *S* seta, *F* foot, *SA* shoot apex). B *Calobryum dentatum*, sporophyte in shoot calyptra. C *Haplomitrium hookeri*, dehiscence of capsule. Ca Through single longitudinal slit, Cb Completely dehisced capsule with double spoon-head, Cc Fixed elaters at apex

(Contd.)

3. **Stegocarpous taxa**. Dehiscence takes place through the operculum or lid, and peristome teeth mainly in mosses like *Pogonatum* and *Funaria* (Fig. 7O-c & P-c), whereas in *Sphagnum* the capsule dehisces by gun-shot/air gun, or splash mechanisms (Fig. 7 S-c).

Spore dispersal is an important aspect to be considered here. After the release of spores from the capsule they are taken to greater distances through wind, water, birds, insects or sometimes by other animals. In most of the hepatics, anthocerotes and mosses wind dispersal is more common. In aquatic forms like *Riella* and *Fontinalis*, dispersal takes place through water, in the case of *Riella*, animal dispersal is also known. In Hepaticae and Musci one time dispersal takes place, while in anthocerotes, due to non-synchronous spore production, spore discharge extends for a longer period (weeks or even months). However, in *Notothylas*, cleistocarpous capsules are also known in some taxa, which have larger spores, usually 40–100 μm in diameter, which can be dispersed over short distances unless they are taken away by birds or water.

Spore Viability. Usually the spores are viable for a short period but, in some cases their viability has been noticed even after a considerable amount of time (several years). Spores, during their viable period when they have access to a suitable substratum moisture, light, temperature and mineral nutrients—germinate to form a new gametophyte, otherwise fail to germinate. Usually thin-walled spores like those of the Lejeuneaceae are viable for a short period and very often germinate precociously (*in situ*) and multicellular spores with fully developed chloroplasts are released which are ready for further development. In contrast, the spores in Marchantiales and Anthocerotales have thick exine and they are viable for a longer period. In extreme cases, viability of spores has been reported up to 20 years in some moss taxa like *Oedopodium* (Chalaud 1932) in *Funaria hygrometrica* up to 11 years (Hoffman 1970) and in *Physcomitrium pyriformae* up to 55 years (Doyle 1970).

Sporeling development in bryophytes has been throughly studied (by various workers), nicely reviewed by Nehira (1983) in his comprehensive account of 'spore germination, protonema and sporeling'. Broadly, it is clear that in foliose bryophytes, protonema includes all the stages from first division of spore to the formation of an apical cell with three cutting faces by which the leafy shoot is formed (Fulford 1956). It may be filamentous, globose, cylindrical, unistratose, disciform, rectangular, ribbon-like. However, in thallose bryophytes the definition of protonema is more difficult. Mehra and Kachroo (1951) define the protonema in Marchantiales as a quadrant stage. Parihar (1959) stated that the protonemal stage of thalloid hepatics is different from the filamentous protonema of Musci or leafy Hepaticae.

Fig. 7 (Contd.)

and base of capsule. Da *Jungermannia tetragona*, perianth and mature capsule, Db *Jungermannia* (Plectocolea) *tetragona*, dehisced capsule. E *Cololejeunea madothecoides*, perianth and four-valved dehisced capsule. Fa *Jackiella javanica*, marsupium. Fb Two-valved dehisced capsule. G *Geocalyx graveolens*, marsupium with four-valved dehisced capsule. H *Frullania*, perianth and dehisced capsule and fixed elaters. Ia *Pellia epiphylla*, involucre and mature sporophyte on thallus. Ib Dehisced capsule with basal elaterophore. Ja *Riccardia levieri*, involucre, calyptra and mature sporphyte. J2 Dehisced four-valved capsule. K *Marchantia* sp., calyptra, perigynium and dehisced capsule. L *Riccia*, embedded sporophyte. M *Notothylas indica*, dehisced capsule with columella. N *Anthoceros erectus*, thallus with involucre, dehisced capsule and columella. Oa *Pogonatum*, female plant with sporophytes. Ob *Pogonatum*, capsule with calyptra lifted. Oc Capsule with detached operculum showing peristome rim. Pa *Funaria* sp., female branch with sporophyte. Pb, c Capsule with operculum removed to show peristome teeth. Q *Andraea rupestris*, portion of plant with sporogonium. Ra Elongated pseudopodium and mature sporogonium. Rb Mature dehisced sporophyte with columella. Rc Sporophyte with calyptra. Sa *Sphagnum* plant apex with mature sporogonium. Sb Cross section of capsule. Sc Dehisced capsule. (A after Udar and Chandra 1965, B after kumar and Udar 1976, Db after Udar and Kumar 1983, E after Udar *et al* 1987, Fb Udar and Kumar 1981, F after Udar *et al*. 1982, H after Kamimura 1961, Ia, 2 after Udar 1976, J-R after Udar 1976).

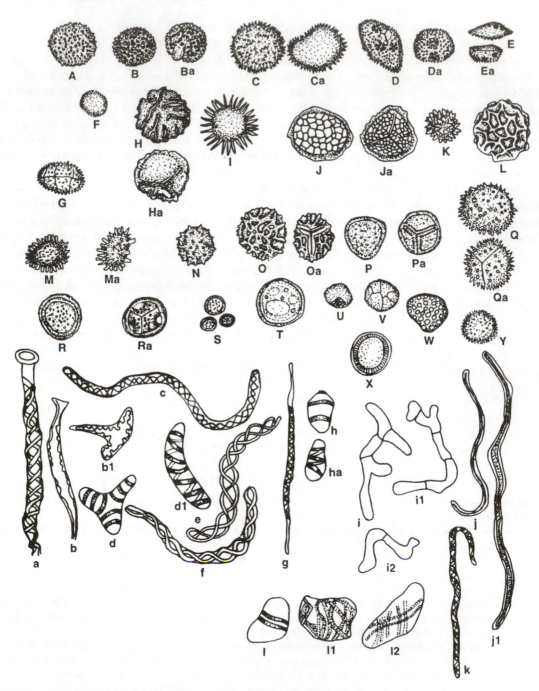

Fig. 8 A-Y. Spores in bryophytes. A *Calobryum dentatum*. B, Ba *Haplomitrium hookeri*. C, Ca *Radula tabularis*. D, Da *Schiffneriolejeunea indica*. E, Ea *Lejeunea indica*. F *Metzgeria indica*. G *Pellia epiphylla*. H, Ha *Fossombronia himalayensis*. I *Calycularia crispula*. J, Ja *Riccia pandei*. K *Athalamia piniguis*. L *Asterella* sp.; M, Ma *Cyathodium indicum*. N *Cyathodium tuberculatum*. O, Oa *Anthoceros erectus*.

(Contd.)

In bryophytes, spore germination is basically of two types, i.e. exosporic and endosporic. In exosporic germination the spore usually ruptures in the early stages and remains at the base of protonema. However, in endosporic types, the spore does not rupture but gets stretched to many times its original size, covering all, or a large part, of the protonema (in the case of intracapsular germination).

Sporeling patterns in Calobryales and Jungermanniales are exosporous as well as endosporous. In exosporous types, massive and filamentous types of protonema are formed. Massive protonema may be globose, e.g. *Haplomitrium* (Fig. 9A and Aa), globose to cylindrical e.g. *Nardia* (Fig. 9F and Fa) or strap-shaped e.g. *Bazzania* (Fig. 9G, Ga, and Gb). A filamentous type of protonema is formed in *Cephalozia*. Under endosporous conditions there is formation of a diamorphic protonema in *Pleurozia* (Fig. 9C, Ca) and *Ceratolejeunea* (Fig. 9D), unistratose disc-shaped protonema in *Radula* (Fig. 9-E-Ec), globose protonema in *Frullania* (Fig. 9H-Hc) and *Lopholejeunea* and thalloid to cylindrical protonema in *Lejeunea* (Fig. 9K, Ka). Nath and Asthana (1993) have also observed the globose protonema in *Frullania physantha* Mitt. Similarly endosporic and exosporic patterns are also found in Metzgeriales. In the exosporic type, a globose to cylindrical protonema is formed in *Pallavicinia*, strap-shaped in *Metzgeria* (Fig. 9I-I3) and filamentous in *Riccardia*. In the endosporic development of protonema, globose to ovoid multicellular protonema are formed inside enlarged exospores in *Pellia* (Fig. 9J).

In Marchantiales, Inoue (1961) recognized seven types of sporeling patterns on the basis of germ rhizoid formation.

1. Targionia type as in *Targionia hypophylla* (Fig. 9Q), *Cyathodium* sp. and *Sauteria* sp. etc.
2. Marchantia type as in *Marchantia* (Fig. 9L-Lb, M).
3. Stephensoniella type as in *Stephenloniella brevipediunculata* (Fig. 9P),
4. Reboulia type as in *Reboulia hemispherica* (Fig. 9N), *Asterella, Plagiochasma, Wiesnerella denudata*, etc.
5. Mannia type as in *Mannia* sp. (Fig. 9O).
6. Conocephalum type as in *Conocephalum* spp., exhibiting a typical endosporous germination.
7. Neohodgsonia type as in *Neohodgsonia* and some species of *Marchantia* (not known in India so far).

In Anthocerotae, the spore germination pattern is also exosporous and endosporous. In exosporous types of germination, e.g. in *Anthoceros* (Fig. 9R) the spore ruptures the exospore at the trilete mark and the spore wall is broken into three pieces, in *Notothylas* (Fig. 9U) the spore coat splits into three or four pieces and a cell mass is developed directly. In *Megaceros* (Fig. 9T), after spore germination, the primary cell divides continuously to form a globose massive protonema with one or two rhizoids. An endosporic germination

Fig. 8 (Contd.)

P, Pa *Phaeoceros laevis*; Q, Qa *Folioceros udarii*; R, Ra *Notothylas dissecta*. S *Pogonatum himalayanum*. T *Oligotrichum* sp.; U *Encalypta* sp. V, Y *Fissidens*, range of sporoderm pattern. a-l Elaters in bryophytes. a *Schiffineriolejeunea indica*. b, b1 *Lejeunea indica*. c *Radula tabularis*. d-f *Fossombronia* sp. g *Riccardia sikkimensis*; h, h1 *Targionia indica*. i, i1 *Anthoceros* sp. i, i1 In *Folioceros* sp. k *Megaceros* sp. l, l2 In *Notothylas* spp. (A after Gupta and Udar 1986, B after Udar and Srivastava 1981, C after Udar and Kumar 1982, D - Ea after Udar and Awasthi 1982, F after Udar and Srivastava 1970, G after Schuster, 1966, H-Ha after Srivastava and Udar 1975b, I after Pande and Udar 1956, J-Ja after Udar 1959, K after Udar 1960, N after Udar and Singh 1987, R, Ra after Asthana and Srivastava 1991, V-Y after Chopra and Kumar 1981, a b, ba after Udar and Awasthi 1981, 1982, c after Udar and Kumar 1982, h, h1 after Udar and Gupta 1983, i-l after Asthana and Srivastava 1991).

pattern is known in Dendroceros (Fig. 9V) where the protonema develops inside the enlarged exospore and multicellular spores are usually present in a mature capsule.

Like Hepaticae and Anthocerotae, the Musci also Show exosporous and endosporous types of germination. On the basis of earlier stages of protonema development, different patterns in different taxa can be easily recognized. In exosporous germination of spores, two types of protonema are produced, i.e. primarily filamentous protonema and primarily massive protonema. In the former case several patterns have been identified (Nehira 1983), like formation of a thalloid protonema at the apex of a short filament in *Sphagnum* (Fig. 10Aa), sex organ formation at the apex of a chloronema branch in *Buxbaumia* (Fig. 10B), formation of vesiculated protonemal cells in *Schistostega* (Fig. 10E), and development of primary rhizoids in *Funaria* sp. (Fig. 10Db). In the case of primarily massive protonema, there is formation of globose mass of cells in *Encalypta* (Fig. 10F) and caulonema and rhizoid development takes place on a massive protonema in *Hedwigia* (Fig. 10Ga). In the endosporous germination pattern a massive protonema develops inside the stretched exospore, as in *Andraea* (Fig. 10H), which later on gives rise to filamentous protonema from the massive one (Fig. 10F-c, H-d).

9.   Hybridization in Nature, by Experimental Methods

Hybridization in Bryophytes takes place by fertilization of an archegonium on the female parent by an antherozoid from the male parent, followed by the development of a hybrid sporophyte on the female parent. In the case of interspecific hybrids, the F1 sporophyte is usually intermediate between parental species. Studies on hybrid sporophytes, spore development and growth of hybrid gametophytes could provide valuable information on degrees of relationship between pairs of hybrid gametophytes.

Khanna (1960) has reported natural hybridization in the genus *Weisia* growing at Pathankot' (Punjab, India). He observed the characteristics of *W. viridula* and *W. crispa,* two closely growing species in *W. exserta*—which is a highly variable species in nature. Subsequently it has also been confirmed that *W. exserta* is a distinct and vigorous species and it is morphologically distinct from its parent species. It has a robust and luxuriant growth with a variation pattern of its own. Among the liverworts natural hybridization is known in *Marchantia polymorpha* as it is a hybrid of *M. alpestris* and *M. aquatica* (Burgeff 1943). The development of this hybrid species has probably taken place during the period of prehistoric man. Apart from this, there have been a number of observations on naturally occurring hybrid sporophytes in mosses. Most of them have been reported in luxuriantly growing monoecious taxa, although a few hybrids have also been reported in dioecious mosses. Absence of hybrid gametophytes, infrequency of recognized hybrid sporophytes and a high percentage of sterility in natural hybrids is the most probable reason of slower evolution in bryophytes.

There have been several reports of the occurrence of hybrids in nature between species of Musci, based upon the morphological characters of the plants in question (Allen 1935). Apart from this, several successful hybridization experiments have included races of *Funaria hygrometrica* and other well recognized taxa of Funariaceae, races of *Sphaerocarpos donnellii* and *Marchantia polymorpha* (Allen 1935). A cross between *Physcomitrium* x *Funaria* (both haploid) resulted into the sporophytes showing dominant *Physcomitrium* characters, but the colour of the capsule is of *Funaria*. In normally developing haploid gametophytes, maternal characters were strongly predominant, however, more or less diploid offspring exhibited a large percentage of paternal characters. In hybridization experiments interspecific moss crosses have been successful between *Funaria hygrometrica* and *F. mediterrnea, Physcomitrium eurystomum* and *P. pyriformae*, etc. (Allen 1935). Sporophytes obtained from reciprocal interspecific crosses have exhibited a remarkable range of variation. In some of the crosses their spores were largely or entirely non-viable. The gametophytic progeny also showed variety of combinations of parental characters. From the results of the various crosses mentioned above, it has been concluded that cytoplasm and chromosomes play an important genetic role.

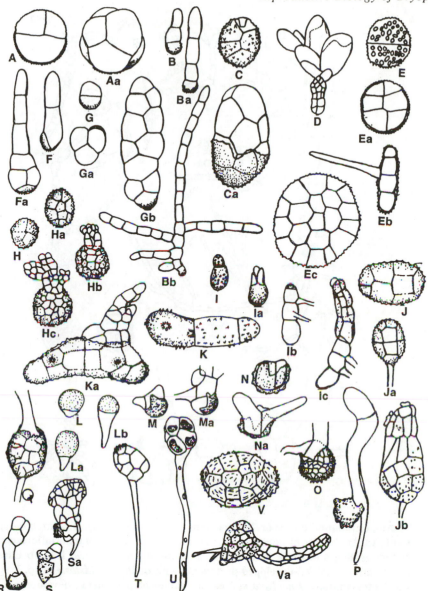

Fig. 9 A-V. Spore germination and sporeling patterns in Hepaticae and Anthocerotae. A *Haplomitrium* sp. A, Aa Early stages of spore germination. B *Nowellia* sp., B, Ba Early stages of spore germination. Bb Filamentous protonema. C *Pleurozia* sp., D *Ceratolejeunea* sp., sporeling. E *Radula* sp. Ea early stage of spore germination, Eb, Ec Discoid protonemma developed inside exospore. F *Nardia* sp., F, Fa Early stages of spore germination. G *Bazzania tridens*. G, G1 Early stages of spore germination. Gc Strap-shaped protonema. H *Frullania physantha* H, Ha Early stages of spore germination, Hb, Hc Sporelings, I *Metzgeria conjugata*, Ia, Ib Early stages of spore germination, Ic Strap-shaped protonema, J *Pellia* sp. K *Lejeunea* sp., K Filamentous protonemma developed inside exospore, Ka Sporeling. L *Marchantia polymorpha*. M *Marchantia* sp. N *Reboulia hemispherica*. O *Mannia* sp. P *Stephensoniella brevipedunculata*. Q *Targionia hypophylla*. R *Anthoceros* sp. S *Phaeoceros laevis*. T *Megaceros* sp. U *Notothylas* sp. V *Dendroceros japonicus*. V Multicellular spore, Va Sporeling (A-V, after Nehira 1983, H after Nath and Asthana 1993).

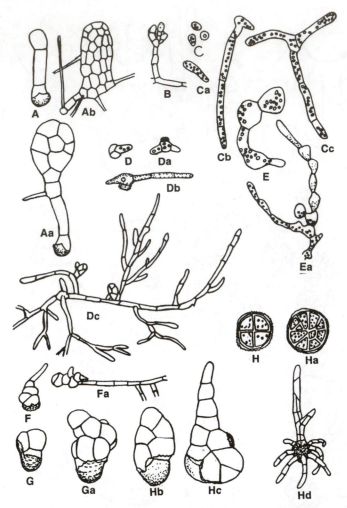

Fig. 10 A-H.   Spore germination and sporeling patterns in mosses (Musci). A *Sphagnum* sp., A Filamentous stage, Aa, Ab thallose protonema. B *Buxbaumia* sp., formation of male inflorescence. C *Polytrichum cummune* stages of spore germination. D *Funaria hygrometrica*. Da, Db Early stages of spore germination. Dc Well-developed protonema with buds. E *Schistostega* sp. E spore germination, Ea Protonemata. F *Encalypta* sp., earlier stages of protonema development. G *Hedwigia* sp., massive protonema, development outside exospore. H *Andraea* sp., H, Ha Early stages of endosporic germination. Hb, Hc Earlier stages of protonemata. Hd Protonema. (A-Ab, Hb, Hc after Nehira 1983; B and Hd after Nishida 1978, C-Cc after Udar 1976. D-Dc after Parihar 1959, E-Ea after Nehira1987, F, Fa, and G, Ga after Nehira, 1983).

Some characteristics of gametophytic progeny are determined by genes (distribution being Mendelian), some are determined by the cytoplasm, the result in maternal inheritance, some by the combined influence of genes and cytoplasm.

10.   Evolutionary Trends in Reproductive Biology
The basic system of reproductive strategies is almost similar in mosses and hepatics involving primitive

features, for example, perennial, dioecious gametophytes, absence of specialized asexual propagules, commonly producing sporophytes with long setae, capsules with a peristome and small spores, as against advanced features like monoecism, production of gemmae and other asexual propagules, immersed capsule, reduced peristome of capsule and larger spores. A specific combination of these primitive and advance features is usually seen in particular situations (Longton and Schuster 1983). Evolution in life history strategies of mosses has been extensively studied by During (1979). He has emphasized habitat stability, colony longevity, age at first reproduction, spore size and effort of reproduction by sexual and asexual means. He defined six main life history strategies in mosses:

1.   **Perennial Stayer** : Which includes perennial and predominently dioecious taxa with lower localized sexual reproduction efforts, low rate of asexual propagules, less than 20 μm spore size. e.g. *Leucobryum glaucum, Hylocomium splendens* etc.;

2.   **Perennial Shuttle**: Characterized by pluriennial (5–10 yrs), monoecious or dioecious taxa having age of first reproduction asexual: 1–2 yrs, and sexual up to 5 yrs, sexual reproduction efforts are variable however, asexual reproduction effort is high in the case of low sexual reproduction effort, 25 μm spore size, e.g. *Orthotrichum* spp., *Macromitrium* spp. etc.;

3.   **Colonist**: Including monoecious or dioecious taxa surviving from one to a few years (pauciennials) with the age of first asexual reproduction more than 1 year, sexual reproduction at the age of 2–3 years, variable effort of sexual and asexual reproduction, spore size 20 μm. e.g. *Bryum bicolor, Grimmia pulvinata*;

4.   **Short-lived Shuttle**: Including many monoecious, pluriennel or pauciennial taxa having first reproduction at 2–3 years of age, high rate of sexual reproduction at 2–3 years of age, high rate of sexual reproduction effort and 25 μm size of spores, e.g. *Bryum marratii, Tetraplodon mnioides*, etc.

5.   **Annual Shuttle**: Including ephemerals (persistence of a given colony <1 year), annuals or pauciennial mostly monoecious taxa having their first sexual reproduction at the age of less than 1 year, which are not restricted to a particular season. Taxa belonging to this category exhibit high efforts of sexual reproduction whereas asexual propagules are rare or absent. Their spore size is more than 25 μm, e.g. *Archidium* spp., *Ephemerum* spp. etc.;

6.   **Fugitives**: These are annual and essentially monoecious taxa having their first sexual reproduction at the age of < 1 year, with a transient habitat. These taxa exhibit a high effort of sexual reproduction while asexual propagules are rare or absent with a spore size less than 20 μm., e.g. *Funaria hygrometrica*.

The evolution of reproductive strategies in hepatics is almost parallel to that of mosses, e.g. trends towards monoecism, immersed, indehiscent capsules and large spores, combined with absence of asexual propagules, assemblages of species from both groups in seasonally arid habitats. For the hepatics, the Bazzania-Anastrophyllum model has been suggested (Longton and Schuster 1983) which is a basic model and includes a wide range of taxa. This model involves all modifications in reproductive modalities involving linkage of (1) unisexuality, (2) rarity of asexual reproduction devices (gemmae, etc.) formed by gametophyte, and (3) perennial gametophytes having the capacity to form clones of indeterminate size and age.

11.   Studies on Reproductive Biology in the 21st Century

The significance of studies on reproductive biology of Bryophytes cannot be fully appreciated till the answers to several fundamental questions are available, e.g. level of ploidy in gametophytes, the role of spores where produced in population maintenance and dispersal, the rate of inbreeding in monoecious taxa, incidence of sexual reproduction by outbreeding to maintain evolutionary effective levels of genetic diversity. Few of these topics are being studied, and there needs to be a serious attempt for this potentially most fruitful avenue to advance the bryological research (Longton and Schuster 1983). Soderstrom (1994) expressed his views in a symposium on 'Reproductive Biology of Bryophytes' that there are two important aspects of study regarding reproduction: first is the role of reproduction in population dynamics and survival, and second in genetic diversity and speciation. As we know that bryophyte populations are distributed in patches and production, dispersal and establishment of asexual propagules play an important role. It has also been observed that genetically more diverse species have a better chance of overcoming environmental changes, compared with stenotypic species. Thus the role of reproduction (asexual and sexual) in the maintenance of genetic variation must be studied. In addition, there are also some aspects like ecological and evolutionary consequences of reproduction which are also hardly studied.

Recent studies have clearly shown that there exists a wide range of genetic variability at interspecific and intraspecific levels in mosses, as shown by electrophoretic patterns. The major challenge for future studies is to discover patterns among various aspects of reproductive biology and genetics which will give an idea of evolutionary processes functioning in different groups of bryophytes (Longton 1994).

In order to derive a clear-cut concept about various reproductive patterns and their evolutionary significance, a comprehensive conceptual framework including application of cytological, electrophoretic and DNA finger-printing methods of study are very much required.

The other most significant aspect of study to be taken up in the 21st century is the 'constraint in reproductive biology' of monotypic, endemic, rare and threatened taxa of bryophytes. There is an urgent need to clearly identify those limiting factors which are instrumental in making a taxon endemic and rare, which is phylogenetically important too. Earlier, in an assessment made by Udar and Srivastava (1983), on rare and endangered liverworts of India, showed that nearly 37 taxa from the eastern Himalayas, 24 taxa from the western Himalayas, 14 taxa from south India and 3 taxa from central India have been listed as rare and threatened. These are either endemic to India or possess a narrow ecological range and restricted distribution in India. Similarly, Pant (1983) has also provided an account of threatened bryophytes of Nainital.

Among the listed rare liverwort taxa, *Aitchisoniella himalayensis* Kash., *Stephensoniella brevipedunculata* Kash. (see also Oder et al. 1983) *Sewardiella tuberifera* Kash. and *Monoselinium tenerum* Griff. are monotypic and possess an immense phylogenetic importance. Out of these, the former three are endemic to India in the northwest Himalayas, however, the last is also distributed outside India, but it is extremely rare in India. The main reason for their frequent depletion is their inability to develop genetic variability in their populations to withstand changing climatic conditions. The second reason seems to be the failure of easy spore germination in *Stephensoniella, Aitchisoniella* and *Sewardiella* as they perenate by means of tubers. Spatial separation of male and female populations also poses a threat to proliferation of progeny and also restricts the required smooth gene flow, e.g. in *Takakia ceratophylla* (Mitt.) Grolle male plants and sporophytes were not known from its discovery in 1958 until 1993. This species was described from Lachen, Sikkim, however, it could not be collected again from this locality. After a gap of decades, antheridia of *T. ceratophylla* were discovered by Davison et al. (1989) in Atka Island. Subsequently, Smith (1990) discovered the first sporophytes in Atka. In the year 1993 Smith and Davison provided a complete account of antheridia and sporophytes of *Takakia ceratophylla*. This particular reproductive behaviour of taxa not only reduces the genetic variability but also restricts the distribution of populations.

It is essential that a critical search is made regularly for the assessment of rare and phylogenetically important bryophytes. the natural habitat of such plants must be declared as protected so that they may get favourable climatic conditions for their luxuriant growth. Those taxa which are facing threat to their survival due to an inefficient reproductive mechanism should be conserved and multiplied under controlled laboratory conditions.

References

Allard RW (1975) The mating system and microevolution. Genetics 79: 115–126

Allen CE (1935) The genetics of bryophytes. Bot Rev 1: 269–291

Allorge V (1955) Muscinces du Pays Basque. Francais et Espangnoi. Rev Bryol Lichén 23: 248–333

Allsopp A, Mitra GC (1958) The morphology of protonema and bud formation in the Bryales. Ann Bot (Lond) 22: 95–115

Anderson LE (1980) Cytology and reproductive biology of mosses. *In*: Taylor RJ, Leviton AE (eds) The mosses of North America. San Francisco California, pp 37–76

Anderson LE, Bryan VS (1956) A cytotaxonomic investigation of *Fissidens cristatus* Wils and *Fissidens adianthoides* Hedw. in North America. Rev Bryol Lichén 25: 254–267

Anderson LE, Lemmon BE (1972) Cytological studies of natural intergeneric hybrids and their parental species in the moss genera *Astomum* and *Weissia*. Ann Mo Bot Gdn. 59: 382–416

Ando H (1972) Distribution and speciation in the genus *Hypnum* in the circumpacific region. J Hattori Bot Lab 35: 68–98

Ashton NW, Cove DJ (1977) The isolation and preliminary characterization of auxtrophic and analogue resistant mutants of moss *Physcomitrella patens*. Mol. Gen Genet 154: 87–95

Asthana AK Srivastava SC (1991) Indian Hornworts (a taxonomic study). Bryophyt: Biblioth 42: 1–230

Bartholomew-Began SE (1986) The sporeling development of *Blasia pusilla* L. J Hattori Bot Lab 59: 255–262

Bartholomew-Began SE (1996) The sporeling ontogeny of *Pellia epiphylla* and *Pellia neesiana* with special reference to protonema. J Hattori Bot Lab 79: 115–128

Bartholomew-Began SE, Crandall Stotler BJ (1994) The sporeling ontogeny of *Monoclea gottscheii* ssp. *elongata*. Bryologist 97: 244–252

Bedford THB (1938) Sex distribution in colonies of *Climacium dendroides* W. and M. and its relation to fruit-bearing. Northwest Nat 13: 312–321

Bergeron T (1944) On some meteorological conditions for the dissemination of spores, pollen, etc., and a supposed wind transport of *Aloina* spores from the region of Lower Yenisey to south-western Finland in July 1936. Sv Bot Tidskr 38: 269–292

Berthier J, Larpent JP, Gourgaud ML (1976) Light action on vegetative propagation in bryophytes. J Hattori Bot Lab 41: 193–203

Bharadwaj DC (1950) Studies in Indian Anthocerotaceae. III. On the morphology of *Anthoceros crispulus* (Mont.) Douin. J Indian Bot Soc 29: 145-163

Blackeslee AF (1906) Differentiation of sex, thallus gametophytes and sporophyte. Bot Gaz 42: 161–178

Boatman DJ, Lark PM (1971) Inorganic nutrition of the protonema of *Sphagnum pappillosum* Lindb., *S. megellanicum* Brid. and *S. cuspidatum* Ehrh. New Phytol 70: 1053–1059

Brown MM (1919) The development of the gametophyte and the distribution of sexual characters in *Funaria hygrometrica* (L.) Schreb. Am J Bot 6: 387–400

Bryan GS (1917) The archegonium of *Catherinea angustata* Brid. (*Atrichum angustatum*). Bot Gaz 64: 1–20

Burgeff H (1943) Genetische Studien an *Marchantia*. Jena.

Burr LL (1939) Morphology of *Cyathophorum bulbosum*. Trans R Soc N Z 68: 437–456

Cameron RG, Wyatt R (1990) Spatial patterns and sex ratio in dioecious and monoecious mosses of genus *Splachnum*. Bryologist 93: 161–166

Campbell DH (1896) The development of *Geothallus tuberosus* Campbell, Ann Bot (Lond) 10: 489–510

Campbell DH (1898) The sytematic position of the genus *Monoclea*. Bot Gaz 25: 272–274

Campbell DH (1913) The morphology and systematic position of *Calycularia radiculosa* (Steph.). Dudley Mem Vol (Stanford Univ) pp 43–61

Campbell DH (1915) The morphology and systematic position of *Podomitrium*. Am J Bot 2: 199–210

Campbell DH (1920) Studies on some East Indian Hepaticae: *Calobryum blumei*. Nab E. Ann Bot (Lond) 34: 1–12

Chalaud G (1932) Germination de spores et phase protonemiqui. *In*: Verdoorn F (ed) Manual of bryology. The Hague, Netherlands pp 89–108

Chopra RN, Gupta A (1991) Effect of some cytokinins on growth and archegonial formation in the liverwort *Riccia discolor* L. grown in vitro. J Hattori Bot Lab 71: 47–54

Chopra RN, Kumra PK (1988) Biology of bryophytes. Wiley Eastern, New Delhi 350 pp

Chopra RN, Rawat MS (1977) Studies on production and behaviour of protonemal gemmae in some Bryaceae. Bryologist 80: 655–661

Chopra RS, Kumar SS (1981) Mosses of the western Himalayas and adjacent plains. Chronica Botanica, New Delhi, pp 1–142

Clarke GC, Greene SW (1970) Reproductive performance of two species of *Pohlia* at widely-separated stations. Trans Br Bryol Soc 6: 114–128

Crandall-Stotler B (1980) Morphogenetic designs and theory of bryophyte origins and divergence. Bioscience 30: 580–585

Crum H (1972) The geographic origins of the mosses of North America's eastern deciduous forest. J Hattori Bot Lab 35: 269–298

Crundwell AC (1981) Reproduction in *Myurium hochstettri*. J Bryol 11: 715–717

Cummins H, Wyatt R (1981) Genetic variability in natural populations of the moss *Atrichum angustatum*. Bryologist 84: 30–38

Davison PG, Smith DK, Mcfarland KD (1989) The discovery of antheridia in *Takakia*. ASB Bull 36: 65

Dixon HN (1924) The student's handbook of British mosses. 3rd edn., V.T. Sumfeld, London 586 pp

Doyle WT (1970) The biology of higher cryptogams, London

Duckett JG, Ligrow R (1992) A survey of diaspore liberation mechanisms and germination patterns of mosses. J Bryol 17: 335–354

Duckett JG, Renzaglia KS (1993) The reproductive biology of the liverwort *Blasia pusilla* L. J Bryol 17: 541–552

During H (1979) Life cycle strategies in bryophytes. A preliminary review. Lindbergia 5: 2–18

Engel PP (1968) The induction of biochemical and morphological mutants in moss, *Physcomitrella patens*. Am J Bot 55: 438–446

Evans AW (1897) A revision of the North American species of *Frullania*: a genus of Hepaticae. Trans Conn Acad Arts Sci 10: 1–39

Fulford M (1951) Distribution patterns of the genera of leafy Hepaticae of South America. Evolution 5: 243–264

Fulford M (1956) The young stages of the leafy Hepaticae. Phytomorphology 6: 199–235

Gayat LA (1897) Recherches sur le developement de l'archegone Chezles Muscinees. Ann Sci Nat Ser 83: 161–258

Gemingham CH, Robertson ET (1950) Preliminary investigations on the structure of bryophytic communities. Trans Br Bryol Soc 1: 330–344

Gemmell AR (1950) Studies in the Bryophyta. 1, The influence of the sexual mechanism on varietal production and distribution of British Musci. New Phytol 49: 64–71

Goebel KV (1930) Organographie der Pflanzen 3. Aufl, II: i–x, 64–1378, Jena

Gupta A, Sarla, Chopra RN (1991) In vitro studies on growth and gametangial formation in *Riccia discolor*. Effect of physical factors. J Hattori Bot Lab 70: 107–118

Gupta A, Udar R (1986) Palynotaxonomy of selected Indian liverworts. Bryophy Bibl 29: 1–202

Hadderson TA, Longton RE (1995) Patterns of life history variation in the Funariales, Polytrichales and Pottiales. J Bryol 18: 639–675

Harvey–Gibson RJ, Miller-Brown D (1927) Fertilization of Bryophyta. Ann Bot (Lond) 41: 190–191

Heald FDF (1898) A study of regeneration exhibited by mosses. Bot Gaz 26: 169–210

Hedwig J (1782) Fundamentum historiae naturalis muscorum frondosorum. Lipsiae

Hoffman GR (1970) Spore viability in *Funaria hygrometrica*. Bryologist 73: 634–635

Imura S (1994) Vegetative diaspore in Japanese mosses. J Hattori Bot Lab 77: 177–232

Ingold CT (1939) Spore discharge in land plants. Oxford (UK), 178 pp

Ingold CT (1959) Peristome teeth and spore discharge in mosses. Trans Proc Bot Soc Edinb 38: 76–88

Inoue H (1961) Studies in spore germination and the early stages of gametophyte development in the Marchantiales. J Hattori Bot Lab 23: 149–191

Inoue H, Schuster RM (1971) A Monograph of the New Zealand and Tasmanian Plagiochilaceae. J Hattori Bot Lab 34: 1–225

Iwatsuki Z (1972) Geographical isolation and speciation of bryophytes in some islands in eastern Asia. J Hattori Bot Lab 36: 126–141

Kamimura M (1961) Monograph of Japanese Frullaniaceae. J Hatton Bot Lab 24: 1–109

Kashyap SR (1929) Liverworts of the western Himalayas and the Panjab Plain. Part I. The University of the Panjab, Lahore, pp 1–129

Khanna KR (1960) Studies on natural hybridization in the genus *Weissia*. Bryologist 63: 1–16

Kimmer RW, Young CC (1995) The role of slugs in dispersal of the asexual propagules of *Dicranum flagellare*. Bryologist 98: 149–153

Krzakowa M, Szweykowski J (1979) Isozyme polymorphism in natural populations of a liverwort, *Plagiochila asplenioides*. Genetics 93: 711–119

Kühn E (1870) Zur Entwicklungsgeschichte der *Andreaeaceen*. Inaugural Dissertation, Leipzig (Germany)

Kumar D, Udar R (1976) *Colobryum dentatum* Kumar et Udar: A new species of *Calobryum* from India. J Indian Bot Soc 55: 23–30

Kumra PK, Chopra RN (1983) Effect of some physical factors on growth and gametangial induction in male clones of three mosses grown in vitro. Bot Gaz 144: 533

Leitgeb H (1869) Wachstum des Stammchens und Entwicklung der Antheridien bei *Sphagnum*. Sitzungsber Kaiser Akad naturwiss Wien 59: 294–320

Lewis KR (1961) The genetics of bryophytes Trans Br Bryol Soc 4: 111–130

Lindberg SO (1882) Europas Och Nord Amerikas huitmosser (Sphagna). Helsingfors

Longton RE (1962) Polysity in the British Bryophyta. Trans B Bryol Soc 4: 111–130

Longton RE (1969) Studies on growth and reproduction in mosses. Def Res Board Can Hazen Ser 38: 6–9

Longton RE (1976) Reproductive biology and evolutionary potential in bryophytes. J Hattori Bot Lab 41: 205–223

Longton RE (1980) Physiological ecology of mosses. *In*: Taylor RJ, Leviton AE (eds) The mosses of North America, San Francisco pp 77–113

Longton RE (1982) The biosystematic approach to bryology. J Hattori Bot Lab 53: 1–19

Longton RE (1988) Life history strategies among bryophytes of arid regions. J Hattori Bot Lab 63: 15–28

Longton RE (1994) Reproductive biology in bryophytes. The challenge and the opportunities. J Hattori Bot Lab 76: 159–172

Longton RE, Greene SW (1966) Relationship between sex distribution and sporophyte production in *Pleurozium schreberi* (Brid.) Mitt. Ann Bot (Lond) 33: 107–126

Longton RE, Greene SW (1967) The growth and reproduction of *Polytrichum alpestre* Hoppe in south Georgia. Philos Trans R Soc B 252: 295–322

Longton RE, Greene SW (1969). The growth and reproductive cycle of *Pleurozium schreberi* (Brid.) Mitt. Ann Bot (Lond) 33: 83–105

Longton RE, Greene SW (1979) Experimental studies on growth and reproduction in the moss *Pleurozium schreberi* (Brid.) Mitt. J Bryol 10: 321–338

Longton RE, Miles CJ (1982) Studies on the reproductive biology of mosses. J Hattori Bot Lab 52: 219–239

Longton RE, Schuster RM (1983) Reproductive Biology. *In*: Schuster RM (ed.) New Manual of Bryology. The Hattori Botanical Laboratory, Japan pp 386–462

Lotsy JP (1909) Vorträge über botanische Stammesgeschichte, vol 2. G. Fischer. Jena

Mehra PN, Handoo ON (1953) Morphology of *Anthoceros erectus* and *Anthoceros himalayensis* and the phylogeny of the Anthocerotales. Bot Gaz 114: 371–382

Mehra PN, Kumar D (1989) Some observations on the embryology of *Calobryum indicum*. J Hattori Bot Lab 67: 239–254

Mehra PN, Kachroo P (1951) Sporeling germination studies in Marchantiales. I. Rebouliaceae. Bryologist 54: 1–16

Meyer SL (1942) Physiological studies in mosses. IV. Regeneration in *Physcomitrium turbinatum*. Bot Gaz 104: 128–132

Miles CJ, Longton RE (1990) The role of spores in reproduction in mosses. Bot J Linn Soc 104: 149–173

Miles CJ, Longton RE (1992) Spore structure and reproductive biology in *Archidium alternifolium* (Dicks. ex Hedw.) Schimp. J Bryol 17: 203–223

Miles CJ, Longton RE (1992) Deposition of moss spores in relation to distance from parent gametophytes. J Bryol 17: 203–222

Miller HA (1979) The phylogeny and distribution of the Musci. *In*: Clarke GCS, Duckett JG (eds) Bryophyte systematics Academic Press, London pp 11–39

Mogenson GS (1978) Spore development and germination in *Cinclidium* (Mniaceae, Bryophyta), with special reference to spore mortality and false anisospory. Can J Bot 56: 1032–1060

Mogenson GS (1981) The biological significance of morphological characters in bryophytes: the spore. Bryologist 84: 187–207

Muller K (1948) Morphologische und anatomische Untersuchungen und Antheridien de Jungermannien. Bot Not 1948: 71–80

Nakosteen PC, Hughes KW (1978) Seasonal life cycles of three species of Funariaceae in culture. Bryologist 81: 307

Nath V, Asthana AK (1993) An observation of spore germination (in situ) in *Frullania physantha* Mitt. J Indian Bot Soc 72: 187–188

Nehira K (1983) Spore germination, protonema and sporeling *In*: Schuster RM (ed) New Manual of Bryology, The Hattori Botanical Laboratory, Japan, pp 343–385

Nehira K (1987) Some ecological correlations of spore germination patterns in liverworts. Bryologist 90: 405–408

Newton ME (1968) Cytotaxonomy of *Tortula muralis* Hedw. in Britain. Trans Br Bryol Soc 5: 523–535

Newton ME (1971) A cytological distinction between male and female *Mnium undulatum* Hedw. Trans Br Bryol Soc 6: 230–243

Newton ME (1972) Sex ratio differences in *Mnium hornum* Hedw. and *Mnium undulatum* Sw. in relation to spore germination and vegetative regeneration. Ann Bot (Lond) 36: 163–178

Nishida Y (1978) Studies on the sporeling types in mosses. J Hattori Bot Lab 44: 371–454

Pant GB (1983) Threatened bryophytes of Nainital *In*: Jain SK, Rao RR (eds) An assessment of threatened plants of India. Bot Surv India, Howrah pp 313–317

Pande SK, Srivastava KP, Misra RN (1955) Studies in Indian Metzgerineae. II. *Sewardiella tuberifera* Kash. Phytomorphology 5: 57–67

Pande SK, Udar R (1956) On two species of *Riccia* new to Indian flora. Curr Sci 25: 232–233

Parihar NS (1959) An introduction to Embryophyta Vol 1. Bryophyta, Central Book Depot Allahabad, pp 380

Proskauer J. (1951) Studies on Anthocerotales. 3. The genera *Anthoceros* and *Phaeoceros*. Bull. Torrey Bot club 78: 331–349

Pujos J (1992) Life history of *Sphagnum*. J Bryol 17: 93–105

Reynolds DN (1980) Gamete dispersal in *Mnium ciliare*. Bryologist 83: 73–77

Richardson DHS (1981) The Biology of Moss, Oxford (UK)

Saxton WT (1931) The life history of *Lunularia* with special reference to the archegoniophore and sporophyte. Trans RSS Afr 19: 259–268

Schuster RM (1953) Boreal Hepaticae: a manual of the liverworts of Minnesota and adjacent regions. Am Midl Nat 49: 257–648

Schuster RM (1959–60) A monograph of the nearctic Plagiochilaceae I-III. Am Midl Nat 62: 1–166, 62: 257–395, 63: 1–130

Schuster RM (1966) The Hepaticae and Anthocerotae of North America, east of the hundredth meridian. Vol 1: Columbia University Press, New York

Schuster RM (1983) Comparative anatomy and morphology of the Hepaticae. *In*: Schuster RM (ed) New Manual of Bryology, pp 760–891. The Hattori Bot Lab, Japan

Shaw AJ (1990) Genetic and environmental effects on morphology and asexual reproduction in the moss *Bryum bicolor* Bryologist 93: 1–6

Smith DK (1990) Sporophyte of *Takakia* discovered. Bryol. Times 57/58: 1, 4

Smith DK, Davison PG (1993) Antheridia and Sporophytes in *Takakia ceratophylla* (Mitt.) Grolle: evidence for reclassification among the mosses. J Hattori Bot Lab 73: 263–271

Soderstrom L (1994) Scope and significance of studies on reproductive biology of bryophytes. J Hattori Bot Lab 76: 97–103

Srivastava SC, Udar R (1975) Taxonomy of Indian Metzgeriaceae- a monographic study. New Bot 2: 1–57

Srivastava SC, Udar R (1975a) Sporeling development in *Fossombronia kashyapii* Sriv and Udar. Geophytology 5: 33–38

Udar R (1957) Culture studies in the genus *Riccia* (Mich.) L. 1. Sporeling germination in *Riccia billardieri* Mont. et N. J Indian Bot Soc 36: 46–50

Udar R (1957) Culture studies in the genus *Riccia* (Mich) L. 2. Sporeling germination and regeneration in *R. crystallina* L. J Indian Bot Soc 36: 580–586

Udar R (1958) Culture studies in the genus *Riccia* (Mich.) L. 3. Sporeling germination in *R. trichocarpa* Howe: a reinvestigation. J Indian Bot Soc 37: 70–74

Udar R (1958a) Studies in Indian Sauteriaceae I. Sporeling patterns in *Athalamia pinguis* Falc. J Indian Bot Soc 37: 300–330

Udar R (1959) Genus *Riccia* in India-IV. A new *Riccia*, *R. pandei* Udar from Garhwal with a note on the species of the genus from the western Himalayan territory. J Indian Bot Soc 38: 146–159

Udar (1960) Studies in Indian Sauteriaceae-II on the morphology of *Athalamia pinguins* Falc. J Indian Bot Soc 39: 56–77

Udar R (1976) An Introduction to Bryophyta. Shashi Dhar Malviya Prakashan, Lucknow, 200 pp

Udar R (1976a) Bryology in India. Ann Cryptogam Phytopathol 4: 1–200, Chron Bot, New Delhi

Udar R, Awasthi US (1981) A new species of *Lejeunea* from India. Cryptogamie Bryol Lichenol 2: 345–348

Udar R, Awasthi US (1982) The genus *Schiffneriolejeunea* Verd. (Hepaticae) in India. Lindbergia 8: 55–59

Udar, R, Awasthi US, Shaheen F (1982) A new *Caudalejeunea* from India. Bryologist 85: 329–331

Udar R, Chandra V (1964) Polyembryony in *Mannia foreaui* Udar et Chandra. Bryologist 67: 55–59

Udar R, Chandra V (1964a) *Exormotheca ceylonensis* Meijer - New to Indian Flora. Curr Sci 33: 436–438

Udar R, Chandra V (1964b) On some anomalous female receptacles of *Reboulia hemispherica* (L.) Raddi. J Indian Bot Soc 43: 521–528

Udar R, Chandra V (1965) On a new species of *Calobryum*, *C. indicum* Udar et Chandra from Darjeeling, India. Rev Bryol et Lichenol. 33: 556–559

Udar R, Gupta A (1977) Development of propagula in *Plagiochila*. J Indian Bot Soc 56: 286–289

Udar R, Gupta A (1982) Natural regeneration in *Lejeunea flava*. New Bot 9: 5–8

Udar R, Gupta A (1983) *Targionia lorbeeriana* Mull. from India. Indian J Bot 6: 215–219

Udar R, Kumar A (1981) *Jackiella javanica* Schiffn. - A rare and interesting taxon from India. J Indian Bot Soc 60: 105–111

Udar R, Kumar A (1983) Studies in Indian Jungermanniaceae II. *Jungermannia* (*Plectocolea*) *tetragona* Lind. from Andaman Islands. J. Indian Bot. Soc 62: 357–360

Udar R, Kumar D (1972) Sporeling development in *Athalamia pusilla*. Phyton 14: 229–237

Udar R, Kumar D (1982) The genus *Radula* Dumort. in India-I. J Indian Bot Soc 61: 177–182

Udar R, Singh VB (1958) Patterns of regeneration in *Notothylas indica* Kash. Curr Sci 27: 23–25

Udar R, Singh DK (1978) In vitro studies on the spore germination of *Cryptomitrium himalayense* Kash. New Bot 5: 15–21

Udar R, Srivastava G, Srivastava SC (1987) on two species of *Cololejeunea* (*Pedinolejunea*) new to India. J Indian Bot Soc 66: 22–26

Udar R, Srivastava SC (1968) Sporeling development in the genus *Exormotheca*. 1. *E. Ceylonensis*. Can J Bot 46: 1009–1012

Udar R, Srivastava SC (1970) Sporeling development in *Preissia quadrata*. Phyton (Argentina) 14: 165–173

Udar R, Srivastava SC (1977) Sporeling development in the leafy liverwort *Lophocolea bidentata* (L.) Dum. New Bot 2: 124–127

Udar R, Srivastava SC (1981) On some noteworthy features in the sporophyte of *Haplomitrium hookeri* from the western Himalayas. J Indian Bot Soc 60: 179–181

Udar R, Srivastava SC (1983) Rare and endangered liverworts of India. *In*: Jain SK, Rao RR (eds) An assessment of threatened plants of India. Bot Surv India, Howrah pp 303–312

Udar R, Srivastava SC (1984) Reproductive biology of some Indian liverworts. Phytomorphology 33: 37–46

Udar R, Srivastava SC, Kumar D (1982) *Geocalyx* Nees: a rare marsupial genus from India. Proc Indian Acad Sci (Plant Sci) 91: 139–143

Udar R, Srivastava SC, Mehrotra L. (1974) Observations on cytokinesis in *Riccia cruciata* Kash. and its spore morphology. New Bot 1: 1–7

Udar R, Srivastava, SC, Srivastava G (1983) Observations on endemic Liverwort taxa from India. 1. Reproductive biology and SEM details of spores in *Stephensoniella brevipedunculata* Kash. J. Hattori Bot Lab 54: 321–330

Une K (1985) Sexual dimorphism in the Japanese species of *Macromitrium* Brid. J Hattori Bot Lab 59: 487–514

Vitt DH (1968) Sex determination in mosses. Mich Nat 7: 195–203.

Wann FB (1925) Some of the factors involved in sexual reproduction of *Marchantia polymorpha*. Am J Bot 12: 307

Watson, EV (1971) The structure and life of Bryophytes. Hutchinson Univ. Library, London.

Zander RH (1979) Patterns of sporophyte maturation dates in Pottiaceae. Bryologist 82: 538–558

9

Reproductive Biology of Pteridophytes

Krishna Kumar

1. Introduction

The pteridophytes were not necessarily the first vascular land plants (see Kenrick and Crane 1997). Linnaeus (1754) included them in the Class Cryptogamae (kruptos = hidden, gamos = marriage). Eichler (1883) regarded cryptogamae as a sub-division of the Plant Kingdom which included 'Club mosses', 'Horse-tails', and 'ferns'. The pteridophytes occupy an intermediate position between the bryophytes and gymnosperms. Engler (1886) suggested the term 'Embryophyta' to include bryophytes, pteridophytes, and spermatophytes. Tippo (1942) considered 'Tracheophyta' to include pteridophytes and spermatophytes.

Due to a more or less similar female reproductive organ the archegonium, the bryophytes, pteridophytes, and gymosperms are also grouped as 'Archegoniatae'. Both bryophytes and pteridophytes show regular alternation of generations—gametophyte (n) and sporophyte (2n). In pteridophytes, the sporophyte is the dominant generation. The gametophytic phase of the life cycle is much reduced. The pteridophytes are seed-less vascular plants which generally grow in moist and shady places. Water is essential for fertilization. Due to delicate and translucent leaf lamina, Hymenophyllaceae are called 'filmy ferns' (Fig. 4H, E). Many ferns are epiphytic. The evolution of a resistant cuticle, complex stomata, vascular tissue, and lignin enabled these plants to dominate habitats on land (Moore et al. 1995). The plant body contains chlorophylls a and b, carotenoids, starch, cellulose and motile flagellated spermatozoids. A true root, vascular tissue (tracheids) and multiflagellate spermatozoids are characteristic features of the pteridophytes.

Vegetative multiplication occurs through fragmentation, gemmae, and bulbils.

2. Synangia

Synangia (three fused sporangia) develop in *Psilotum* (Fig. 1A-C, E, F). In a transection the synangium appears trillocular (Fig. 1J, K, M); a bifid appendage (Fig. 1R) may bear a bilocular synangium (Fig. 1S, T). The leaves which bear synangia in *Tmesipteris* may be murconate (Fig. 1Q) rather than lanceolate. *Lycopodium* and *Equisetum* bear cones containing similar spores (homosporous) (Figs. 2I, J, K, 3F, I). *Selaginella, Marsilea, Salvinia, Azolla, Regnellidium, Pilularia, Stylites, Isoetes* and *Platyzoma* are heterosporous (produce two types of spores—microspores and megaspores; Figs. 4A, D, C, L, M, 5A, F, G). Mukhopadhyay and Bhandari (1999) found microspores in polyads in *S. intermedia*. Only the above nine taxa show heterospory which leads to a 'seed' habit in higher plants. The spores (Fig. 2K) are contained in sporangia which are borne on sporophylls (Fig. 2H, J) or sporangiophores (Fig. 3E, I). These become aggregated into cones or strobilli (Figs. 2A, C, L, 3A, B, 5A).

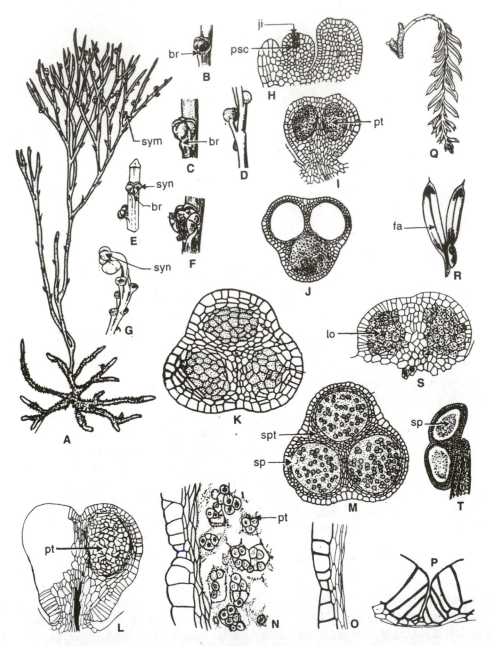

Fig. 1 A-T. Synangia. A-P *Psilotum*, Q-T *Tmesipteris*. A Plant with synangia (*syn*). B, C Bifid bract (*br*) bears trilocular synangium. D Lateral branch with terminal synangium. E Cauline synangium (*syn*), bract (*br*) bifid. F Dehisced synangium. G Terminal synangium. H Eusporangium shows jacket initials (*ji*) and primary sporogenous cells (*psc*). I, L Plasmodial tapetum (*pt*). J, K Transection synangium. M Trilocular synangium (t.s.) with spores (*sp*), note sterile partitions (*spt*). N Synangium wall, spores, and remnants of plasmodial tapetum. O Three- to four-layered wall of synangium. P Part of mature wall. Q-T *Tmesipteris*. Q Plant with mucronate leaves bear synangia. R Fertile appendage (*fa*). S Two locules (*lo*) of synangium. T Synangium (l.s.) with spores (*sp*) (A, M from Parihar 1965, B, C, E, R from von Wettstein 1935, D from Bierhorst 1956, E, Q from Pritzel 1900, G from Rouffa 1967, H, I from Smith 1955, J from Rashid 1976, K, N from Bower 1908, L, O, P, S from Bierhorst 1971, T from Foster and Gifford 1974).

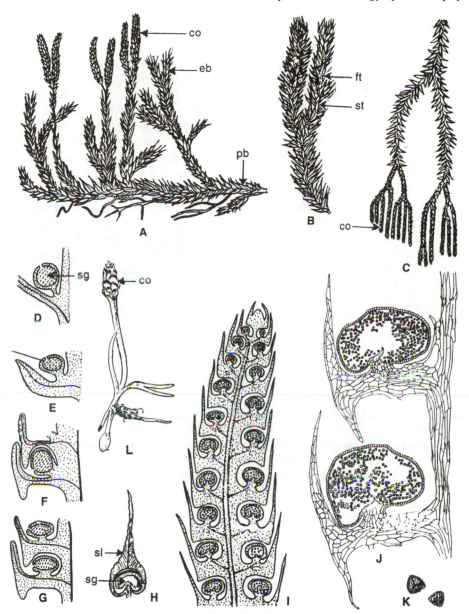

Fig. 2 A-L. Strobilus (Cone) A-K *Lycopodium, L. phylloglossum.* A *L. clavatum*, prostrate (*pb*) and erect branches (*eb*), and cones (*co*) (Rhopalostachya). B *L. selago* with alternate fertile (*ft*) and sterile (*st*) region (Urostachya). C. *L. phlegmaria*, dichotomously branched cones (*co*). D *L. leucidulum*, axillary and exposed sporangium (*sg*). E *L. squarrosum*, subfoliar and exposed sporangium. F *L. inundatum*, axillary and protected sporangium. G *L. cernuum*, foliar and protected sporangium. H *L. clavatum*, sporophyll (*sl*) with sporangium (*sg*). I *L. clavatum*, cone (*l.s*). J Abaxial extension of sporophyll protects the lower sporangium, note one-layered wall. K Spores with triradiate ridge. *L. Phylloglossum*, plant with terminal strobilus (A from Strasburger 1894, B, H from von Wettstein 1935, C from Smith 1955, D from Eames 1936, E-G from Sykes 1908a,b, I from Parihar 1965, J, K from Bracegirdle and Miles 1971, *L* from Foster and Gifford 1974).

3. Cone or Strobilus

A strobilus is essentially a stem tip with several closely-packed leaves (= sporophylls; Fig. 3C), or branches that bear sporangia in their axil. The strobilus is one of the significant developments in the reproductive organization of Lycophyta (*Lycopodium* spp.-club mosses, *Selaginella* spp.- spike mosses and Arthrophyta *Equisetum*-horse-tails; Figs. 3C, F, G, 5F, CE). Of the ca. 13,000 species, Ca. 12000 are ferns Ferns 11,000 (Raven et al 1999). Strobili do not develop in Psilophyta (*Psilotum*-whisk fern, and *Tmesipteris*), and Pterophyta (ferns). Pterophyta (the most successful and largest group of ferns) are both homosporous and heterosporous. Among ferns, the homosporous ferns (the largest group) have three special characters: (1) a cordate prothallus, (2) the sporangium has a special wall which develops annulus and stomium (lip cells) concerned in dehiscence, and (3) multiflagellate spermatozoids. The spore-producing organs (sporangia) in homosporous ferns are usually borne on the under-surface of leaves. They may be fertile spikes (Fig. 7E, F), tassels (Fig. 7A, B), and sori (*Adiantum, Dryopteris*) (Fig. 9I, K). The heterosporous ferns develop special organs, the sporocarps—Marsileales (Fig. 4A, K, L, M) and Salviniales (Fig. 4C, D, F, G, I).

The sporangia, in ferns, within a sorus arise from fertile tissue (receptacle; Fig. 10A). Sometimes it is elongated (Fig. 4B, J). The sporangia may or may not be covered by a flap of tissue arising from the receptacle, the true indusium (Fig. 10A). Sometimes, a true indusium may be absent and the 'false' indusium is formed by inrolling of the leaf margin (Fig. 10B, C).

Lycopodium selago bears alternate fertile and sterile regions (selago condition) on the stem (Fig. 2B). Such alternate fertile and sterile regions are also distinguishable in *Psilotum* (Bower 1908) and *Selaginella* (Fig. 5B). The cones are dichotomously branched in *Lycopodium phlegmaria* (Fig. 2C) and *L. lucidulum*. In *L. clavatum* and *Equisetum arvense* the cones are borne on fertile shoots which may be unbranched and bear pale leaves (Figs. 2A, 3A).

3.1 Sporophylls

The sporophylls (Fig. 2H, I) are compactly arranged to form cones in *Lycopodium, Selaginella* and *Equisetum* (Figs 2A, 3A, 5A, B). In *Equisetum* the sporangia are borne on the under surface of a special structure (peltate sporangiophore) (Fig. 3E). In ferns the sporophylls are photosynthetic. In *Lycopodium* spp. the sporophylls may be pale-yellow in the Rhopalostachya (distinct from vegetative leaves; Fig. 2A), but in Urostachya they are similar to vegetative leaves (Fig. 2B).

3.2 Sporangia

The pteridophytes are polysporangiate. The sporangia are cauline in *Psilotum* (Fig. 1D; Bierhorst 1971) and *Equisetum* (Eames 1936). Sometimes the sporangia are terminal as in *Psilotum* (Fig. 1G; Rouffa 1971). The sporangia are mostly associated with leaves. If the sporangium develops from many initials, it is called a eusporangium and the method of development eusporangiate but, on the other hand, if the entire sporangium develops from one initial only, it is called leptosporangium and the method of development leptosporangiate (Goebel 1881).

The pteridophytes can be classified on the basis of development of the sporangium only, as Eusporangiopsida and Leptosporangiopsida. In homosporous leptosporangiate ferns, the development of a special sporangium (leptosporangium) (Figs. 8, 10D) is a unique feature in the Plant Kingdom (Foster and Gifford 1974). Sometimes, the sporangia may fuse and form synangia (Figs. 1E, 9N, P; Psilophyta, Marattiales—Wolf 1997). The sporangia may occur in a group, the sorus. The sorus may be covered by an indusium. The indusium may be true or false. Synangia (Fig. 9J, P) also occur in Marattiales (*Marattia, Danaea* and *Christensenia*).

3.3 Eusporangium

The sporangial initials are superficial and the periclinal divisions (Fig. 1H) result in an outer layer of jacket

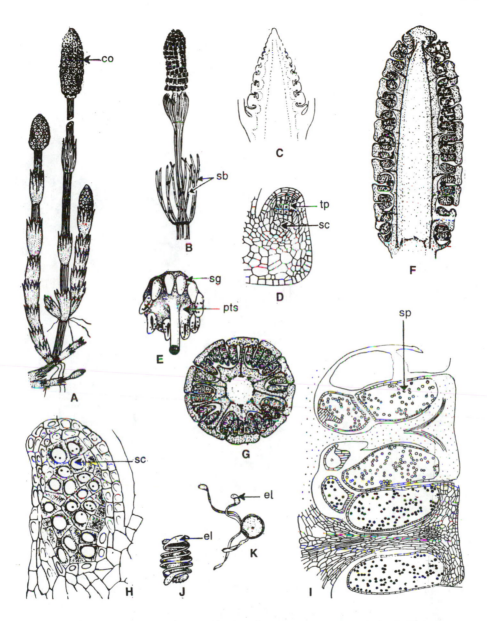

Fig. 3 A-K. *Equisetum,.*Strobilus. **A** Branches bear scale leaves and cones (*co*). **B** Branch with secondary branches (*sb*) and terminal cone. **C** *E. arvense*, young strobilus (l.s.). **D** Peltate sporangiophore (l.s.) shows tapetum (*tp*), wall and sporogenous cells (*sc*). **E** Peltate sporangiophore (*pts*) (ventral surface) with 10 sac-like sporangia (*sg*). **F** *E. arvense*, strobilus (l.s.). **G** Strobilus (t.s.). **H** Peltate sporangiophore (l.s.), shows sporogenous cells (*sc*). **I** Part of cone (l.s.), with sporangia containing spores (*sp*). **J** Spore with coiled (*el*) elaters (under moist conditions). **K** Spore; when dry, elaters (*el*) uncoil. (A from Wossidilo-cited in Strasburger 1894, B, I from Bracegirdle and Miles 1971, C from Hofmeister 1851, D from Bower 1894, E, K from Dodel-Port—cited in von Wettstein 1935. F, G from Parihar 1965 H from Foster and Gifford 1974, J from Sadebeck 1902)

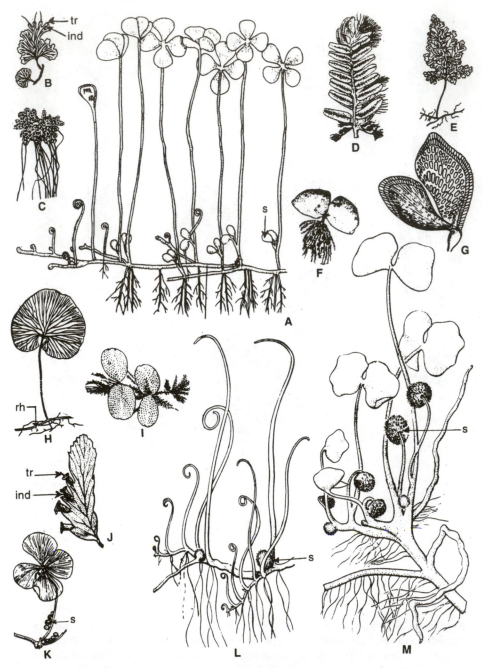

Fig. 4 A-M Ferns. A *Marsilea*, sporocarps (*s*). B *Trichomanes cuspidatum*, sporangia on elongated receptacle, enclosed by indusium (*ind*); note trichomes (*tr*). C *Azolla*. D *Salvinia oblongifolia*. E *Hymenophyllum multifidum*. F, I *Salvinia natans*, leaves. G *Azolla microphyllum*. H *Trichomanes reniforme*, rhizome (*rh*) with a leaf. J. *T. australicum*, portion of fertile frond, note indusium (*ind*) and trichome (*tr*). K *Marsilea polycarpa*, note sporocarps (*s*). L *Pilularia* with sporocarp (*s*). M. *Regnellidum*, with sporocarps (*s*). (A, E, H, L, M from Eames 1936, B from Christ 1897, C from Foster and Gifford 1974, D, G from Martinus—cited in Rashid 1976, F, I-K from Bierhorst 1971).

initials and an inner layer of 'primary sporogenous cells': *Psilotum, Tmesipteris, Lycopodium, Selaginella, Isoetes, Equisetum* (Fig. 3H), and homosporous ferns (Fig. 11A, B). Leptosporangiate ferns (homosporous or heterosporous) have a leptosporangium. Although the development is similar, leptosporangia in heterosporous ferns do not show an annulus and stomium (Figs 10E, F; 11H-K). The eusporangia may be protected (Fig. 2F, G) or exposed (Fig. 2D, E; *Lycopodium*). In Psilophyta (many layers), *Lycopodium* (one layer; Fig. 2J), *Selaginella* and Equisetophyta (2 layers), the wall of the mature sporangium consists of one or more than a single layer of cells (Fig. 1L-P) and tapetum (Fig. 3D). The sporogenous cells undergo meiosis to form spores. The spore dispersal is due to dry elaters (Fig. 3J, K). The spore output of eusporangia is variable and is also correlated with a homosporous or heterosporous condition. The massive microsporangia of *Isoetes* produce the maximum number of spores (Fig. 5I, J).

4. Leptosporangium

The spore output is much less from a leptosporangium compared with a eusporangium (Fig. 10B). The leptosporangium originates from a single superficial cell. The mature sporangium has a wall of a single layer of cells (Fig. 10E). The sporangial initial divides by a periclinal wall and the entire sporangium is derived from the outer cell (Fig. 11H). Occasionally, the inner cell divides, and the daughter cells contribute to the sporangial stalk. The outer cell forms an apical cell (Fig. 11C) which produces a filament of cells (Fig. 11 I). The lower segments form a 3-rowed stalk (Fig. 11D-G, I, J), and the upper segments contribute to the sporangium wall. The apical cell divides periclinally to form an outer wall initial and an inner primary archesporial cell. The wall initial gives rise to the wall, which is a single layer of cells (Fig. 11D, K). The primary archesporial cell cuts off four tapetal initials by diagonal walls, and the innermost primary sporogenous cell (Fig. 11D-F). The tapetal initials give rise to a two-layered tapetum, and the primary sporogenous cell to a fixed number of sporocytes (=spore mother cells, *SMC*) (4, 8, 16) which undergo meiosis and form spore tetrads (Fig. 11 E-G).

4.1 Transitional Type
In Osmundaceae (Bower 1926) the sporangium has a massive stalk (Fig. 11N, O). Sometimes the initial cell is truncated at the base (Fig. 11L) and sometimes pointed (Fig. 11M). The former is like eusporangium, and the latter (pointed at the base) like a leptosporangium initial. The spore output (512) is also much more than in a leptosporangium, and less than in a eusporangium.

5. Tapetum

Goebel (1905) recognised a secretory tapetum which provides nutrition to sporocytes in Lycophyta and Pterophyta (except Ophioglossales), and plasmodial tapetum in the Psilophyta (Fig. 1I, L, N), Equisetophyta and Ophiolossales. In *Lycopodium* the tapetum is an innermost wall layer of the sporangium, and in *Selaginella* it is the outermost sterile cells of sporogenous tissue. In leptosporangiate forms, the tapetum is derived from the primary archesporial cell (Fig. 11D)

6. Sorus in Ferns

In homosporous ferns the megaphylls serve both vegetative and reproductive functions. When reproductive, the megaphylls are called sporophylls (Fig. 9F) as in other groups of pteridophytes. They may bear sporangia; when in groups the sporangia are designated sori (Fig. 9H, K, O). Sometimes the sori are further protected in a special fructification called the sporocarp (Fig. 4A, L, M). Occasionally, the leaves are dimorphic-photosynthetic sterile megaphylls, and fertile sporophylls (=megaphylls) as in *Osmunda cinnamomea* (Fig. 7D), *O, javanica* and *Matteucia struthiopteris*. There are transitional forms in which the megaphylls

182 KUMAR

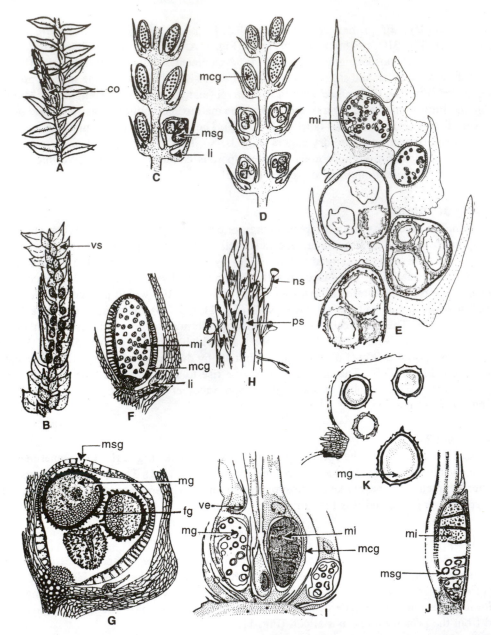

Fig. 5 A-K   Heterosporous sporangia. A-H *Selaginella,* I-K *Isoetes.* A *S. kraussiana,* leafy branch with cone (*co*). B Cone, with vegetative shoots (*vs*) both above and below the sporangia. C-E Strobilus (l.s.), shows disposition of mega (*msg*)- and microsporangia (*mcg*). F Microsporangium (*mcg*) with microspores, note ligule (*li*). G Megasporangium (*msg*), of the three megaspores (*mg*), two show female gametophytes (*fg*). H New sporophyte (*ns*) on the parent strobilus (*ps*); vivipary. I-K *Isoetes.* I, J Sporophylls (l.s.) with micro (*mi*)- and megaspores (*mg*); note velum (*ve*). K Megaspores (*mg*). (A, C, D from Parihar 1965, B from Mitchell 1910, E, I, K from Bracegirdle and Miles 1971, F-H from Lyon 1901, J from Rashid (1976).

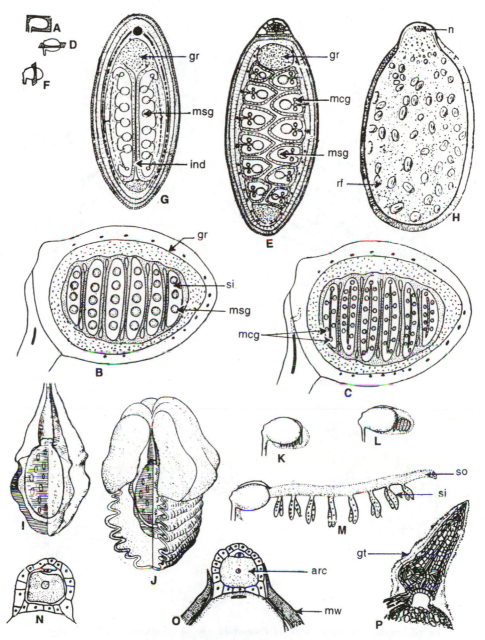

Fig. 6 A-P   Heterospory, *Marsilea*. A, D, F Types of section of B, C, E, G. B Sporocarp (vertical longisection, vls); sori (*si*) with megasporangia (*msg*); the gelatinous ring (*gr*) surrounds the sori. C Sori (*are*) with microsporangia (*mcg*), the gelatinous ring surrounds the sori. E Horizontal longisection shows alternate sori with mega (*msg*)- and microsporangia (*mcg*), the gelatinous ring (*gr*) is much better developed on the dorsal than the ventral side. G Sori (t.s.) show megasporangia (*msg*) separated by indusium (*ind*). H two-nucleate female gametophyte in megaspore (l.s.), nucleus (*n*) in apical portion, and reserve food (*rf*) throughout the megaspore. I, J, Gelatinous ring (*gr*) surrounds female gametophyte. K-M Sporocarp bears sorophore (*so*) with sori (*si*). N Single archegonium (*arc*) (l.s.) with neck canal cell, ventral canal cell, and egg cell. O Single archegonium (*arc*) surrounded by megaspore wall (*mw*) (l.s.). P Embryo surrounded by sheath of gametophytic tissue (*gt*); note rhizoids (*ri*). (A-G from Sharp 1914, H, N, O from Campbell 1918, I, J from Machlis and Rawitscher-Kunkel 1967, K-M from Eames 1936, P from Parihar 1965).

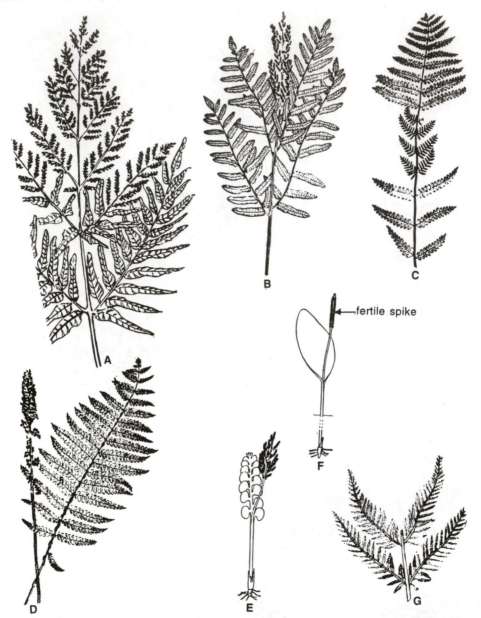

Fig. 7 A-G. Fertile region in ferns. A-D, G Osmundaceae. A *Osmunda javanica*, segment of fertile frond. B *O. regalis*, leaf with fertile pinna. C *O. claytoniana*, part of fertile pinna. D *O. cinnamomea*, sterile and fertile fronds. E, F Ophioglossaceae. E *Ophioglossum vulgatum*, fertile spike. F *Botrychium lunaria*, part of lamina, sterile and partly fertile. G *Todea barbara*, abaxial sorus. (A-D, G from Hewitson 1962, E, F from Luerssen—Cited in Sporne 1970).

are partly fertile and partly sterile as in *Osmunda regalis* (Fig. 7B), *O. claytoniana* (Fig. 7C), *O. javanica* (Fig. 7A), *Anemia* and *Botrychium* (Fig. 7E), and *Todea barbera* (Fig. 7G).

6.1 Diversity of Sori

In the majority of the homosporous ferns, the sporangia (Fig. 9E) are grouped in sori (Fig. 9D, K). Sometimes the sporangia are united to form synangia (singular synangium) as in Marattiales (Fig. 9P, J). The sori may be circular (Fig. 9Q), cup-shaped (Fig. 9L), reniform or linear. The larger sori fuse to form a coenosorus (*Pteris*). Coenosori may be broken up into segments as in *Blechnum* (Fig. 9C), and *Woodwardia*. The sori may occur on a vein (Fig. 9H) or at the end of a vein (Fig. 9B). The raised or elevated leaf surface bearing a sporangium is termed a receptacle (Fig. 10A-C). The receptacle may project as trichomes after dispersal of sporangia, hence the name, *Trichomanes* (Fig. 4B, J). The sporangia originate from superficial cells of the receptacle. The sori are classified, according to Eames (1936), as : (1) marginal, (2) intramarginal, and (3) abaxial (Fig. 10I).

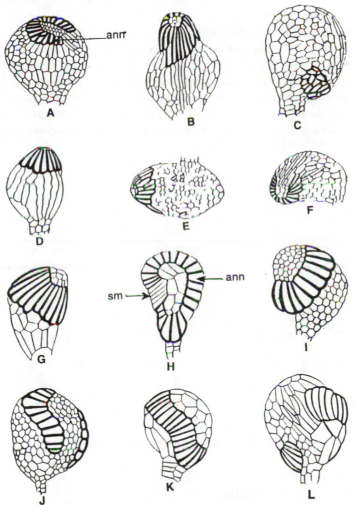

Fig. 8 A-L. Position of annulus and stomium in homosporous leptosporangiate ferns. A *Mohria caffarum.* B *Schizaea dichotoma.* C *Osmunda cinnamomea*, sporangia. D *Actinostachys oligostachys*, annulus (*ann*) terminal. E *Lygodium reticulatum.* F *Anemia phyllitidis.* G *Gleichenia bracknirdgi.* H *Cyathea capensis*, note stomium (*sm*). I *Gleichenia caudata.* J *Matonia pectinata.* K, L *Hymenophyllum dilatatum.* (A to L from Bower 1935).

6.2   Fern Spores

The spores are tetrahedral (trilete) or bilateral (monolete). In addition to exospore (exine) and endospore (intine), there may be an additional outer envelope called perispore-perine (Foster and Gifford 1974, see also (Santha) Devi 1977).

Fern Spore Output. In Ophioglossales the sporangia produce more than 2000 spores, in *Botrychium* and *Ophioglossum vulgatum* upto 15000 spores. In Marattiales it varies from 1440 in *Angiopteris*, 2500 in *Marattia*, and 7000 in *Christensenia*.

6.3   Indusium

The sori are usually covered by two laterally-developed indusial flaps (Fig. 9G, M, O). Both the flaps originate from the receptacle (true indusium) as in *Dryopteris* (Fig. 10A). When the sori are protected by inwardly-turned margins of the leaflets, the protective structure is called a 'false indusium', e.g. in *Pteris* and *Adiantum* (Fig. 9I). The indusium is reniform in *Dryopteris* (Fig. 9K), circular in *Polystichum lobatum*, funnel-shaped in *Davallia* (Fig. 9A), elongated and curved in *Asplenium lanceolatum*. In *Blechnum* the sori become confluent and are covered by a common indusium (Fig. 9C). In *Athyrium filix-foemina* the indusium is reniform with lacerated margins. In *Matteucia struthiopteris*, the indusium is cup-shaped with dentate margins, thin and papery, the leaflet margins become strongly inrolled and afford additional protection to sori. In *Adiantum* the sporangia develop on the underside of special marginal flaps of the lamina that become reflexed and protect the sorus (Fig. 9I).

7.   Sporocarp

In Marsileales and Salviniales the sporangia develop in sori that are borne within a sporocarp. In *Marsilea* the sori contain both micro- and megasporangia (Fig. 6A-G). The apoptosis (programmed death) occurs due to genes. Out of 64 spores one survives (Bell 1996). The sporocarp contains a gelatinous ring (sorophore) which bears sori. If the sporocarp is injured, the sorophore comes out bearing sori which contain sporangia (Fig. 6K-M). In Salviniales the smaller sporocarp contains many microsporangia, and the larger sporocarps contain one or more megasporangia.

8.   Classification of Sori

Bower (1935) classified sori in ferns according to the three modes of development of sporangia in a sorus.

Simple Sorus. All the sporangia in the sorus develop and mature simultaneouslly as in *Ophioglossum* and *Osmunda* (Fig. 10L).

Gradate or Basipetal Sorus. The earliest sporangium differentiates close to the apex of the receptacle, the successive younger sporangia develop towards the base (Fig. 10J).

Mixed Sorus. The sporangia are intermingled in different stages of development and are the most advanced evolutionary type (Fig. 10K).

9.   Dehiscence of Leptosporangium

The annulus is associated with dehiscence and the forceful ejection of the spores. As the sporangium matures, water is lost from the cells of the annulus. The annulus has thin outer tangential walls which stretch inwards and the bases of the radial walls are pulled towards each other. This results in tearing open of the sporangium on the weaker side (stomium or 'lip cells') (Fig. 10F). The annulus bends back, carrying with it all the spores (Fig. 10G). The spores are all thrown off (catapulted) to a distance of 1 cm, when the

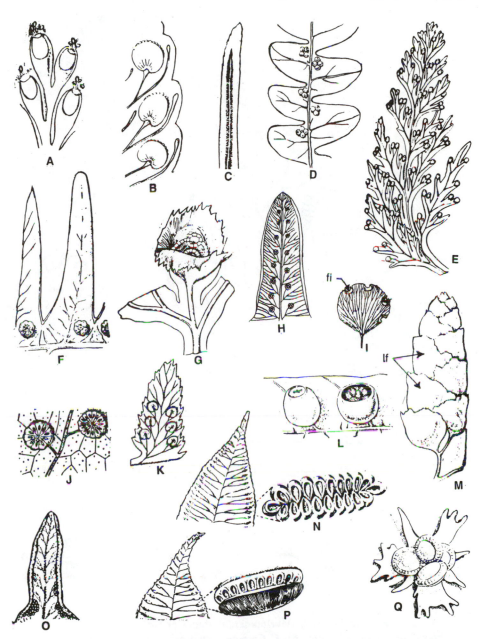

Fig. 9 A-Q.   Sori and indusia of Homosporous Leptosporangiate Ferns. A *Davallia*, marginal sori. **B** *Nephrolepis*, indusium attached on one side. **C** *Blechnum*, pinna shows two coenosori. **D** *Gleichenia*, superficial position of sori. **E, G** *Hymenophyllum*. **E** *H. dilatatum*, portion of frond with sori. **F** *Matonia*, peltate indusium. **G** *H. tunbrigense*, portion of sorus, note indusial flaps. **H** *Gleichenia*, fertile pinnule (ventral view), sori on veins. **I** *Adiantum*, marginal sori covered by false indusium (*fi*). **J** *Christensenia*, sori. **K** *Dryopteris*, reniform indusia. **L** *Cyathea*, cupshaped indusium. **M** *Lygodium*, each sporangiam covered by a laminal flap (*lf*). **N** *Angiopteris*, sori. **O** *Pteridium*, coenosorus and false indusium. **P** *Marattia*, synangia. **Q** *Woodsia*, basal membranous indusial segments. (A, G, M, Q from Foster and Gifford 1974, B from Bower 1928, C, O from Rashid 1976, D-F, N, P from Bower 1926, H, L from Eames 1936, I to K from Bitter 1900).

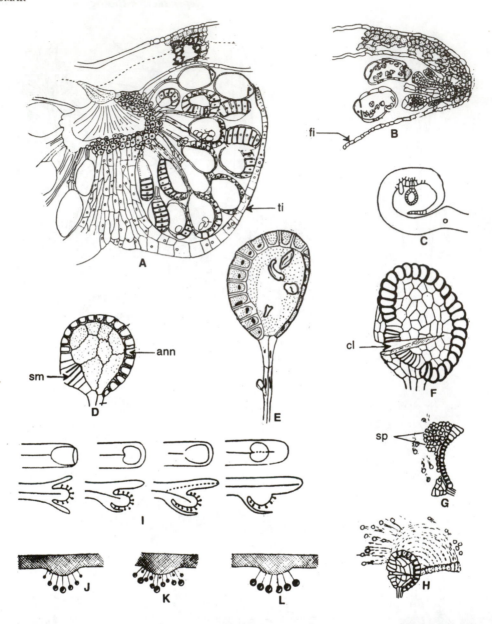

Fig. 10 A-L.   A-H Leptosporangium, **I-L** Developmental stages of sorus. **A** *Dryopteris,* true indusium (*ti*). **B** *Pteridium,* false indusium (*fi*). **C** *Cryptogramma crispa,* intramarginal sorus protected by leaf margin. **D, E** Leptosporangium shows annulus (*ann*) and stomium (*sm*). **F** Cleft sporangium (*cl*) at dehiscence. **G** Annulus bent backwards, note the spores (*sp*). **H** Spore liberation (as from a catapult); the sporangium snaps back into position. **I** Marginal sorus, indusium abaxial. Three types (J-L) of sorus—gradate (J) mixed (K) and simple sorus (L). (A, B from Bracegirdle and Miles 1971 (A), 1973 (B), C from Wardlaw 1961, D, E from Bracegirdle and Miles 1971, F from Bierhorst 1971, G from Bower 1928, H from Meyer and Anderson 1939, I-L from Foster and Gifford 1974).

sporangium wall snaps back into position (Fig. 10H), as the annulus is under tremendous tension due to continuous loss of water. The tension prior to dehiscence is equivalent to 300 atmospheres.

10. Gametophyte

The gametophyte generation in the pteridophytes is relatively inconspicuous. According to Wagner (1963), some of the pteridophytes like *Vittaria* remain permanently in the gametophytic phase. In the homosporous pteridophytes, the gametophyte is much reduced, exosporic (Fig. 12B, E, G, J), not enclosed by the spore wall, terrestrial and green (Fig. 12J), or subterranean (Fig. 12B, E, H, G). If it is green, it grows on the ground (terrestrial) exposed to light, as in some *Lycopodium* spp., e.g. *L. cernuum,* and *Equisetum.* In homosporous leptosporangiate ferns (*Pteris*), the gametophyte is a free-living structure (Figs 12P, 15H). If it is subterranean, as in Psilotales, *Ophioglossum* and some species of *Lycopodium*, e.g. *L. clavatum* and *L. annotinum,* it is nourished through a mycorrhizal fungus (Fig. 12D, L, R). When there is excess nutrition as in a green prothallus of *L. cernuum,* which also has mycorrhiza (in *Lycopodium* species, green prothalli also have mycorrhiza), the nutrients get stored in a new structure from the embryo, the protocorm (Fig. 12F, M-O). The exosporic gametophytes are monoecious and develop both types of sex organs (Fig. 13F), the antheridia and archegonia. The heterosporous pteridophytes are endosporic and are completely enclosed by the wall of the microspores or megaspores. The micro- and megaspores are enclosed in micro- and megasporangia, respectively, as in *Selaginella* and *Isoetes* (Fig. 5K). The microgametophyte develops up to the 13-celled stage in microspores, and then the microspores are liberated from the microsporangia. The microsporangia are enclosed by sporophylls. The megaspores develop female gametophytes (megagametophytes) followed by differentiation of archegonia. Such megaspores with archegonia are discharged from the strobilus. In a strobilus the megaspores have the megagametophytes at the same stage of development. With female gametophytes in megaspores, the megaspore is dispersed by a megasporangium, but in *S. rupestris* vivipary has been reported (Fig. 5H). The endosporic gametophyte is strictly dioecious (Figs. 16N, Fig. 19C). The microspores produce micro- (male)-gametophytes and the megaspores produce mega- (or female)- gametophytes which survive in a much wider range of habitats since they are well-protected by the parent plant. The male gametophyte of the water ferns *Salvinia* and *Azolla* may be washed away if shed singly, and are therefore shed in a mass, the massula (Fig. 16V). In some species of *Azolla* the massulae have an anchoring structure, the glochidia (Fig. 16V). The female gametophyte is surrounded by an irregular special wall, the perispore, to which the glochidia get attached, which facilitates fertilization (Fig. 18N).

10.1 Exosporic Gametophyte

Radial gametophytes are primitive (Fig. 12I) and dorsiventral ones are advanced, and occur in some *Lycopodium* species, Psilotales and *Ophioglossum sp.* The archegoniophores of filamentous gametophytes of *Trichomanes* are also radial and change from radial to the dorsiventral condition, as in *Lycopodium selago*—a transitional form. This change occurs due to adaptation to light (see Rashid 1976, 1999).

The exosporic gametophyte has a structural arrangement like the thallus, commonly called the prothallus. The prothallus is a multicellular, green, self-nourishing body of simple organization like thallose liverworts. When it grows on the soil, one-celled rhizoids (Fig. 12C) act chiefly as the rooting and absorbing organs. Its form varies considerably. It may be branched, filamentous and green as the protonema of mosses, exemplified by *Schizaea* (Fig. 14G, L) and *Woodsia* (Fig. 14E). In homosporous leptosporangiate ferns, the gametophyte simulates a thallose liverwort: thin, flat, cordate, dorsiventrally differentiated, and prostrate, with the sex organs borne on the ventral surface as in *Dryopteris filix-mas* (Figs. 14H, O, 15HI). The gametophyte may be cone-shaped or cylindrical, elongate, or radial, as in Psilotales (Figs 12B, 13B), Ophioglossales (Fig. 14A, B, F), and in some species of *Lycopodium*. In *Equisetum* and in some species

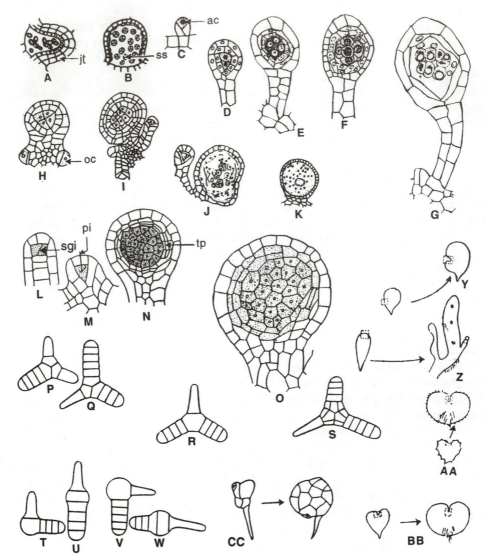

Fig. 11 A-BB. Two types of sporangial development (A-K). Transitional type (L-O); three patterns of spore germination (P-X); different types of gametophytes (Y-BB). A, B Eusporangium shows jacket (*jt*) and spore sac (*ss*). C-G Leptosporangium development in homosporous ferns. C Apical cell (*ac*). D Wall layer, tapetal layer, and spore mother cells. E Later stage, note two-layered tapetum. F, G Later stages of development of sporangium. H-K Leptosporangium development in a heterosporous fern (*Marsilea*). H Two microsporangia initials (lateral), and a megasporangium (central) note outer cell (*oc*). I Microsporangia have longer stalks. J Degenerated spores in a megasporangium (one spore survives after meiosis of 16 spore mother cells (smc). It enlarges up to the size of megasporangium. K Megasporangium contains the megaspore. L-O Transitional type. L the sporangium initial in *Osmunda* is like a eusporangium (*sgi*). M Initial cell pointed (*pi*) like leptosporangium initial. N, O Wall layer, tapetum (*tp*) and sporogenous cells. P-X Spore germination; note patterns of walls in sporeling. P-S Tripolar germination. T-W Bipolar germination. X Amorphous germination (*Angiopteris*). Y-BB Gametophyte development. Y Ceratopteris type. Z Kaulinia type. AA Aspidium type. BB Marattia type (*Adiantum, Drynaria, Osmunda*) give rise to cordate, CC Marattia type of prothallus). (A, B, E-G, from Foster and Gifford 1974, C, D from Parihar 1965, H-J from Marschall-Corelia 1925, K-M from Bower 1926, N, O from Bower 1899, P-BB from Nayar and Kaur 1971).

Fig. 12 A-R. Gametophytes and embryos. A, B *Psilotum*, C *Tmesipteris*, D-R *Lycopodium*. A Exoscopic embryogeny, epibasal cell gives rise to embryo (*emb*) (towards neck of archegonium), hypobasal cell develops as haustorial foot (*hf*). B Gametophyte. C Gametophyte bears antheridia (*an*), archegonia (*arc*) and rhizoids (*ri*). D *Lycopodium annotinum*, gametophyte development, shows origin of mycorrhizal association. E *L. clavatum,* gametophyte. F *L. laterale,* subaerial or terrestrial gametophyte (*pr, ppl*). G *L. phlegmaria,* note gemmae (*gm*) on branches of gametophyte. H *L. camplanatum*, gametophyte. I *L. selago*, gametophyte. J, P *L. cernuum*, lobed gametophyte. K, Q *L. clavatum*, embryo with large foot (subterranean type). L Gametophyte (l.s.) shows antheridia in the centre and archegonia on the margins. M Protocorm, note protophylls. N Protocorm with protophylls. O Protocorm with shoot apex. R *L. clavatum*, gametophyte (l.s.) shows embryo, archegonia (on the sides) and antheridia (in the centre). (A, B, C from Bierhorst 1956, D, E, H-L, Q, R from Bruchmann 1885, F, P from Chamberlain 1917, G, N, O from Treub 1884, M from Eames 1936).

of *Lycopodium*, on the dorsal surface the prothallus develops vertical lobes, somewhat resembling the leaves of Hepaticae (Figs. 12J, 13F-I).

10.2 Endosporic Gametophyte

The endosporic gametophytes of heterosporous forms are reduced and sexually differentiated as in *Selaginella*, *Isoetes, Marsilea* and *Salvinia* (Fig. 19E-I). The gametophytes consist of a few cells each and persist within the spore wall, or may slightly protrude from it. The reduction is extreme and the microgametophyte (= male gametophyte) consists of only one antheridium as in *Selaginella* (Fig. 16 O), *Isoetes* and *Azolla* (Fig. 16S). There are two antheridia in *Marsilea* (Fig. 16P, Q), and a few or even only one vegetative cell.

The megagametophytes, though multicellular, are also much reduced (Fig. 19A). In *Marsilea* the megagametophyte shows extreme reduction, and consists of only a single archegonium (Fig. 6O) and a few vegetative cells. In all these taxa, the development of the prothalli and the sex organs takes place mainly at the expense of the reserve food material deposited in the spores (from the sporophyte).

10.3   External Factors

The heterosporous gametophytes are independent of external control. In *Marsilea* and *Salvinia*, the germination of spores, development of gametangia and gametes, and even embryo development may take place in the dark. Rhizoids are absent in the male prothalli of all the heterosporous ferns, and also in the female prothalli of *Salvinia* and *Azolla* (Fig. 18M-O). The female gametophyte in heterosporous forms shows poor development, and growth ensues only after fertilization.

10.4   Life-Span of Gametophytes

Some taxa, like *Vittaria,* remain permanently in the gametophytic phase. In homosporous pteridophytes the gametophytes are relatively long-lived, and in certain species of *Lycopodium* (*L. clavatum*) they may require 6 to 8 years to attain maturity; they remain 'alive' for 8–15 years. In contrast, the gametophytic life span in heterosporous forms ranges from a few hours to a few days.

Vegetative Propagation of Gametophytes. In homosporous forms, vegetative propagation by adventitious branching is frequent. Gemmae occur in *Psilotum, Lycopodium* (Figs. 12G, 13C; Urostachya group), Hymenophyllaceae and Vittariaceae (Gifford and Foster 1989). The prothalli of *Gymnogramma* form tubers which perennate.

11.   Gametangia

In the monoecious exosporic gametophytes, the distribution pattern of antheridia and archegonia varies considerably. In the radial, fleshy, non-photosynthetic gametophytes of Psilotales, they are irregularly distributed (Figs. 12C, 13A, B). In a number of species of *Lycopodium,* with subterranean gametophytes— *L. clavatum, L. complanatum, L. annotinum, L. inundatum*—the sex organs differentiate in distinct patches on the flattened and hollow upper surface (Fig. 12J, L, R, H). Antheridia differentiate first and are borne centrally. The archegonia appear later and are closer to the margins. In ·*Marattia* and *Angiopteris,* the archegonia are restricted to the ventral surface of the midrib (Fig. 14P). Once the archegonia have been initiated, the antheridia may also develop on the dorsal surface of the mid-rib. In homosporous leptosporangiate ferns, the antheridia and archegonia are usually restricted to the ventral surface of the prothallus. The archegonia are restricted to the thick central cushion behind the apical notch, whereas the antheridia occur near the base as well as on the wings (Fig. 15I).

11.1   Antheridia

The mature antheridium is globular. It has a wall which encloses a group of androcytes (Fig. 16I, J). The

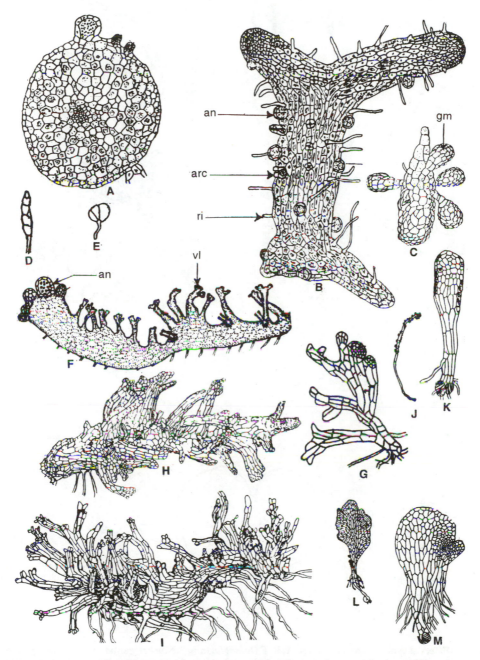

Fig. 13 A-M. **Gametophytes. A, B** *Psilotum*, **C** *Lycopodium*, **D-I** *Equisetum*. **J, L** *Platyzoma*, **K, M** *Ceratopteris*. **A** Gametophyte (transection) with antheridia and archegonia. **B** Gametophyte, antheridia (an), archegonia (arc) and rhizoids (ri) are scattered all over the surface. **C** *L. phlegmaria*, gametophyte bears gemmae (*gm*). **D, E** Young gametophytes. **F** Prothallus with vertical lobes (vl) and antheridia and archegonia. **G** Male prothallus bears emergent (projecting) antheridia. **H, I** Sectional view of prothallus with vertical lobes, antheridia, archegonia and rhizoids. **J** *Platyzoma* (heterosporous, but forms exosporic instead of endosporic gametophytes), male prothallus bears antheridia. **K** *Ceratopteris* (homosporous, but forms two types of prothalli, male and female). Male prothallus bears antheridia on the sides. **L** Female prothallus. **M** Female prothallus has a lateral meristem. (A from Holloway 1939, B from Lawson 1917, C from Treub 1884, D-G from Walker 1931, H from Duckett 1970, I from Walker 1937, J, L from Tryon 1964, K, M from Rashid 1976).

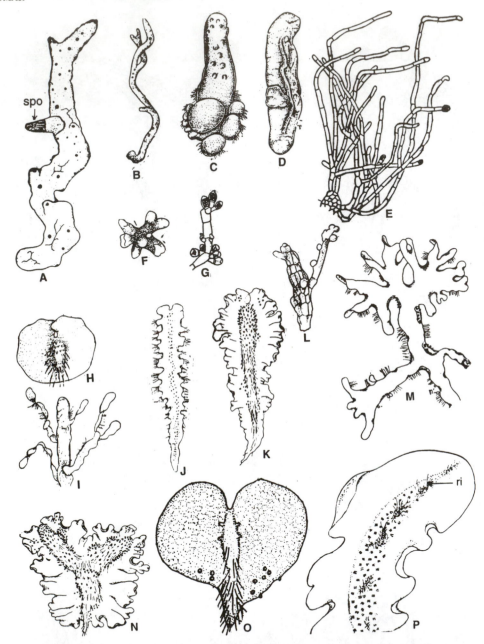

Fig. 14 A-P.   Different types of gametophytes. A, B, F *Ophioglossum.* **A** Mature gametophyte with a young sporophyte (*spo*). **B** Branched gametophyte. **C** *Helminthostachys,* tuberous prothallus. **D** *Botrychium,* tuberous prothallus. **E** *Woodsia* (Aspidiaceae), highly branched portion of a gametophyte. **F** *Ophioglossum pendulum,* gametophyte. **G** *Schizaea pusilla,* antheridial branch. **H** *Adiantum cordata,* cordate prothallus. **I** *Paraleptochillus decumbens* (Polypodiaceae), ribbon-shaped prothallus. **J** *Elaphoglossum,* prothallus. **K** *Gleichenia bifida,* gametophyte. **L** *Schizaea dichotoma,* branched gametophyte. **M** *Hymenophyllum blumeanum,* prothallus. **N** *Gleichenia glauca,* gametophyte. **O** *Osmunda regalis,* cordate prothallus with antheridia and archegonia, note rhizoids. **P** *Marattia douglasii,* archegonia and rhizoids (*ri*) on the projecting ventral midrib (antheridia not shown). (A, B from Bruchman 1910, C, D, I, J from Britton and Taylor 1901, H, M from Stokey and Atkinson 1956, K, N from Stokey 1950, L from Bierhorst 1971, O from Stokey and Atkinson 1956, P from Foster and Gifford 1974).

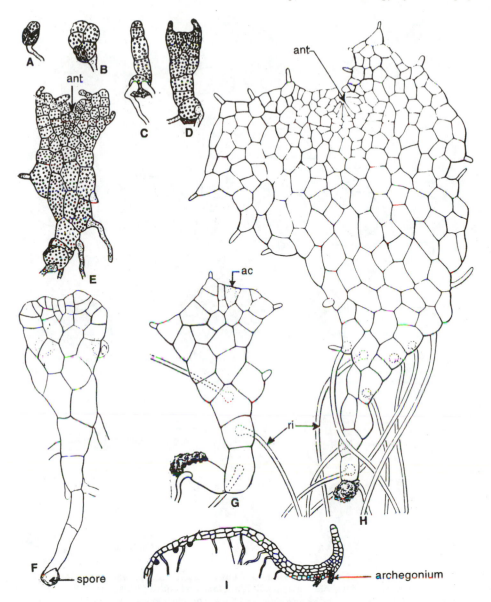

Fig. 15 A-I.   Gametophyte of homosporous leptosporangiate ferns. A-E, I *Dryopteris filix-mas*. A Germinated spore. B, C Gametophyte. D, E Gametophyte, one-cell thick, apical notch (*ant*) in *E*. F-H Development of cordate prothallus, note rhizoids (*ri*) in H. I Vertical transection of prothallus to show archegonia, antheridia and rhizoids, restricted to the ventral surface of the prothallus. (A-H from Kny 1875, F-F from Foster and Gifford 1974, I from Parihar 1965).

later give rise to motile antherozoids. The number of antherozoids may be four as in *Isoetes* (Fig. 16T), several hundred in *Gleichenia* and Marattiaceae, and several thousand in *Ophioglossum pendulum*. The antheridium may be embedded partly, or entirely sunk in the tissue of the gametophyte, or remain free, or become emergent (project from the prothallus). The embedded type occurs in *Lycopodium* (Fig. 16A),

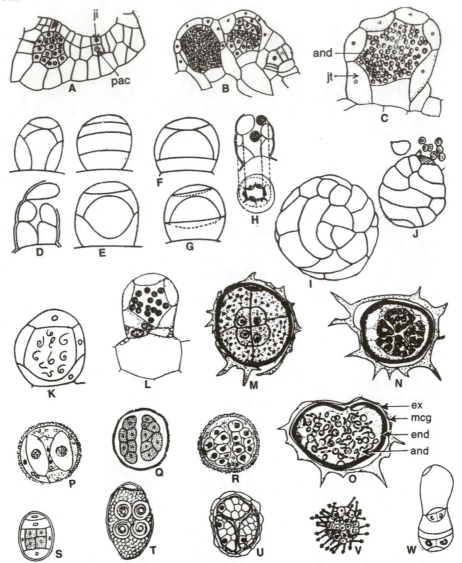

Fig. 16 A-W.   Development of antheridia. A-L Homosporous forms. A *Lycopodium,* antheridium shows jacket initial
(*ji*) and primary androgenial cell (*pac*). B *Equisetum*, embedded antheridia. C *Psilotum*, mature embedded
antheridium, note one-celled jacket (*jt*) and a mass of androcytes (*and*). D-G Antheridium (transection),
sequence of wall formation. H *Woodsia silvensis* (*Polypodiaceae*), dehisced antheridium, I *Osmunda
javanica,* antheridium (top view). J *Matonia pectinata*, dehisced antheridium (top view), an opercular
cell is blown off to release androcytes. K Polypodiaceae, mature antheridium with two ring cells and
a cap cell. L *Gammitis billardieri* (Polypodiaceae) shows a funnel-shaped cell, a ring cell, and a cap cell.
M-W Heterosporous forms. M-O *Selaginella.* M Endosporic male gametophyte, four jacket cells and
four spermatogenous cells. N Four masses of androcytes. O Mature male gametophyte shows microspore
spinous wall and mass of androcytes; the prothallial cell and eight jacket cells disintegrate. P-R *Marsilea.*
P Two spermatogenous cells. Q, R Two masses of spermatogenous cells. S, T *Isoetes.* S Four spermatogenous
cells (androcytes). T Four sperms. U, V *Azolla.* U Microspores. V Massulae with glochidia. W *Salvinia*,
two spermatogenous cells separated by two sterile cells. (A, B from Bruchmann 1898, C from Lawson
1917, D from Campbell 1940, E from Davie 1951, F from Stone 1962, G from Verma and Khullar 1966,
H from Bower 1928, I from Bierhorst 1971, K from Rashid 1976, L from Stone 1960, M-O from Slagg
1932, P-R from Sharp 1914, S from Liebig 1931, T from Belajiff 1885, U from Smith 1955, V from
Bernard 1904, W from Belajiff 1898).

Marattia, and Ophiolossaceae, and also in all the heterosporous pteridophytes—*Selaginella, Marsilea.* When the antheridia are embedded, the adjoining cells of the prothallus protect the antherozoids. An opercular layer develops, which not only protects, but is also specialized as an opening mechanism for the discharge of antherozoids. The opercular mechanism is usually a single layer, except in *Botrychium* and *Helminthostachys* where it consists of two layers, except in the centre. One or more centrally-located cells of this opercular layer separates (disintegrates), forming a singular triangular pore through which the antherozoids escape. In *Lycopodium, Ophioglossum* and *Marattia,* there is usually a single opercular cell, while in *Equisetum* there may be one or more opercular cells.

The free or emergent type of antheridia usually occur in homosporous leptosporangiate ferns. They either project free as a spherical body upon the surface of the gametophyte, or are seated among the rhizoids on the ventral surface of prothallus as in *Dryopteris filix-mas.* In the emergent type, the wall always comprises a single layer of cells. The number of cells of the wall may vary from 3 to 12. When the cells are fewer, they are regularly arranged, and consist of two ring cells and one or two opercular cells. In *Gleichenia* (Stokey 1950), there may be 10 or 12 wall cells and one of these functions as an opercular cell, this is also the case for *Matonia pectinata* (Fig. 16J).

Antherozoids. In pteridophytes, the male gamete or antherozoid, or spermatozoid, is usually a motile flagellated cell. Water is essential for reaching the passive, non-motile egg (for fertilization). The newly-formed antherozoid in *Selaginella* (Fig. 17C), *Equisetum* (Fig. 17F, M) and Filicales is a slender, spirally-coiled band of two to five counter-clockwise turns bearing the flagella on the outer surface of the anterior portion (Fig. 17F, J, L, M), and enclose one or two droplets of cytoplasm in the hollow posterior of the coil. The antherozoids of *Selaginella* and *Lycopodium* have two (Fig. 17G, H), *Equisetum* many (Fig. 17M), and *Marsilea* 40 to 50 flagella (Fig. 17J). The antherozoids swim by the whip-like lashing of their flagella. In some genera the strokes are synchronous, in others, the flagella act in series (waves). The motion is at first 'trembling' and made irregular by the weight of cytoplasm. As soon as excess cytoplasm is cast off , the antherozoid becomes narrower and it moves straight ahead with a cork screw-like motion, rotating counter-clockwise on its axis. The archegonium secretes (into the external medium) a gradient of chemotactic substances which exert an attractive influence on antherozoids. These attractive substances are malic acid and citric acid.

11.2 Archegonia

In pteridophytes the archegonia are flask-shaped and consist of a basal embedded portion which encloses the egg and a ventral canal cell in the venter, and a neck enclosing the neck canal cells (Fig. 18C, D, F-L). The wall of the venter is formed from the prothallial cells, and is embedded in the tissue of the prothallus. The neck comprises four vertical rows of cells and protrudes above the prothallus. There is an axial row of cells, the lowermost being the egg, topped by the ventral canal cell; and 1 to 16 neck canal cells (*L. complanatum*). At maturity the neck opens at the tip by the separation of apical cells. The disintegration of the canal cells leaves a tubular passage for the motile antherozoids to reach the egg cell (Fig. 18J-L). In *Psilotum* the opening of the neck is due to the sloughing off of the upper tiers of the neck (Fig. 18A, B, E).

The largest neck is reported in *Lycopodium complanatum* with 12 to 16 neck canal cells. Very short, scarcely projecting necks are known in homosporous leptosporangiate ferns. The archegonia of rapidly maturing prothalli of *Lycopodium, Isoetes* and *Selaginella* also have short necks. The neck canal cells are usually a superposed row of cells. Some species of *Equisetum* have two laterally-placed (boot-shaped) neck canal cells.

The longest neck occurs in *Lycopodium* and the shortest in homosporous leptosporangiate ferns. This cannot be the reductional series' as Psilophyta (*Psilotum* and *Tmesipteris*) have only one binucleate neck

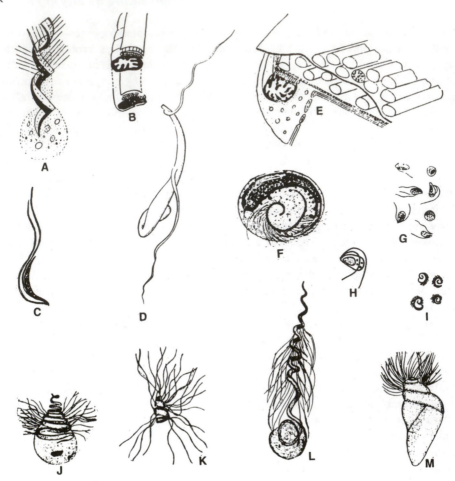

Fig. 17 A-M.   **Antherozoids/spermatozoids. A Multiflagellate spermatozoid of fern. B Biflagellate sperm shows elongated nucleus, mitochondria, microtubular band, parts of spermatozoid based on electron micrograph. C *Selaginella*, biflagellate spermatozoid. D Biflagellate spermatozoid E, F, M, *Equisetum*. E 3-D multiflagellate spermatozoid. F Spermatozoid nucleus is helicoid. G, H *Lycopodium*, biflagellate antherozoids. I *Psilotum*, multiflagellate antherozoids. J, L *Marsilea*, multiflagellate antherozoids. J Corkscrew-like antherozoid with closely packed coils. K *Pteridium aquilinum*, spirally-coiled, multiflagellate antherozoids. L Coils of antherozoids loosen when moistened. M Spirally-coiled mature antherozoid, flagella restricted to anterior end. (A, B from Rashid 1976, C from Parihar 1965, D from Gifford and Goster 1989, E from Duckett 1973, F, M from Sharp 1912, G, H from Bruchmann 1910, I from Lawson 1917, J, L from Sharp 1914, K from Lagerberg 1906).**

canal cell like the higher ferns. Ventral canal cells are also variable. They are well-defined cells in lower forms, but are evanescent in Ophioglossales.

After fertilization, the zygote becomes surrounded by a wall. It divides and the daughter cells also divide (Figs. 20S-V, X, Y, 21A-C) to form the pro-embryo (sporophyte).

Except for the gametophyte of Psilotales, which is similar to the sporophyte in its dichotomous branching, presence of vascular tissue, and mycorrhiza, in most of the pteridophytes, the gametophyte is entirely

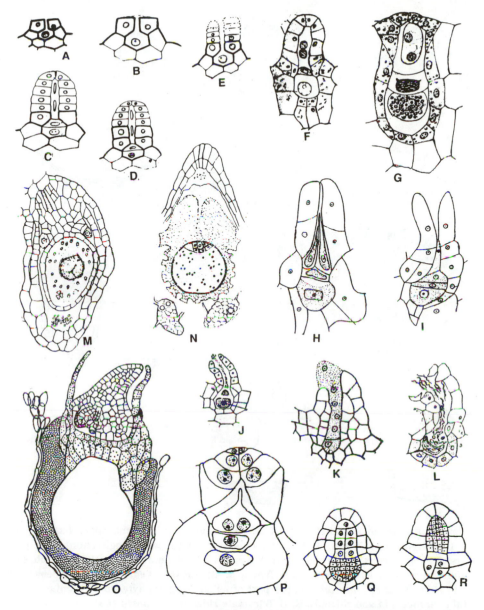

Fig. 18 A-R. Archegonia. A-E *Psilotum,* F, G, J-L fern (Polypodiaceae). H-I *Equisetum,* M, N *Azolla,* O, P *Salvinia,* Q, R *Lycopodium lucidulum.* A, B, E Archegonia with most of the neck having fallen off; the remaining cells of the neck become cutinized. C, D, F, G Binucleate neck canal cells. H Two boot-shaped neck canal cells which lie side by side. I Two neck canal cells one above the other; the uppermost neck cells fall apart. J, K Ventral canal cell and neck canal cells have degenerated. L Antherozoids (spirally coiled) enter archegonial neck canal. M Sporocarp with megasporangium which contains one megaspore. N Megagametophyte with apical cushion and massulae. O Megagametophyte. P Archegonium, Q, R Abnormal archegonia. Q Venter with many cells (like spermatogenous cells). R Neck region with many cells (like antheridium). (A-E from Bierhorst 1954, F, G from Haupt 1940, H from Sethi 1928, I-K from Parihar 1965, L from Shaw 1898, M, N from Smith 1955, O from Pringsheim 1863, P from Yasui 1911, Q, R from Spessard 1922).

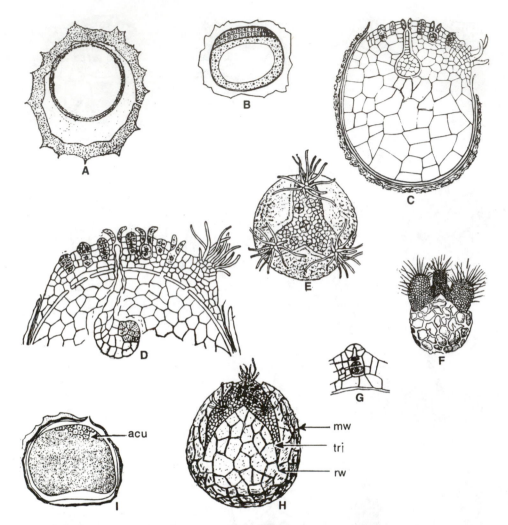

Fig. 19 A-I.   Development of female gametophyte of heterosporous Lycophyta (endosporic female gametophyte). A-H *Selaginella,* **I** *Isoetes.* **A, B,** *S. kraussiana.* **A** Megaspore (longisection), note protoplasmic vesicle, and spinous wall of megaspore. **B** Apical cushion, note many nuclei in female gametophyte and central vacuole. **C** *S. denticulata,* female prothallus shows embryo with suspensor, archegonia and rhizoids. **D** *S. kraussiana,* upper part of female gametophyte shows diaphragm, embryonal tube, archegonia and rhizoids. **E,·H** Triradiate ridge (tri) exposed to show archegonia, rhizoids, and megaspore wall (mw). **F** Three female gametophytes project from the megaspore. **G** Archegonium (egg cell). **I** *Isoetes,* female gametophyte, note upper cellular strip (acu). (A from Campbell 1902, B from Rashid 1976, C from Bruchmann 1912, D to H from Bruchmann 1909, I from Smith 1955)

different from the sporophyte. The gametophyte may be ribbon-shaped (Fig. 14I). It is long-lived in *Marattia douglasii,* and is nourished by the endophytic fungus. The sex organs develop on the ventral surface which also bears rhizoids (Fig. 14P).

Incipient heterospory occurs in *Platyzoma microphyllum* (Tryon 1964). Initially it produces unisexual gametophytes (Fig. 13J, L). The variability of gametophytes of ferns has been discussed by Atkinson and Stokey (1964).

12. Spore Dormancy

There is a period of low metabolic activity between maturation and germination of spores. In the majority of pteridophytes the spores remain viable for several years. In contrast, the spores of *Ophioglossum, Osmunda, Helminthostachys, Botrychium* and *Equisetum* are viable for a few hours to a few days. The chlorophyllous spores have a high metabolic rate and low viability. They can be freeze-preserved ($-10°C$) in glycerine for about 3 years (Jones and Hook 1970).

In *Lycopodium* spp., the spores which have a thin wall germinate readily and give rise to green gametophytes, but thick-walled spores germinate after 3 to 8 years. The spores of *Alsophila australis* remain dormant for 1 year.

The subterranean gametophytes germinate in the dark, e.g. some *Lycopodium* species, *Psilotum, Osmunda regalis, Pteridium aquilinum, Polypodium crassifolium* and *Ceratopteris thalictroides*. The spores of *Psilotum* and *Botrychium dissectum* have an obligate requirement for light.

12.   Spore Germination (Homosporous ferns)

The germination of spores and establishment of gametophytes in Psilotales has been studied by Holloway (1918, 1939).

Nayar and Kaur (1968) have classified spore germination of homosporous ferns into:

1. **Bipolar germination** is of the Osmunda, Gleichenia, Vittaria, Cyathea and Anemia types (Fig. 11T V, W). It is also characteristic of Lycophyta and Equisetophyta. In *Equisetum*, crowded and starved prothalli form a filament of cells after spore germination (Fig. 13D) but well-nourished prothalli show longitudinal divisions only (Fig. 13E). A rhizoidal cell and a prothallial cell are formed. The prothallial cell gives rise to the gametophyte in homosporous Lycophyta and Equisetophyta but in ferns (the particular types mentioned above), the prothallial cell leads to a uniseriate filament of cells.

2. **Tripolar germination** is characteristic of Hymenophyllaceae. The initial division results in an equatorially-expanded plate of three equal cells. Further divisions give rise to the gametophyte (Fig. 11P-S).

3. **Amorphous germination** of spores is restricted to *Angiopteris* (Fig. 11 X) and *Marattia*. A mass or plate of cells is formed by irregular divisions and irregular growth. Later, a marginal cell becomes meristematic and further growth of the prothallus is in the direction of the meristematic cell.

Nayar and Kaur (1971) discuss seven patterns of gametophytic development in homosporous ferns:

1. **Osmuda type.** A plate of four cells is formed and a symmetrical cordate prothallus develops (Fig. 11 AA, BB).

2. **Marattia type.** Amorphous germination of spore gives rise to a cordate prothallus.

In rest of the ferns, 1-D-growth gives way to 2-D-growth with a change in the plane of cell division. Exceptionally, in the Grammitidaceae, the uniseriate stage is more extensive, and in *Schizaea* and *Trichomanes* there is no change in 1-D-uniseriate growth, resulting in an adult filamentous prothallus (Fig. 14E, M, G, L).

3. **Adiantum type** is characterized by a wedge-shaped meristematic cell, its activity leading to a one-cell-thick obovate prothallial plate. Later, a notch at the meristematic region gives rise to a cordate stage (Fig. 11AA).

4. **Drynaria type** has delayed formation of apical cell. A broad spatulate prothallial plate develops into a cordate prothallus.

5. **Kaulinia type** has an ameristic prothallial plate and a meristem is not initiated. The thallus elongates and becomes ribbon-shaped (Fig. 11Z). When mature, the thallus develops irregularly-scattered small circular cushions, two to four cells thick, in its middle region.

6. **Ceratopteris type** The juvenile prothallus is a broad ameristic plate of cells (Fig. 11Y), and the growth of the male prothallus ceases soon after the initiation of antheridia (Fig. 13K). The female prothallus continues to grow and a group of laterally-situated cells show meristematic activity (Fig. 13M). This lateral meristematic region has pleuricellular meristem, the prothallus becomes asymmetric and one wing is larger than the other (Fig. 13M).

7. **Aspidium type** has early hair formation, and the development is variable in the same species. A broad lopsided prothallial plate is formed by the posterior cells, as the terminal cell of the protonemal filament produces a unicellular papillate hair. This terminal apical cell and one or two lower cells remain inactive. Later development is similar to that in the Adiantum type.

Nayar and Kaur (1971) have described five types of mature gametophyte in homosporous leptosporangiate ferns. Spore germination is of three types. Development of the prothallus is of seven types (see Sec. 13). The five types of mature gametophytes are:

1. **Cordate** prothallus as in Polypodiaceae (Fig. 15), sometimes with fimbriated or convoluted margins (Fig. 14K, N).
2. **Filamentous** prothallus like branched algal filaments or moss protonema, as in *Schizaea, Woodsia* (Fig. 14E, M), *Hymenophyllum* (Fig. 14M) and *Trichomanes*.
3. **Strap-shaped,** long, fimbriate and with margins (breadth between cordate and ribbon-shaped) as in Grammitidaceae, Lomariopsidaceae and Polypodiaceae (Fig. 14J).
4. **Ribbon-shaped,** one-cell-thick, profusely branched, as in Vittariaceae, Polypodiaceae, Hymenophyllaceae and Loxogrammaceae (Fig. 14M)
5. **Tuberous,** as in *Ophiolossum, Botrychium and Helminthostachys* (Fig. 14A-D).

Transitional Forms. The heterosporous pteridophytes form endosporic gametophytes; *Platyzoma* is no doubt heterosporous but it shows latent homospory. It forms exosporic gametophytes like homosporous pteridophytes.

The microspores produce male gametophytes (Fig. 13J) and macrospores female gametophytes (Fig. 13L). The female gametophytes produce antheridia if fertilization is delayed.

Ceratopteris is a homosporous fern but shows incipient heterospory, and produces male and female gametophytes (Fig. 13K, M). If fertilization is delayed, the female gametophytes become hermaphrodite.

14. Endosporic Gametophytes in Heterosporous Pteridophytes

Male Gametophyte. On germination, the microspore (except in *Platyzoma*) produces a lenticular cell and a larger cell. The latter, by a series of divisions, differentiates into a number of spermatogenous cells and a variable number of sterile wall cells. In *Isoetes* (Fig. 16S, T) and *Azolla* (Fig. 16U), there is one spermatogenous cell which produces four or eight spermatozoids. *Marsilea* has two spermatogenous cells (Fig. 16 P, R) which produce 8 or 32 spermatozoids. In *Selaginella* with four spermatogenous cells (Fig. 16M, N), 128 or 256 spermatozoids are produced. The entire male gametophyte in *Selaginella, Isoetes, Marsilea,* and *Azolla* is represented by an antheridium, but in *Salvinia* the antheridium is represented by two spermatogenous cells separated by sterile cells (Fig. 16W).

Female Gametophyte In *Selaginella* and *Isoetes*, under the tri-radiate ridge, the megaspore nucleus undergoes free-nuclear divisions followed by wall formation (slower In *Isoetes*). In some species of *Selaginella* the gametophytic tissue (apical cushion) is separated from the free-nuclear portion below by a diaphragm (Fig. 19D). Archegonia develop on the cellular portion of the gametophyte (apical cushion) in all the species of *Selaginella* (Fig. 19B, D). In *Marsilea, Salvinia,* and *Azolla*, the megaspore divides and forms a small papillate cell, and the larger cell occupies the rest of the spore which stores abundant food reserves (Fig. 6H). The papillate cell forms a tissue and it is in this region that the archegonia differentiate. In

Marsilea, there is extreme reduction, as there is only one archegonium per gametophyte (Fig. 6N). The female gametophyte has a papillar envelope (Fig. 6I, J).

Sex Organs (Gametangia). The exosporic gametophyte of homosporous forms are monoecious, while endosporic gametophytes of heterosporous forms are dioecious.

The axial gametophytes of Psilotales and Ophioglossales develop antheridia and archegonia all over their surfaces (Figs. 12C, 14A, B). In *Lycopodium lucidulum* the sex organs are intermingled, but in other species distinct patches are present (*L. clavatum*). On dorsiventral gametophytes of homosporous ferns, sex organs occur on the ventral surface. The antheridia appear towards the base among the rhizoids, and the archegonia are confined to the meristematic region on the central cushion. In primitive forms, such as *Marattia* and *Angiopteris*, the archegonia are restricted to the central cushion (Fig. 14P) and antheridia occur on both surfaces.

Antheridia. Embedded antheridia develop in *Lycopodium, Equisetum,* Ophioglossaceae, Marattiaceae, and the heterosporous ferns, also in *Selaginella* and *Isoetes.*

Emergent (projecting) antheridia are reported in Psilotales (Fig. 16C), in the short-lived male gametophytes of *Equisetum* (Fig. 13A, F, G), and leptosporangiate ferns (Fig. 16C, D). The antheridia project as spherical bodies on the prothallus, and have a sterile wall or jacket. In embedded antheridia the jacket is one cell thick (Fig. 16B), but in *Botrychium* and *Helminthostachys*, it is two cells thick except in the centre. During dehiscence one or more opercular cells of the jacket breakdown, and help release the spermatozoids (Fig. 16H). Various patterns of dehiscence are known in different species of *Equisetum*, and it is variable even within the same species. Dehiscence is preceded by the swelling of mucilaginous substances stored in the antheridium.

In emergent antheridia of homosporous ferns the jacket is always one cell thick, and there is only one opercular cell (Fig. 16D).

The emergent antheridia of *Psilotum, Tmesipteris,* and the lower leptosporangiate ferns *Gleichenia, Matonia* (Fig. 16J), and *Osmunda* (Fig. 16I) are similar in strucutre to the embedded type. In higher leptosporangiate ferns the antheridium is a 'more' evolved and highly-reduced organ, the jacket consisting of two ring cells, and one terminal opercular cell (Fig. 16D).

The total number of cells in the jacket is variable in different families. In the massive antheridia of Gleicheniaceae, the jacket consists of 10–12 cells. In polypodiaceous ferns, the cap cell is bodily thrown off (Fig. 16D).

Spermatozoids. The number of sperms is lowest (four) in *Isoetes*. Normally the sperm output is 100 and, exceptionally, a few thousand in *Ophioglossum*. There are two types of spermatozoids: biflagellate in *Lycopodium* and *Selaginella* (Fig. 17C, D, G, H), and multiflagellate in *Psilotum, Tmesipteris, Isoetes, Equisetum* and ferns (Figs. 16K 17A, F, I, J-M).

Archegonium. The female reproductive organ, the archegonium, has a neck which protrudes while the lower portion (venter) remains embedded in the gametophytic tissue. The venter contains the egg and a ventral canal cell. The neck has neck canal cells. In lower pteridophytes, the prothallial cells which enclose the venter are not very well demarcated, whereas in ferns they form a discrete jacket (Fig. 18G).

The archegonial neck comprises four vertically-arranged rows of cells. The number of neck canal cells varies. In subterranean types of gametophytes of *Lycopodium*, there are 12–16 neck canal cells, whereas in leptosporangiate ferns there is a single binucleate neck canal cell (Fig. 18F, G, P). In terrestrial gametophytes (which are green), the short neck has a few neck canal cells. In subterranean gametophytes of *Psilotum* and *Tmesipteris* (Psilophyta), there is only one binucleate neck canal cell (Fig. 18C, D), compared with *Lycopodium* which has 12-16 neck canal cells (mentioned earlier). In subterranean gametophytes, the neck is short (Psilophyta), it falls off (sloughs off) and one or two thick cutinized neck cells persist to facilitate fertilization.

Ontogeny of Gametangia. Both types of gametangia (antheridia and archegonia) develop from a single cell of the gametophyte.

Embedded Antheridia. The antheridial initial divides to form an outer jacket initial, and a primary spermatogenous cell, or primary androgonial cell. Numerous spermatocytes (androcytes) which give rise to spermatozoids (antherozoids) develop from the spermatogenous cells formed by the primary spermatogenous cell (Fig. 16L)

Emergent Antheridia. In higher ferns, homosporous leptosporangiate ferns, according to earlier workers, the antheridial initial divides using three successive walls—the first and third are funnel-shaped and the second is hemispherical (Fig. 16D). According to Davie (1951), all three walls are transverse (Fig. 16E), but Stone (1962) reported that in many ferns the first wall is transverse, the second hemispherical, and third funnel-shaped (Fig. 16F). In polypodiaceous ferns, according to Verma and Khullar (1966), The first and third walls are transverse and the second hemispherical (Fig. 16G).

Archegonia. A single initial undergoes a periclinal division to form an outer primary cover cell, and a central cell. The outer cell after two successive anticlinal divisions at right angles to each other, forms four neck initials which give rise to the neck, comprising four rows of neck cells. The central cell gives rise to the primary neck canal cell and the primary ventral cell. The former gives rise to a varying number of neck canal cells. The primary ventral cell forms a ventral canal cell and the egg cell.

15.   Homology of Gametangia

Spessard (1922) observed abnormal archegonia in *Lycopodium lucidulum* in that the neck (Fig. 18R) and venter (Fig. 18Q) region showed cells which simulated spermatogenous cells as found in an antheridium. Abnormal archegonia in subterranean gametophytes of *L. complanatum* have 12–16 binucleate neck canal cells, and the archegonia simulate antheridia.

In ferns, archegonia with multinucleate neck canal cells and two ventral canal cells have been reported by Rashid (1976).

Gametes. The male gametes (spermatozoids) are formed from spermatocytes (androcytes) when the latter come in contact with water. The female gamete (egg cell) remains in the archegonium.

Spermatogenesis. The last mitosis of spermatogenous cells gives rise to the spermatids (spermatocytes). Their protoplasts differentiate (within the cell) and form mature sperms. The protoplast recedes from the cell wall (Fig. 17F), the nucleus assumes a crescent-shape, elongates, and to coil.

Yusuwa (1933) observed in the spermatozoids of Polypodiaceae the nucleus, cilia-bearing band, border, brim, lateral bar, cilia and plasma fragment. Duckett (1973, 1975) states that an elongated nucleus, a continuous mitochondrial band, and a series of microtubules form a spermatozoid (Fig. 17B, E). The cilia are characteristic of a 9 + 2 arrangement.

Oogenesis. Bell (1959) mentions that the cytoplasmic contents of the spore and zygote are very characteristic. Bell (1963), working on *Pteridium aquilinum*, observed DNA in the cytoplasm of the mature egg. The egg contains large amounts of RNA and basic proteins, compared with somatic cell cytoplasm. Therefore, the egg is seen as a special cell. Extranuclear DNA was also present in the primary cell of the axial row of neck canal cells (Sigee and Bell 1971). There is a five-fold increase in the volume of the primary cell. The central cell remains active for about 24 h and the level of extranuclear DNA is at its maximal level. During the formation of the egg, there is degeneration of plastids and mitochondira derived from the spore. At the periphery of the egg there is an enormous increase in endoplasmic reticulum. The remnants of the earlier organelles appear to be excreted, and accumulate on the surface of the egg. These remnants provide the egg with an additional membrane, not present in any other cell of the gametophyte. In the maturing egg cell, there is a period of nuclear evagination resulting in a new population of mitochondira. This unusual behaviour of the nucleus becomes clear very early in the life of an egg, and is concerned with synthetic activity in the cytoplasm (Sheffield and Bell 1987).

15. Fertilization

Ward (1954) described opening of the antheridium and archegonium, and the process of fertilization in *Phlebodium aureum*.

Opening of Antheridium. The antheridium dehisces within 3–7 min when immersed in water. As the mature antheridial wall absorbs water, the sperm mass becomes more distinct and is suddenly discharged. The coiled bodies escape through an opening made by bodily blowing off of cap cell. The sperms are inactive at first, their rounded bodies lie in a mass near antheridium and, in less than a minute, they expand and assume a helical form for swimming. The sperms move spirally with marked rapidity, and a rotary motion through the water.

Opening of the Archegonium. The archegonia open in water within an hour or so. The distal portion enlarges from an internal pressure due to degeneration in neck canal cells (Fig. 18K), and the four cover cells separate from each other. According to Campbell (1940), the cover cells are thrown off which has not been confirmed by Ward (1954). The separation of cover cells (apical cells) is followed by exudation of a colourless mucilaginous mass (disintegrated neck canal cells and ventral canal cell) from the apex, in three or four rounds. The uppermost neck cells, next to the cover cells, turn outward (Fig. 18I), and the four rows of neck cells split apart.

16.1 Attraction of Sperms Towards Archegonium

The archegonium exudes substances which at first repel the sperms (due to excessive concentration). Sperms are attracted chemotactically to the canal of the archegonium, (Fig. 18J) when water diffuses into the open canal of the archegonium and dilutes these chemotactic substances (citric acid, malic acid).

Entry of Sperm into Archegonium. The sperms circle about and orient at the mouth of archegonium and are ultimately drawn in. Accelerated rotary motion is characteristic of sperms reaching the venter. They have freedom of movement which is vigorous and at random in the space above the egg. The sperms soon cease side-to-side gyrations, and stand with the flagellated end against the egg. The movement stops within 4 to 20 min.

Syngamy. Yamonouchi (1908) described the entry of the sperm into the egg in *Nephropodium*. Although several sperms may enter the neck of the archegonium (Fig. 18L) and reach the exposed surface of the egg protoplast—the receptive spot, only one spermatozoid penetrates the egg. When the spermatozoid reaches the nucleus, the coils become shortened and the anterior region penetrates the nucleus first, followed by the posterior region.

16.2 Mating Systems

Free-living gametophytes have two types of mating systems in pteridophytes: (1) intergametophytic, and (2) intragametophytic.

The first one is characteristic of both homosporous and heterosporous forms, and the second of homosporous forms only (especially ferns). The majority of pteridophytes undergo intragametophytic selfing. There are morphological, populational and genetic adaptations to lessen the selfing. The first one is formation of antheridia followed by the bisexual phase. Simple polyembryony (more than one sporophyte from a gametophyte) is known in Osmundaceae and Equisetaceae (Fig. 22D, F), and favours intergametophytic mating.

In Equisetaceae, there is a populational adaptation. The gametophytes are classified as monogametophytic; either male (although homosporous, the crowded, starved, small, short-lived prothalli are male in *Equisetum* which bear a few projecting antheridia and very soon) or female. The female prothalli bear arehegonia first. They are larger prothalli which develop under favourable conditions of growth (environmental heterospory of *Equisetum*). If they remain unfertilized, they bear antheridia.

The prothalli are bigametophytic (male and female, male and hermaphorodite, female and hermaphoridite) or trigametophytic (male, female and bisexual). The third adaptation (genetic) is self-incompatibility.

17. Embryogeny

The division of the zygote is either transverse (Fig. 20A) (lower pteridophytes) or parallel (leptosporangiate ferns) (Fig. 20B) to the long axis of the archegonium (Fig. 20E). Exoscopic embryogeny is characteristic of *Psilotum* (Fig. 12A), *Tmesipteris, Stromatopteris, Equisetum* (Fig. 20S-U) and some members of Ophioglossales. In endoscopic embryogeny, the shoot apex is directed towards the neck of the archegonium. Endoscopic embryogeny is characteristic of *Lycopodium, Selaginella, Isoetes* and certain eusporangiate ferns (Fig. 20C, D). The shoot-apex is directed away form the neck of the archegonium. The embryos of leptosporangiate ferns tend to be endoscopic but do not fall into any of the above categories. Endoscopic embryos may be with or without a suspensor. The former type occurs in *Lycopodium* (Fig. 12Q, K), *Selaginella* (Fig. 21D, E), *Phylloglossum* and certain eusporangiate ferns. *Isoetes* (Fig. 21F) and some members of Marattiaceae show examples of endoscopic embryos without suspensors. The suspensor always originates from the upper cell of a two-celled proembryo, and the embryo proper is derived from the lower embryonic cell (Fig. 20F-H). Most commonly, the division of the embryonic cell is transverse, forming an epibasal cell and a hypobasal cell. Two successive vertical divisions in each of these cells results in eight-celled prombryo (Fig. 20E) with epibasal and hypobasal tiers. However, many species of *Lycopodium* deviate and the first divison is vertical, as well as the second division, and forms a group of four cells. Transverse division in each cell produces an eight-celled proembryo divisible into epibasal and hypobasal tiers.

17.1   Organogenesis

The epibasal tier gives rise to the shoot apex and first foliar appendage. The foot arises from the hypobasal tier (*Psilotum*). The foot is conspicuous in *Selaginella, Equisetum, Ophioglossum* and leptosporangiate ferns (Fig. 20V, X, Y). It serves as the anchoring and absorbing organ. The green gametophytes of *Lycopodium* (*L. cernuum*) have a suspensor, small foot, and develop a protocorm (Fig. 20p) due to storage of reserve food (Wardlaw 1968). It is green, has endophytic mycorrhiza, and synthesizes its own 'food'. The subterranean gametophytes of *Lycopodium* have a large foot (*L. clavatum, L. annotinum*) (Fig. 20Q, R, W, 12K). The foot in Psilotales is extensive (Fig. 13A), lobed and truly a haustorial structure (like *Anthoceros*). *Psilotum* resembles *Anthoceros*, has embedded sex organs, exoscopic embryogeny, and a lobed foot.

The suspensor places the embryo deeper into the gametophyte for obtaining nutrition. It is present in *Helminthostachys* and some species of *Botrychium*, whereas it is absent in *Ophioglossum*. The root is absent in Psilotales. It may arise from the epibasal half in *Isoetes* (Fig. 20L), Marattiaceae and Ophioglossaceae. In *Selaginella* the root arises from the hypobasal half. The root is a lateral organ (Fig. 20M-O), and variable in *Selaginella* (Fig. 20 I-K). In *S. denticulata* the root is on the same side as the suspensor of the pro-embryo. In *S. poulteri*, it is opposite the suspensor, whereas in *S. galeotti*, it lies between the suspensor and the shoot-apex.

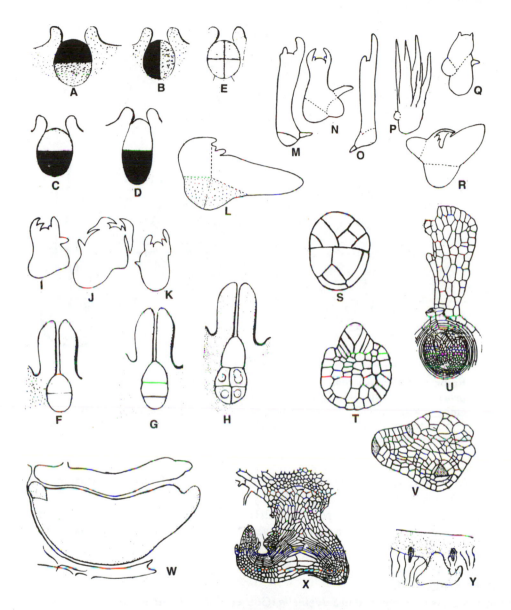

Fig. 20 A-Y.   Development of embryo (A-D diagrammatic). A Exoscopic embryogeny, as in some members of *Psilotum, Tmesipteris, Equisetum* **and Ophioglossaceae. B Orientation of two-celled proembryo as in leptosporangiate ferns. C Endoscopic proembryo as in** *Isoetes,* **some members of Marattiaceae and Ophioglossaceae. D Endoscopic embryo with suspensor, as in** *Lycopodium, Selaginella, Phylloglossum,* **and some Marattiaceae. E Octant embryo. F Two-celled proembryo. G Suspensor. H Suspensor with octant embryo. I-K** *Selaginella,* **note three types of orientation of embryo. L-R Embryo, root lateral. S-V** *Equisetum.* **S Young embryo. T Exoscopic embryo. U Embryo with shoot apex. V, X, Y, Fern embryos (homosporous leptosporangiate ferns). W** *Lycopodium selago,* **embryo with suspensor. (A-H from Foster and Gifford 1974, I-R, V from Bower 1935, S-U from Sadebeck 1902, W from Bruchmann 1910, X from Hofmeister 1851)**

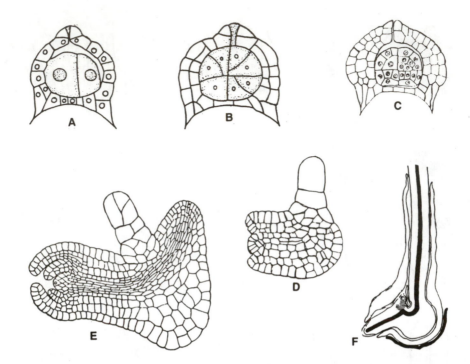

Fig. 21 A-F.   Embryo of heterosporous pteridophytes. A-C *Marsilea*. A Vertical division of zygote. B Young proembryo. C Later stage of embryo. D, E *Selaginella*. D Young embryo with suspensor. E Older embryo with rhizophore, foot, suspensor. F *Isoetes*, young embryo. (A, B from Campbell 1898, C from Haupt 1953, D from Bruchmann 1909, E from Bruchmann 1912, F from Baldwin 1933).

18.   Sporelings

The gametophyte bears a young sporophyte (Fig. 22A, B), sometimes both ends of rhizome bear aerial axes in *Tmesipteris*. The primary root is soon replaced by an adventitious root (Fig. 22C). The first leaves are scale-like (Fig. 22C) or highly developed (Fig. 22H). There may be many sporophytes per gametophyte (Fig. 22D-F). The circinate vernation may occur very early in homosporous leptosporangiate ferns (Fig. 22G).

19.   Parthenogenesis

Parthenogenesis has been reported in *Selaginella* (Goebel 1905), *Marsilea drummondii* (Strasburger 1907), *Athyrium filix-foemina* var. *clarissima,* and *Scolopendrium* sp. (Farmer and Digby 1907).

20.   Apogamy

The origin of a vascular sporophyte from a thalloid gametophyte, without fertilization, is of considerable interest (Farlow 1874, in *Pteris cretica*). It has been induced in *Lycopodium* and *Equisetum*. The factors which induce apogamy are: Presence of glucose or sucrose in the medium, if the thallus is planted erect on the medium, exposed to bright light and polyploidy.

Whittier et al. (1970) have described details of developmental aspects of apogamy in *Pteridium* and *Cheilanthes*. It is known in 20 species of Himalayan ferns. Apogamous life cycles can be meiotic or

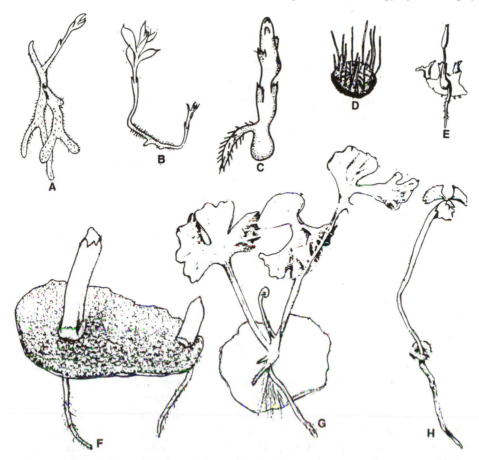

Fig. 22 A-H. Sporelings. A, B *Tmesipteris tannensis.* A Gametophyte bears young sporophyte. B Young sporophyte, both ends of rhizome (with rhizoids) have formed aerial axes. C *Lycopodium clavatum*, young sporophyte bears scale leaves, a lateral primary root, and bulbous foot. D-F *Equisetum.* D *E. debile*, prothallus with many sporophytes (polyembryony). E, F *E. arvense.* E Vertical section of lobed prothallus shown young sporophyte with leaf sheath, note primary root. F prothallus shows two young sporophytes, each has its own primary root. G Fern prothallus bears young sporophyte, note circinate vernation. H *Botrychium,* prothallus bears young sporophyte. (A, B from Holloway 1918, C from Bruchmann 1910, D from Kashyap 1914, E from Hofmeister 1851, F from Walker 1921, G from Foster and Gifford 1974, H from Eames 1936)

ameiotic (Doyle 1970). The differentiation of stomata (Smith 1955), tracheids or sporangia in gametophyte indicate that spogamy has occurred.

21. Apospory

Druery (1884), for the first time, reported the origin of a gametophyte from sporophyte, without the formation of spores, in *Athyrium filix-foemina* van. *clarissima*. The formation of rhizoids, antheridia, archegonia and cordate prothallus indicate apospory. The factors which cause apospory are absence of sucrose or glucose in the medium (starvation), dim light and small (size of) segments of sporophyte. In *Phlebodium aureum*, Ward (1963) observed that from the rhizome segments aposporous gametophyte and sporophyte regenerated. It was independent of any cultural conditions, or it was out of control. The modern concept of development of sporophyte or gametophyte is based upon two hypotheses:

(a) The gene block hypothesis (Mehra 1984) which postulates the presence of 4-gene-blocks in fern systems. A gene-block controlling the formation of root, leaf or gametophyte can be stimulated into action in any cell, whether of a sporophyte or gametophyte. Every gene-block has a 'master gene' for that block, and it is necessary to activate this master gene in order to trigger into action any sub-division of a gene block. The gene-block hypothesis is entirely theoretical and needs to be demonstrated.

(b) The second hypothesis (The Genetic Theory of Alternation of Generations) concerns 'alternation of generations' (Bell 1970, Smith 1979).

The 'Antithetic' and 'Homologous' theories are also not convincing. Bell (1970) confirmed the totality of evidence leads us to perceive the life cycle as an integral sequence of events, each key event is coded by a specific set of genome—The action of genes in the organization of a living system, but what is not clear is the location of 'gene switches'.

According to Smith (1979), certain metabolic changes may be responsible for determining the developmental pathways during the germination of spore, rather than specific gene-action. One of the major mysteries in cell biology is the 'missing link' between the macromolecules and visible structural organization (Loyal 1982).

22.   Concluding Remarks

The pteridophytes are not the first group of land plants. In a very thought-provoking and interesting article on *The Origin and Early Evolution of Plants on Land*, Kenrick and Crane (1997) point out that fossil spores are known which belong to land plants which must have existed before the pteridophytes. Future studies may reveal the identity of the ancient plants to which these spores belong.

The pteridophytes are indeed a very heterogeneous group of plants which occur in diverse conditions as epiphytes, aquatic (floating) plants, inhabit moist and shady areas, and are even found in xeric environments.

Male gametes some bi- and others multiflagellate, depend on water for fertilization. The gametophytic phase is much shorter than the sporophytic phase. The sori without and with indusia (true or false), and sporocarps in *Marsilea* and *Regnellidium*, and several features of embryogenesis make the pteridophytes a very distinctive group.

Acknowledgments

I am grateful to Professor B.M. Johri for encouragement and guidance, am highly indebted to my colleague Dr (Mrs) Bharati Bhattacharyya of the Botany Department, Gargi College, for help in the preparation of this manuscript, especially the illustrations, and to Dr (Mrs) Chhaya Biswas, formerly Principal of Gargi College, University of Delhi (South Campus), for her sustained interest in this work.

Glossary

Amorphous germination. Spore germination by irregular divisions and irregular direction of growth

Bipolar germination. First division in spore germination forms two cells, rhizoidal cell and a prothallial cell

Blepheroplast. A minute extranuclear granule which gives rise to motile cilia of antherozoid

Cauline sporangia. Sporangia are borne on stem (*Psilotum, Equisetum*)

Diaphragm. Thick-walled cells separate apical cushion from the lower food-filled portion of megaspore. Embryonal tube passes through diaphragm

Dimorphism of leaves. Ferns have two types of leaves, photosynthetic and reproductive

Dormancy of spores. Period before spore germination

Exoscopic embryogeny. Shoot-apex directed towards neck of archegonium

Endoscopic embryogeny. Shoot-apex is directed away from the neck of archegonium

Epibasal cell. Gives rise to upper part of embryo

Exosporic gametophyte. Spore germinates to form a green dorsiventral prothallus (homosporous forms), monoecious gametophyte

Endosporic gametophyte. It is characteristic of heterosporous ferns. Gametophyte is dioecious and remains contained in the spore (mega-/micro-)

Exindusiate. Without indusium

Eusporangiate. When there are many sporangial initial cells. Leptosporangiate development of sporangium (Leptosporangium). When one initial cell gives rise to the sporangium

Glochidia. Having outgrowths on microspores to anchor massulae in *Azolla* (otherwise they will be washed off)

Lip cells (stomium). Leptosporangium dehisces at the cleft (of lip cells or stomium)

Massula. Microspores are studded with hairy outgrowth called massulae (in *Azolla*)

Intramarginal sorus. When sori occur on the abaxial surface and leaf margin protects them

Parthenogenesis. Development of embryo without fertilization

Periplasmodium. Tapetal cells break-down and provide nutrition to developing spores.

Tripolar germination. Initial division of spore germination which results into an equatorially-expanded plate of three equal cells

Urostachya. Primitive features of *Lycopodium* spp.: protostele, haplostele, sporangia naked (not protected)

Rhopalostachya. Advanced features of *Lycopodium* spp.

Trichomes. Hairy trichomes are present in *Trichomanes*

Velum. Membranous outgrowth in sporangia of *Isoetes*.

Vivipary. The new sporophyte develops when spores germinate while still contained in the strobilus

References

Atkinson LR, Stokey AG (1964) Comparative morphology of the gametophyte of the homosporous ferns. Phytomorphology 14: 51–70

Baldwin WK (1933) The organization of the young sporophyte of *Isoetes englemanni* A. Br. Philos Trans R Soc Lond B 27: 1–19

Balajeff W (1885) Antheridien und Spermatozoiden der heterosporen Lycopodiaceen. Bot Zeitg 43: 793-802, 808–819

Belajeff W (1898) Über die mannlichen Prothallien der Waserfarne. Bot zeit 56: 141–194

Bell PR (1959) The experimental investigation of the pteridophyte life cycle. J Linn Soc London (Bot) 56: 188–203

Bell PR (1963) The cytochemical and ultrastructural peculiarities of the fern egg. J Linn Soc London (Bot) 58: 353–359

Bell PR (1970) The archegoniate revolution. Sci Progress (Oxford, England) 58: 27–45

Bell PR (1996) Megaspore abortion: A consequence of selective apoptosis? Intl J Plant Sci 157: 1–7

Bernard C (1904) A propos d'*Azolla*. Rcc Trav Bot Neer 1: 1–13

Bierhorst DW (1954) The gametangia and embryo of *Psilotum nudum*. Am J Bot 41: 274–281

Bierhorst DW (1956) Observations on the aerial appendage in the Psilotaceae. Phytomorphology 6: 176–184

Bierhorst DW (1971) Morphology of Vascular Plants, pp 1-560. Macmillan, New York

Bitter G (1900) Marattiaceae. *In*: Engler A, Prantl K (eds) Die natürlichen Pflanzenfamilien. 1(pt 4): 422–444

Bower FO (1899) Studies in the morphology of spore-producing members IV. The leptosporangiate ferns. Philos Trans R Soc London B 192: 29–138

Bower FO (1908) The Origin of a Land Flora, pp 1–727. MacMillan, London.

Bower FO (1926), (1928) The Ferns. 2: 1-344, 3: 1–306. Cambridge Univ Press, New York

Bower FO (1935) Primitive Land Plants, pp 1–658. MacMillan, London

Bracegirdle B, Miles PM (1971) Atlas of Plant Structure. Vol 1: 1-123. Heinemann Educational Books, London

Bracegirdle B, Miles PM (1973) Atlas of Plant Structure. Vol II: 1-106. Heinemann Educational Books, London

Britton EG, Taylor A (1901) The life history of *Schizaea pusilla*. Bull Torrey Bot Club 28: 1–19

Bruchmann H (1885) Das Prothallium von *Lycopodium annotinum*. Bot Zbl 21: 23-28, 309–313

Bruchmann H (1898) Über die prothallien und die Keimpflanzen mehrerer Europaischer Lycopodien, pp 1–111. Gotha

Bruchmann H (1909) Von prothallium der grossen sporen und von der Keimes-entwicklung einizer *Selaginella* arten. Flora 99: 12–51

Bruchmann H (1910) Die Kiemung der Sporen und die Entwicklung der prothallien von *Lycopodium clavatum* L., *L. annotinum* L., und *L. selago* L. Flora 101: 220–267

Bruchmann H (1912) Zur Embryologie der Selaginellacean. Flora 104: 180–224

Campbell DH (1898) Einige hotizen über die Keimung von *Marsilea aegyptiaca*. Ber Deutch Bot Gesells 6: 340–345

Campbell DH (1902) Studies on the gametophytes of *Selaginella*. Ann Bot (London) 16: 419–428

Campbell DH (1918) The Structure and Development of Mosses and Ferns, pp 1–708. 3rd edn. MacMillan, New York

Campbell DH (1940) The Evolution of the Land Plants (Embryophyta), pp 1–731. Stanford, California

Chamberlain CJ (1917) Prothalli and sporelings of three New Zealand species of *Lycopodium*. Bot Gaz 63: 51–65

Christ H (1897) Die Frankrauter der Erde, pp 1–388. Jena

Davie JH (1951) The development of the antheridium in the Polypodiaceae. Am J Bot 38: 621–628

Devi, Santha (1977) Spores of Indian Ferns, pp 1–228. Today & Tomorrow's Publ, New Delhi

Doyle WT (1970) The Biology of Higher Cryptogams. Mcmillan, London, pp 1–163

Druery CT (1884) Observations on a singular development in the lady fern. J Linn Soc Bot Lond 21: 354–538

Duckett JG (1970) Sexual behaviour of the genus *Equisetum* subgenus *Equisetum*. Bot J Linn Soc 63: 327–352

Duckett JG (1973) An ultrastructural study of the differentiation on the spermatozoid of *Equisetum*. J Cell. Sci 12: 95–129

Duckett JG (1975) Spermatogenesis in Pteridophytes. *In*: Duckett JG, Racey PA (eds) The biology of male gamete. Biol J Linn Soc 7 (Suppl 1): 97–127

Eames AJ (1936) Morphology of Vascular Plants: Lower Groups, pp 1–433. McGraw-Hill, New York

Eichler AW (1883) Syllabus des von les ungen über spezielle und medicinisch-Pharmaceutische. Botanik. 3rd edn. Berlin

Engler A (1886) Fuherer durch den Koniglichen botanischchen Garten der Universitat Zu Breslau, Breslau

Engler A, Parantl K (eds) (1898-1902) Die natürlichen Pflanzenfamilien 1(4): Engleman, Leipzing

Farmer JB, Digby L (1907) Studies in apogamy and apospory in ferns. Ann Bot (London) 21: 161–199

Foster AS, Gifford EM Jr (1974) Comparative Morphology of Vascular Plants, pp 1–751. 2nd edn. WH Freeman, San Francisco (USA)

Gifford EM Jr, Foster AS (1989) The Morphology and Evolution of Vascular Plants, pp 1–626. 3rd edn. WH Freeman, San Francisco (USA)

Goebel K (1880-81) Development of sporangia. Bot Z 38: 545–552, 561-571. 39: 681–694, 697-706, 713–720

Goebel K (1905) Organography of Plants. English Translation by IB Balfour, pp 1-707. Oxford (UK)

Haupt AW (1940) Sex organs of *Angiopteris erecta*. Bull Torrey Bot Club 67: 125–129

Haupt AN (1953) Plant Morphology. McGraw Hill, New York, pp 1–464

Hewitson W (1962) Comparative morphology of the Osmundaceae. Ann Missouri Bot Gard 49: 57–93

Hofmeister W (1851) Vergleichende Untersuchungen der Keimung, Entfaltung und Frucht bildung hoherer Kryptogamen (Moose, Farn, Equisetaceae, Rhizocarpeen, Lycopodiaceen) und der samenbildung der coniferen, pp 1–179. F Hofmeister, Leipzig

Holloway JE (1918) The prothallus and young plant of *Tmesipteris*. Trans NZ Inst 50: 1–44

Holloway JE (1939) The gametophyte, embryo and young rhizome of *Psilotum triquetrum*. Ann Bot (London) 3: 313–336

Jones JL, Hook PW (1970) Growth and development in microculture of gametophytes from stored spores of *Equisetum*. Am J Bot 57: 430–435

Kashyap SR (1914) Structure and development of the prothallus of *Equisetum debile*. Ann Bot (London) 28: 163–181

Kenrick P, Crane PR (1997) The origin and early evolution of plants on land. Nature 389: 33–39

Kny L (1875) *Ceratopteris*. Nova acta K. Leopcarol Deutch Ak d Naturf 37: 1–66

Lang WH (1898) On apogamy and development of sporangia on fern prothalli. Philos Trans R Soc London B 190: 182–239

Lagerberg T (1906) Zur Entwicklungsgeschichte des *Pteridium aquilinum* (L) Kuhn. Arkiv Bot 6: 1–28

Lawson AA (1917) The gametophytic generation of the Psilotaceae. Trans R Soc Edinburgh 52: 93–113

Liebig J (1931) Erganzungen zur Entwicklungsgeschichte von *Isoetes lacustris* L., Flora 125: 321–358

Linnaeus C (1754) Genera Plantarum, Salvi, Stockholm, 5th edn. pp 1–485

Loyal DS (1982) Some aspects of recent advances in phylogeny and life cycle of the Pteridophyta with particular reference to ferns. Aspects of Plant Sciences 6: 31–93. Today & Tomorrow's Printers & Publishers, New Delhi.

Lyon FM (1901) The evolution of sex organs of plants. Bot Gaz 37: 280–293

Machlis L, Rawitscher-Kunkel H (1967) The hydrated megaspore of *Marsilea vestita*. Am J Bot 54: 689–694

Marschall-Corelia C (1925) Differentiation of sporangia in *Marsilea quadrifolia*. Bot Gaz 79: 85–94

Meyer BS, Anderson DD (1939) Plant Physiology. Van Nostrand, New York

Mehra PN (1984) Some aspects of differentiation in flowering plants. SL Hora Gold Medal Lecture Award. *In*: INSA Award Lectures (1984-1993), Diamond Jubilee Publication (1994) 4: 1362–1401

Mitchell Gertrude (1910) Contributions towards a knowledge of the anatomy of *Selaginella* Spr. Pt V. The strobilus. Ann Bot (London) 24: 19–33

Moore R, Dennis Clark W, Stern KR (1995) Botany. Wm C. Brown Communications, Inc, USA, pp 1–868

Mukhopadhyay R, Bhandari JB (1999) Occurrence of microspore polyads in *Selaginella intermedia* (Bi.) Spring. Phytomorphology 49: 75–78

Nayar BK, Kaur S (1968) Spore germination in homosporous ferns. J Palynol 4: 1–14

Nayar BK, Kaur S (1971) Gametophytes of homosporous ferns. Bot Rev 37: 295–396

Parihar NS (1965) An Introduction to Embryophyta. Vol 2: 1–331. Pteridophytes. Central Book Depot, Allahabad, (India)

Pringsheim N (1863) Zur Morphologie der *Salvinia natans*. Jahrb Wiss Bot 3: 484–541

Pritzel E (1900) Lycopodiales. *In*: Engler A, Prantl K (eds) Die naturlichen Pflanzenfamilien. 1(Abt 4): 563–606

Rashid A (1976) An Introduction to Pteridophyta, Vikas Publishing House, New Delhi, pp 1–283

Rashid A (1999) An Introduction to Pteridophyta, 2nd edn. Vikas Publishing House, New Delhi, pp 1–426.

Raven PH, Evert RF, Eichhorn SE (1999) Biology of Plants. 6th edn. WH Freeman, USA, pp 1–944

Rouffa AS (1967) Induced *Psilotum* fertile appendage aberrations. Morphogenetic and evolutionary implications. Can J Bot 45: 855–861

Rouffa AS (1971) An appendageless *Psilotum*. Introduction to aerial shoot morphology. Am Fern J 61: 75–86

Sadebeck R (1902) Isoetaceae. *In*: Engler A, Prantl K (eds) Die naturlichen Pflanzenfamilien. Vol 1. Wilhelm Engleman, Leipzig

Sethi ML (1928) Contributions to the life history of *Equisetum debile*. Ann Bot (London) 42: 729–738

Sharp LW (1912) Spermatogenesis in *Equisetum*. Bot Gaz 54: 89–119

Sharp LW (1914) Spermatogenesis in *Marsilea*. Bot Gaz 58: 419–431

Shaw WR (1898) The fertilization of *Onoclea*. Ann Bot (London) 12: 261–285

Sigee D, Bell PR (1971) The cytoplasmic incorporation of tritiated thymidine during oogenesis in *Pteridium aquilinum*. J Cell Sci 8: 467–487

Slagg RA (1932) The gametophyte of *Selaginella kraussiana* 1. The microgametophyte. Am J Bot 19: 106–127

Sheffield E, Bell PR (1987) Current studies of the pteridophyte life cycle. Bot Rev 53: 442–490

Smith GM (1955) Cryptogamic Botany. 2nd edn. Vol 2: 1–399. McGraw Hill, New York

Smith GM (1979) Biochemical and physiological aspects of gametophyte differentiation and development. *In:* Dyer AF (ed) The Experimental Biology of Ferns. Academic Press, London

Spessard EA (1922) Prothalli of *Lycopodium* in America 2. *L. lucidulum* and *L. obscurum* var. *clendroideum*. Bot Gaz 74: 392–413

Sporne KR (1970) Morphology of Pteridophytes. Hutchinson, London, 3rd edn. pp 1–192

Stokey AG (1950) The gametophyte of the Gleicheniaceae. Bull Torrey Bot Club 77: 323–339

Stokey AG, Atkinson LR (1956) The gametophytes of Osmundaceae. Phytomorphology 6: 19–40

Stone IG (1960) Observations on the gametophytes of *Grammitis billardeiri* Willd, and *Crenopteris heterophylla* (Labill.) Tindala (Grammitidaceae). Austral J Bot 8: 11–37

Stone IG (1962) The ontogeny of the antheridium in some leptosporangiate ferns. Austral J Bot 10: 76–92

Strasburger E (1894) On the periodic reduction of the chromosomes in living organisms. Ann Bot (London) 8: 281–316

Sykes MG (1908b) Notes on the morphology of the sporangium bearing organs of the Lycopodiaceae. New Phytol 7: 41–60

Tippo 0 (1942) A modern classification of plant kingdom. Chron Bot 7: 203–206

Treub M (1884) Etudes sur les Lycopodiacees. Ann Jard Bot Buitenzorg 4: 107–138

Tryon A (1964) *Platyzoma*—A Queensland fern with incipient heterospory. Am J Bot 51: 939–942

Vashishta PC (1992) Botany for degree students, Pteridophyta pp 1-509. S Chand & Co. Ltd N. Delhi

Verma SC, Khullar SP (1966) Ontogeny of the polypodiaceous fern antheridium with particular reference to some Adiantaceae. Phytomorphology 16: 302–314

Wagner WH Jr (1963) A remarkably-reduced vascular plant in the United States. Science 142: 1483–1484

Walker ER (1921) The gametophytes of *Equisetum laevigatum*. Bot Gaz 71: 378–391

Walker ER (1931) The gametophyte of 3 species of *Equisetum*. Bot Gaz 92: 1–22

Walker ER (1937) The gametophyte of *Equisetum scirpoides*. Am J Bot 24: 40–43

Ward M (1954) Fertilization in *Phlebodium aureum* J. Sm. Phytomorphology 4: 1–17

Ward M (1963) Developmental patterns of adventitious sporophytes in *Phlebodium aureum* J. Sm. J Linn Soc Lond Bot 58: 377–380

Wardlaw CW (1968a) Morphogenesis in plants. A contemporary study. Methuen, London.

Wardlaw CW, Sharma DN (1961) Experimental and analytical studies of Pteridophytes. Morphogenetic investigation of sori in leptosporangiate ferns. Ann Bot (Lond) 25: 477–490

Wettstein R von (1935) Handbuch der Systematichen Botanik, Franz Deuticke, Leipzig. 4th edn. pp 1–1152

Whittier DP (1962) The origin and development of apogamous structures in the gametophytes of *Pteridium* in sterile culture. Phytomorphology 12: 10–20

Whittier DP (1970) The initiation of obligate apogamy in the fern *Cheilanthes castanea*. Am J Bot 57: 1249–1254

Wolf PG (1997) Evaluation of ATP B nucleotide sequences for phylogenetic studies of ferns and other pteridophytes. Am J Bot 84: 1429–1440

Yamanouchi S (1908) Apogamy in *Nephrodium*. Bot Gaz 45: 280–318

Yasui K (1911) On the life history of *Salvinia natans*. Ann Bot (Lond) 25: 469–483

10

Reproductive Biology of Gymnosperms

Chhaya Biswas and B.M. Johri

1. Introduction

The gymnosperms (an ancient group of plants) date back to the Devonian (395–359 my BP—million years before present). These plants constituted most of the world's dominant vegetation throughout the Late Palaeozoic (350–250 my BP), Mesozoic (141–65 my BP) and steadily declined thereafter. In the Tertiary (65 my BP) while the gymnosperms declined, the angiosperms evolved steadily. Thereafter, the angiosperms became dominant and the gymnosperms occupied a secondary position. At present there are ca. 72 genera and 760 species. These are grouped into seven orders: Cycadales, Ginkgoales, Coniferales, Taxales, Ephedrales, Welwitschiales and Gnetales.

While the plant body is quite diverse, most—except *Welwitschia* and *Gnetum*—gymnosperms show a more or less similar mode of reproduction.

The gymnosperms are:

Monoecious: Pinaceae (according to Troup 1921, *Cedrus deodara* is dioecious, occasionally male and female cones occur on the same tree, while it has been reported to be monoecious as a rule—see Maheshwari and Biswas 1970), Taxodiaceae, Cupressaceae, and Araucariaceae (except *Araucaria*); or

Dioecious: Cycadaceae, Ginkgoaceae, Podocarpaceae, Cephalotaxaceae, Araucariaceae (except *Agathis*), Ephedraceae, Welwitschiaceae and Gnetaceae.

The members of Taxaceae are usually dioecious, occasionally monoecious. In teratologic cases, the same strobilus of *Araucaria* is male at the base and female upwards. In plants with naked ovules, ca. 30% of the species are dioecious (Favre-Duchartre 1984).

2. Male Cones

All gymnosperms have male reproductive structures in cones/strobili (Fig. 1A-E). Each cone has a central axis with microsporophylls arranged spirally (cycads, conifers; Figs. 1A; 3D). The male cone may be pendant and catkin-like (*Ginkgo*; Fig. 1B), or stalked globose head and microsporophylls with peltate pollen sacs (taxads; Figs. 2G; 3F). In *Ephedra* the strobili are in clusters and compound (Fig. 1C), while in *Welwitschia* the strobilus has opposite and decussate scales (Fig. 1D). In *Gnetum* there are six to eight superposed cupules or collars around an axis. Each collar bears male flowers basipetally, and each male flower has a stalk bearing two unilocular anthers (Figs. 1E, F; 2J).

Microsporophyll. The microsporophylls bear: (1) a large number of microsporangia on the entire lower surface of the sporophylls (cycads; Figs. 2A, B; 3A), (2) two—occasionally three to seven—pendulous microsporangia (*Ginkgo*; Fig. 2C, D), (3) two sporangia on the lower/abaxial side (conifers; Figs. 2E, F; 3E), and (4) peltate pollen sacs (taxads). In *Ephedra* a sporangiophore bears at its tip two–six bilobed

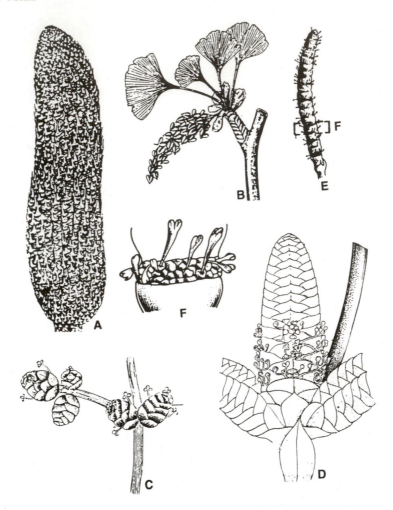

Fig. 1 A-F.   Male Cone. A *Cycas*, **B** *Ginkgo*, **C** *Ephedra*, **D** *Welwitschia*, **E, F** *Gnetum*. **F Inset from E (A after Pant 1973, B after Gangulee and Kar 1982, C after Tiagi 1966, D after Chamberlain 1935, E, F after Vasil 1959).**

sessile microsporangia which dehisce apically (Fig. 2H). In *Welwitschia* each microsporangiophore bears three fused sporangia (Fig. 2I), while in *Gnetum* the male floral stalk bears two unilocular anthers (Fig. 2J).

 Microsporangia. A mature microsporangium comprises: (1) epidermis, (2) a multilayered microsporangial wall, the innermost layer functions as tapetum. The epidermal cells of the microsporangium develop fibrous thickenings (except along the line of dehiscence), and form an exothecium. *Ginkgo* is an exception and fibrous thickenings in the hypodermal layer form an endothecium (cf. angiosperms; Fig. 3C) The middle layers are crushed during maturation of the microsporangium (Fig. 3B, C). The tapetum encloses the sporogenous tissue, and comprises a single (occasionally double) layer of large, richly-cytoplasmic multinucleate cells. The tapetum is mostly of the secretory (glandular) type. The tapetal cells show maximal activity during meiosis in microspore mother cells, and degenerate after the spores are released from the tetrads. The main function of the tapetum is to nourish the sporogenous cells and young microspores. In addition, the tapetum is concerned in the production and release of callase, transmits PAS-positive material into the loculus, forms an acetolysis-resistant membrane, and the exine of the microspores.

The primary sporogenous cells, on repeated divisions, form a mass of sporogenous cells which mature into microspore mother cells (mimc).

The mimc, after meiotic divisions, forms tetrads of microspores. The latter are liberated from tetrads with a thin exine (Favre-Duchartre 1984).

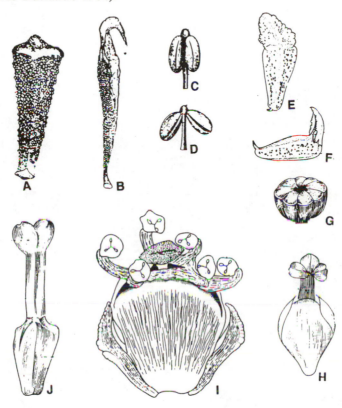

Fig. 2 A-J. **Microsporophyll. A, B** *Cycas*, **abaxial (A) and lateral (B) view of microsporangia arranged in sori. C, D** *Ginkgo*, **two pendant sporangia. E, F** *Cedrus*, **dorsal (E) and lateral (F) view of abaxial microsporangia. G** *Taxus*, **transection radially-symmetrical and peltate microsporangia. H** *Ephedra*, **fertile shoot with three bilocular microsporangia at the tip. I** *Welwitschia*, **male flower shows lateral bracts and perianth: each microsporangiophore fused at the base, bears a synangium of three fused microsporangia. J** *Gnetum*, **male flower shows elongated stalk and anthers (A, B after Pant 1973, C, D after Gangulee and Kar 1982, E, F after Johri 1936, G after Oliver 1982, H after Tiagi 1966, I after Martens 1961, J after Sanwal 1962).**

Male Gametophyte. The development of male gametophyte follows an uniform pattern (except in *Welwitschia* and *Gnetum*).

A mature pollen grain shows one (cycads, *Welwitschia, Gnetum*) or two (Pinaceae, *Ginkgo, Ephedra*) prothallial cells which are usually inconspicuous and ephemeral. The number of prothallial cells varies from three or four (*Podocarpus*), five–seven (*Agathis*) to 18–20 (*Araucaria*). The prothallial cells are absent in Taxodiaceae, Cupressaceae, Cephalotaxaceae and Taxaceae. The pollen grain functions directly as the antheridial initial.

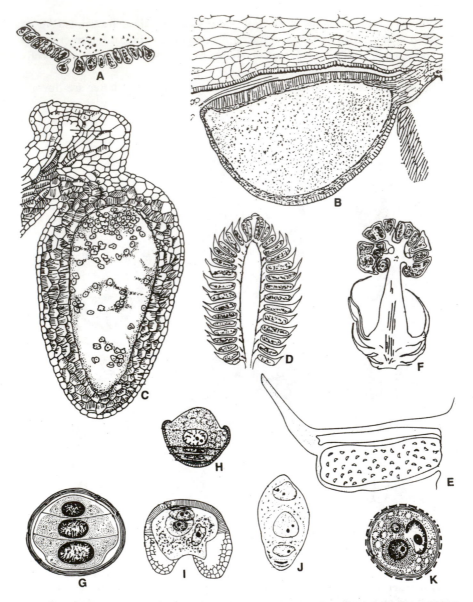

Fig. 3 A-K.   Microsporangium and mature pollen. A, B, G *Cycas*. A Longisection microsporophyll shows abaxial arrangement of microsporangia. B Sporangium. C, H *Ginkgo*, longisection (C) microsporangium shows multilayered endothecium (shaded) and microspores (H). D, E, I *Cedrus*. D Male cone longisection, microsporophylls arranged spirally. E Longisection microsporophyll shows abaxial arrangement of microsporangium. F Longisection male cone. G-K Mature pollen grains at shedding. J *Ephedra*. K *Gnetum* (A, B after Pant 1973, C after Jeffrey and Torrey 1916, D after Roy Chowdhury 1961, E, I after Johri 1936, F after Bracegirdle and Miles 1973, G Original, H after Chamberlain 1935, J after Singh and K. Maheshwari 1962, K after Sanwal 1962).

In cycads and *Welwitschia*, the only prothallial nucleus degenerates precociously. In Pinaceae and *Ephedra* both prothallial nuclei degenerate early and become imbedded in the intine, while in *Ginkgo* only the first prothallial nucleus degenerates, and the second nucleus persists. In Araucariaceae and Podocarpaceae, the multiple prothallial nuclei become free in the common cytoplasm.

The logical evolutionary tendency is the suppression or non-production of the prothallial cells. However, it is interesting that there are two prothallial cells in *Ephedra* and one each in *Welwitschia* and *Gnetum*—the three most advanced taxa in gymnosperms.

The antheridial initial divides to form a small antheridial cell and a large tube cell. The antheridial cell divides to form the stalk cell and the body cell. The latter has the major part of the reproductive cell cytoplasm. The stalk cell has a distinct wall which eventually breaks down and the cytoplasm merges with that of the tube cell. The body cell enlarges, has dense cytoplasm, and a large nucleus. It divides and gives rise to two male gametes. Occasionally, the body cell divides several times to produce multiple male gametes (*Cupressus, Juniperus*). The male gamete is a flagellate spermatozoid only in cycads and *Ginkgo*, and is an unflagellated sperm cell, equal/unequal in the other taxa.

Pollen Grain. The mature pollen, at the time of shedding, differs in various taxa (Fig. 3G-K). The pollen grain is one-celled in *Chamaecyparis, Callitris, Cryptomeria, Cupressus* and *Taxus*; two-celled in *Cephalotaxus, Athrotaxis, Torreya* and *Taxodium*; three-called in most cycads, *Ginkgo* (Fig. 3G, H), *Welwitschia* and *Gnetum* (Fig. 3K), four-celled in *Pinus*, five-celled in *Cedrus* and *Ephedra* (Fig. 3I, J), and multicelled (due to a large number of prothallial cells) in *Podocarpus* and *Araucaria*.

The pollen grains have wings in several taxa, e.g. members of the Pinaceae (except *Larix*), *Microcachrys, Pherosphera, Podocarpus*. The shape, structure and the number of wings vary. The wings are absent in members of the Taxodiaceae, Cupressaceae, Araucariaceae, Cephalotaxaceae, Taxaceae, and *Ephedra, Welwitschia* and *Gnetum*.

The pollen is produced in large quantities, is dispersed by wind, and the surrounding areas become clouded by a yellow dust, the 'sulfur shower'.

3. Female Cones

All gymnosperms, except *Cycas*, bear ovules in cones/strobili (Fig. 4A). In *Cycas* the ovules are borne on loose megasporophylls (Fig. 4B). Some of the cycad cones are enormous. A ripe cone of *Dioon spinulosum* is ca. 61 cm long and weighs up to 28 kg (see Biswas and Johri 1997).

In *Ginkgo* the cone is reduced. The ovules occur in pairs, one each at the tip of a forked peduncle borne on the dwarf shoot (Fig. 4C, D).

The cone in Pinaceae consists of seed-scale-complexes (bract scale and ovuliferous scale) spirally arranged on the cone axis (Fig. 4E). Each complex bears two ovules on the abaxial side (Fig. 4F). In Taxodiaceae the spirally-arranged fertile scales bear three or four adaxial ovules. The cone in Cupressaceae consists of three or four pairs of decussate scales which bear one to three ovules on the adaxial side. In Podocarpaceae the cone bears bracts and only the uppermost is fertile. It bears a single ovule encircled by an epimatium (Fig. 4G).

The cone in Araucariaceae consists of spirally-arranged bracts and ovuliferous scales which are partially fused to form a ligule. a single ovule occurs on each cone scale.

The small cones of Cephalotaxaceae consist of five to seven pairs of opposite and decussate bracts. Each bract bears two ovules in its axil. Generally, only one ovule of a pair matures.

In taxads, the female reproductive organ consists of a terminal ovule (Fig. 4H) on the short secondary shoot.

The female 'inflorescence' of *Ephedra* consists of a short shoot which bears two to four pairs of bracts. At the apex of the shoot there are one to three 'female flowers' (Fig. 4I).

In *Welwitschia* the female strobilus has broad opposite and decussate cone bracts (Fig. 4J). These cone bracts are broad and ovate with tapering thin lateral wings. It is quite thin near the base and thick towards the apex. The flattened seed lies in a pocket formed by the thin area.

Fig. 4 A-K.   Female cones and ovules. A, F *Pinus wallichiana*, E *P. roxburghii*. A Young female cone. B *Cycas revoluta*, megasporophylls with ovules. C, D *Ginkgo biloba*, dwarf shoot (C) bearing young strobili. D Longisection female strobilus. E Longisection young female cone. F Seed-scale complex, lateral view, bearing ovule. G *Podocarpus*, longisection anatropous ovule with epimatium (*epi*). H *Taxus* sp., longi-section fertile shoot with terminal orthotropous ovule. I *Ephedra foliata*, female strobilus shows two ovules and pairs of opposite and decussate bracts (*br*). J *Welwitschia mirabilis*, stalk bearing several ovulate strobili. K *Gnetum gnemon*, cone bearing two seeds (A, E, F after P. Maheshwari and Konar 1971, B after Pant 1973, C, D after Gangulee and Kar 1982, G after Konar and Oberoi 1969, H after Bracegirdle and Miles 1973, I after Tiagi 1966, J after Chamberlain 1935, K after P. Maheshwari and Vasil 1961).

The 'female flowers' in *Gnetum* are borne in whorls on spike-like axes (Fig. 4K). Each whorl is (apparently) subtended by a fleshy collar.

The individual sporophytes produce male or female cones. The cycads require several decades before sporulating—*Ginkgo* and *Cedrus* ca. 30 years, *Tsuga* and *Abies* ca. 20 years, *Larix* and *Taxus* ca. 15 years and *Juniperus* ca. 5 years. Within the same genus, the species demonstrate diversity such as *Pinus cembra* ca. 25 years, *P. nigra, P. halepensis* ca. 15 years, *P. pinaster* ca. 10 years and *P. sylvestris, P. strobus* ca. 5 years.

In most species the male cones are initiated one year earlier than the female cones, on the same tree.

Ovule. The young ovule has a central nucellus covered by a single integument. *Ephedra* has two coverings and *Gnetum* three. Of these, only the innermost covering is the integument. All along the central region of integument, up to the nucellus, is a narrow passage, the micropyle. The ovule is mostly orthotropous; in *Podocarpus* (Fig. 4G) and *Dacrydium* it is anatropous.

A ring-shaped collar is present at the base of the ovule in *Ginkgo* (Fig. 5A). In podocarps an epimatium (equivalent to ovuliferous scale) covers a part or the entire ovule. An aril (Fig. 5B, C) is initiated as a ring of meristematic tissue at the base of the integument in taxads, further development occurs only after pollination.

Most gymnospermous ovules are vascularized, except in conifers. However, in *Cedrus* (Pinaceae) and *Cephalotaxus* (Cephalotaxaceae) the ovules are vascularized (Fig. 5D, E).

Megasporogenesis. One or more sporogenous cells (in an ovule) may function directly as the megaspore mother cell (mgmc). The latter is usually large, elongated, has prominent nuclei and dense cytoplasm. It undergoes meiosis and produces a triad (due to an undivided upper dyad cell) or linear tetrad. Generally,

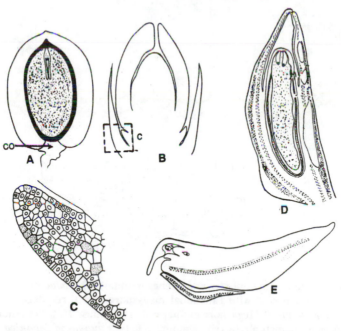

Fig. 5 A-E. Ovule. A *Ginkgo*, longisection mature seed, the collar (*co*) is inconspicuous. B, C *Taxus* sp. B Longisection ovule with a short aril at the base. C Inset marked C in B. *D, E Cedrus*, longi- and transection ovule shows vasculature of integument (A after Chamberlain 1935, B, C after Loze 1965, D, E after Roy Chowdhury 1961).

the lowermost megaspore functions (Fig. 6A). In *Welwitschia* and *Gnetum* a wall is not laid down after meiosis I and II, in mgmc. This results in the formation of a four-nucleate coenomegaspore (Fig. 6B, C). Investigations have revealed the presence of a megaspore wall in all extant gymnosperms. The structure is more elaborate in the primitive members—cycads and *Ginkgo*—where motile male gametes bring about fertilization.

Female Gametophyte. The development of female gametophytes (in gymnosperms) is usually monosporic, but tetrasporic in *Welwitschia* (Fig. 6B, C) and *Gnetum*. The number of free nuclei (in the female gametophyte), varies before wall formation (Fig. 6D). It may be ca. 250 in taxads and ca. 8000 in *Ginkgo*. The female

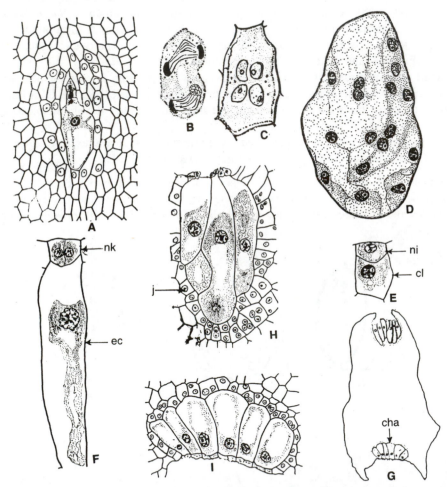

Fig. 6 A-I. Megaspore, female gametophyte, and archegonium. A, D *Podocarpus gracilior*. A Longisection central part of nucellus with functional megaspore and three degenerated megaspores. B, C *Welwitschia mirabilis*. B Megaspore mother cell at telophase II. C Four-nucleate coenomegaspore. D Free-nuclear gametophyte (wholemount). E-I *Cryptomeria japonica*. E, F Development of archegonium, *cl* central cell, *ec* egg cell, *ni* neck initial, *nk* neck. G Outline diagram of gametophyte shows micropylar and chalazal (*cha*) archegonial complex. H, I Micropylar (H) and chalazal (I) archegonial complex, *J* jacket (A, D after Konar and Oberoi 1969, B, C after Martens 1963, E-I after Singh and Chatterjee 1963).

gametophyte becomes completely cellular before the differentiation of the archegonium. *Gnetum* is an exception, the female gametophyte is partly cellular and becomes completely cellular only at the time of fertilization or thereafter.

In *Welwitschia* the coenomegaspore undergoes eight to ten free-nuclear divisions. Walls are laid down by free-cell formation, the cytoplasm becomes divided into irregular multinucleate compartments with cell walls. Two zones differentiate: (1) a smaller fertile micropylar zone (*mz*), and (2) a larger sterile chalazal zone (*cz.*, Fig. 11A).

Archegonia. The archegonia develop in all gymnosperms, except *Welwitschia* and *Gnetum*. They differentiate mostly at the micropylar pole (Fig. 5D); lateral and chalazal archegonia are rare. The archegonia occur singly in cycads, *Ginkgo*, *Ephedra*, members of Pinaceae, Podocarpaceae, Cephalotaxaceae, Araucariaceae and Taxaceae. In Cupressaceae and Taxodiaceae (except *Sciadopitys* where four or five archegonia occur singly), the archegonia are grouped into one or more complexes which have a common jacket layer (Fig. 6G-I). In *Athrotaxis* and *Callitris* the archegonial complexes are lateral.

The archegonium develops from an apical archegonial initial (Fig. 6E, F). A mature archegonium has a short neck and a large egg cell. The latter has an ephemeral ventral canal cell (Pinaceae) or only the nucleus (*Cephalotaxus*, *Ephedra*, Taxodiaceae, Cupressaceae). In *Taxus*, *Torreya taxifolia*, *Widdringtonia cupressoides* the central cell functions as the egg. The egg nucleus moves to a slightly lower site and enlarges. The cytoplasm of the egg accumulates numerous proteid vacuoles which are only cytoplasmic formations of several types: large inclusions, small inclusions, microbodies and vesicular bodies. The cytoplasm of the egg also shows a conspicuous ER, plastids, mitochondria, Golgi bodies and ribosomes. The egg nucleus becomes filled with nucleoplasm.

One to three layers of densely-cytoplasmic cells, the jacket layers, surround the central cell/egg/proembryo. There is a thick wall, with simple pits, between the jacket cells and the central cell of the archegonium. The jacket layer is concerned in the nutrition of egg/proembryo. The food reserve in the gametophytic cells becomes available through plasmodesmata on the inner tangential walls of jacket cells. These cells also secrete enzyme(s).

The gametophytic tissue around the archegonia grows upward and forms a depression, the archegonial chamber, so that the archegonia appear sunken.

In *Welwitschia* the micropylar cells along with the contents form a cluster of embryo sac tubes or prothallial tubes. These tubes pierce the megaspore memberane and grow in all directions in the nucellus. Each tube grows independently of the others, and normally ends in an enlarged swelling, the 'fertilization bulb' (Fig. 11A-D). All the nuclei in the bulb are potentially female nuclei.

In *Gnetum*, the pollen tube enters the free-nuclear micropylar zone of the female gametophyte, one or more groups of adjoining cells become densely-cytoplasmic. Only one (rarely two) cells from each group function as the 'egg'.

4. Pollination

Most gymnosperms are wind-pollinated. There is effective insect pollination in *Ephedra aphylla* (Bino et al. 1984), *Gnetum* sp. (Kato et al. 1995) and *Welwitschia* (Kubitzki 1990).

At the time of pollination, a sugary exudate (pollination drop) is secreted at the micropyle in the gymnosperms so far investigated. The exceptions are: *Abies*, *Cedrus*, *Larix*, *Pseudotsuga* and *Tsuga* which have a 'stigmatic' micropyle. In *Araucaria*, *Agathis* and *Tsuga dumosa*, the pollen grains do not land on the micropyle.

The pollination drop serves as a receptor of the wind-borne pollen as well as a means for transporting it to the nucellus where it germinates. Secretion of the drop appears to be a cyclic (24-h cycle) phenomenon

(McWilliam 1958). It is secreted in the night or early in the morning, evaporates/retracts during the day, and is secreted again the following morning. The cycle continues for a few days or until ovule is pollinated. The presence of the pollen on the pollination drop effects a rapid and complete withdrawal of the fluid (see Singh 1978).

A chemical analysis of the pollination drop in *Pinus nigra* (McWilliam 1958) shows three sugars as the main constituents: D-glucose (40 m moles), D-fructose (40 m moles) and sucrose (2.5 m moles). Several amino acids, peptides, malic and citric acids in the pollination drops of *Ephedra* and *Taxus* have been reported (Ziegler 1959), as well as inorganic phosphates and sugars. The sucrose concentration in the pollination drop of *Ephedra* is high (25%). It has been observed that for in vitro germination of pollen in *Ephedra*, a high concentration of sugar is required (Mehra 1938).

In different gymnosperms the ovules receive pollen at various stages of development: (1) sporogenous cells or megaspore mother cell (*Ginkgo*, conifers and taxads), (2) free-nuclear gametophyte (many cycads, *Gnetum*), (3) young archegonia (*Macrozamia*), and (4) mature archegonia (*Ephedra*).

Pollen Germination. The pollen grains are caught in the pollination drop. Opinions differ whether the pollen grains are sucked in or transported through active absorption into the micropyle. The pollen finally lands on the nucellus and germinates.

According to Biswas and Johri (1997), a fresh study is required (by using modern tools and techniques) of the pollination drop, its secretion and chemical composition, likewise the mode of transport of pollen within the ovule.

Role of Pollen Tube. The pollen tube grows towards the female gametophyte. The short and unbranched tube is quite prominent. At its tip the tube contains the body cell/male gametes. The pollen tube finally ruptures the neck cells of the archegonium and delivers its contents into the egg cytoplasm.

In cycads and *Ginkgo* the pollen tube arises from the distal (upper) end of the pollen grain. The tube may branch (*Cycas*, *Ginkgo*) or may remain unbranched (*Zamia*). These pollen tubes (or its branches) are enucleate and are not concerned with the transport of male gametes for fertilization. They absorb nutrition from the surrounding nucellar tissue. In *Zamia furfuracea* (Choi and Friedman 1991) the pollen tube penetrates the nucellar cells by destroying them enzymatically. Intracellular haustorial growth ultimately leads to a complete destruction of each penetrated cell. The pollen tube growing in the nucellar tissue is interpreted as a vegetative structure since it does not transport the male gametes. The pollen tube establishes only a haustorial sytem. The proximal lower pole of the pollen grain (*Microcycas*, *Zamia*) produces a pollen tube ca. 1.5–2 mm wide and 3–5 mm long. It contains the prothallial, stalk and body cells/male gametes (see Johri 1992). The proximal lower pole of the pollen grain expands rapidly and breaks through the inner nucellar epidermis into the space of the archegonial chamber. The proximal ends of numerous pollen grains hang into the archegonial chamber. Finally it bursts and forms a short pollen tube (*Cycas*) through which it discharges the male gametes into the archegonial chamber.

5. Fertilization

The release of male gametes varies in different taxa of gymnosperms. In cycads and *Ginkgo* the flagellated gametes are released in the archegonial chamber which usually contains a fluid. In taxa where the archegonia occur singly (*Pinus*, *Cedrus*, *Podocarpus*, *Cephalotaxus*, *Taxus*, *Ephedra*), the neck cells of the archegonium degenerate, the pollen tube penetrates the egg cell, and releases the male gametes. In taxa where the archegonial complexes are present at the micropylar end or are laterally placed (*Athrotaxis*, *Callitris*), the pollen tubes grow adpressed to the neck of several archegonia. The male gametes, while still inside the pollen tube, become closely attached to the neck cells which eventually degenerate. The male gamete then passes into the egg cell leaving behind its cytoplasmic sheath. Two male gametes from a pollen tube may enter the same archegonium or two different archegonia. On entering the egg cytoplasm, the male gamete (along with some cytoplasm) comes into contact with the egg nucleus. the non-functional male nucleus in the egg cytoplasm usually persists for some time. The nucleoplasm of both male and female nuclei—forming the ground substance of neocytoplasm—alone take part in the formation of embryonal cytoplasm, while the remaining cytoplasm of the zygote degenerates.

Double Fertilization. Double fertilization has been reported in *Ephedra* spp. (Khan 1943: Moussel 1978; Friedman 1990a, b, 1991). The pollen tube pushes its way between the neck cells of the archegonium. The tip ruptures and discharges the two male gametes into the archegonium. One of the gametes fuses with the egg. At an advanced stage of fusion of egg and male gamete, the ventral canal nucleus (sister of egg nucleus) moves down, followed by the second male gamete, and both the nuclei fuse. Thus, 'double' fertilization is accomplished.

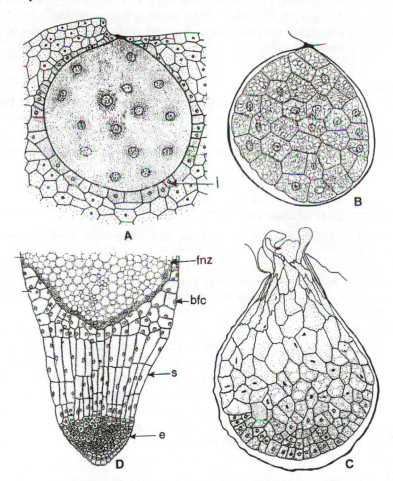

Fig. 7 A-D. Proembryogeny. A-C *Ginkgo biloba*. A Longisection free-nuclear proembryo shows jacket layer (j) and gametophytic cells. B Cellular proembryo. C Upper cells of cellular proembryo elongate to form suspensor. D *Zamia*. Longisection proembryo shows embryonal mass (*e*), massive suspensor (*S*), a narrow zone of buffer cells (*bfc*), and upper free-nuclear region (*fnz*) (A-C after Lyon 1904, D after Bryan 1952).

In *Gnetum gnemon* the pollen tube discharges two male gametes which fuse with nearby undifferentiated female nuclei within the partially coenocytic female gametophyte and form two zygotes. According to Carmichael and Friedman (1995), this is an expression of a rudimentary pattern of double fertilization. It has been proposed that double fertilization evolved before the origin of angiosperms. The original manifestation of double fertilization in seed plants may have led to the formation of two embryos (Friedman 1992b, 1994, 1995).

In angiosperms the zygote (diploid) and the primary endosperm nucleus (triploid) are formed as a result of double fertilization. Of the two male gametes from a pollen tube, one gamete fuses with the egg and the other with the two polar nuclei or fused polars. Thus, the zygote (2n) produces the embryo, and the primary endosperm nucleus (3n) gives rise to endosperm. If there is only one polar, the zygote and the primary endosperm nucleus are both diploid and genetically identical, but their products are quite different. There is no such parallel in gymnosperms (Biswas and Johri 1997). According to Khan (1943): "The type of double fertilization as seen in *Ephedra* may have no phylogenetic significance at all, and may simply be the natural outcome of a tendency towards fusion between any two nuclei of opposite sexual potencies that happen to lie in a common chamber."

6.   Embryogeny

In the development of the embryo in gymnosperms, there is a free-nuclear phase as the division of the zygote is not followed by a wall. *Sequoia*, *Gnetum* sp., and *Welwitschia* are exceptions since the division of the zygote is followed by wall formation cf. angiosperms).

The development of the embryo is in three distinct phases: Proembryogeny, and early and late embryogeny.

Proembryogeny

This begins with the division of the zygote and lasts up to the stage prior to elongation of the suspensor. Four distinct variations have been observed:

1. Cycads and *Ginkgo*. The nuclear division of the zygote is followed by repeated synchronous divisions in quick succession. Later divisions are not synchronous, some of the nuclei may even fail to divide. In most cycads the nuclei migrate to the base of the proembryo where further divisions occur. Only a few nuclei persist in the upper portion which has scanty cytoplasm. Subsequently, the nuclei degenerate.

The free nuclei are evenly distributed in the proembryo in *Ginkgo* (Fig. 7A). Wall formation takes place when there are ca. 256 free-nuclei. In cycads the number of nuclei is variable. In *Dioon edule* walls are laid down after ten mitoses (1024 nuclei, the highest in gymnosperms); in *Zamia floridana* after more than eight free-nuclear divisions, while in *Bowenia scurrulata* after six free-nuclear divisions (lowest in Cycads).

The newly-formed cells fill the entire proembryo in *Ginkgo*. The cells at the base divide and function as the embryonal cells, while the upper cells elongate to form a massive suspensor (Fig. 7B, C).

In most cycads, cells are formed in the lower portion of the proembryo, the nuclei in the upper region remain free. Eventually, these free nuclei degenerate and form a plug. The cells at the base function as the embryonal cells. The cells just above the embryonal cells differentiate into a suspensor, and the uppermost cells function as buffer cells (Fig. 7D). The buffer cells may direct the pressure of the elongating suspensor to downward growth.

2. Conifers and Taxads. The division of zygote is intra-nuclear, the resulting nuclei remain within the neocytoplasm. The nuclear membrane of the zygote disappears at the end of mitosis. The two nuclei with the neocytoplasm migrate to the base of the archegonium where further synchronous mitoses occur. The number of free nuclei before the initiation of walls varies. It is constant for each taxon.

The free nuclei become arranged in two tiers. Walls are laid down resulting in a lower group of cells—

the primary embryonal tier (p^E), and an upper group of cells—the primary upper tier (p^U). The cells of the p^U tier remain open towards the archegonium. The cells of both the tiers undergo division. In the lower tier p^E, the divisions result in doubling the number of cells (in podocarps the cells become binucleate). This group of cells is designated the embryonal tier (E). In the p^U tier the transverse division is followed by wall

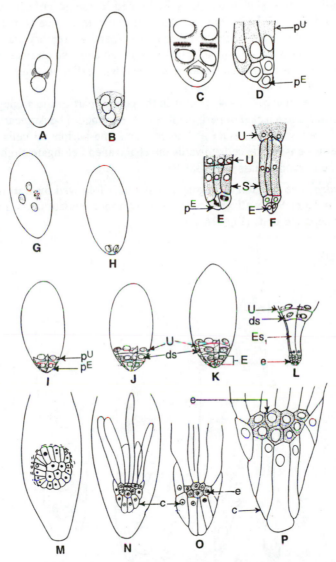

Fig. 8 A-P. **Proembryogeny. Diagrammatic representation of proembryo development in Taxodiaceae (A-F), Pinaceae (G-L) and Araucariaceae (M-P). A-D, G-I Wall formation takes place at the eight-nucleate stage to form p^U and p^E tiers. Internal division in both tiers form a three-tiered (E, F) or a four-tiered (J-L) proembryo. *ds* disfunctional suspensor, *E* embryonal tier, *e* embryo tier, *Esl* embryonal suspensor, *S* suspensor tier, *U* upper tier. M-O *Agathis australis*. M Longisection proembryo after wall formation. N, O proembryos show embryonal cells in the center, cap cells (C) and elongated suspensor. P *Araucaria brasiliensis*, lower part of mature embryo (A-F after Dogra 1966, G-L after Buchholz 1929, M-O adapted from Eames 1913, P adapted from Burlingame 1915).**

formation, and gives rise to an upper open tier (U) and a lower suspensor tier (S). The mature proembryo in conifers consists of three tiers of cells–U: S: E arrangement, termed the *basal plan* (Fig. 8A-F).

There are variations in the basal plan. In Pinaceae the four nuclei move to the base of the proembryo. They become arranged in one layer and divide. The resultant eight nuclei are arranged in two tiers of four each. After wall formation the usual p^U and p^E tiers are formed. These tiers of cells undergo further division and four tiers are formed (Fig. 8G-L). A four-tiered proembryo with four cells in each tier is characteristic of Pinaceae. the lowest two tiers comprise the E group, i.e. E and *esi* (embryonal suspensor) followed by S and U tiers. The cells of the S tier do not elongate but undergo abortive meristematic activity. Therefore' this tier has been designated as the *disfunctional suspensor (ds)*. Earlier, this tier was known as *rosette tier* and the resultant group of cells as *rosette embryos*.

In Araucariaceae the free nuclei (32–64) persist in the central part of the archegonium, even after wall formation. A central group and a jacket of peripheral cells are formed. The Araucarias thus lack the internal division phase, and do not show the usual p^U and p^E tiers. The peripheral cells towards the micropyle elongate and form the suspensor. The cells towards the chalazal end elongate slightly and form a cap. The middle cells form the embryonal mass (Fig. 8M-P).

3. *Sequoia* and *Ephedra*. The division of zygote in *Sequoia* is followed by wall formation. The resultant daughter nuclei have independent walls and function as two independent units. Further division leads to the formation of four separate units (Fig. 9A-C).

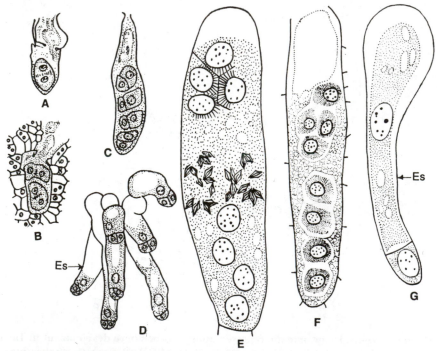

Fig. 9 A-G.   Proembryogeny. A-D *Sequoia*. A Division of zygote followed by wall formation. B Young proembryo, with surrounding gametophytic cells show second division. C Two-celled four units derived from one zygote. D Young embryos from two adjacent zygotes show embryonal cells divided and the embryonal suspensors (*Es*) elongated. E-G *Ephedra*. E Eight-nucleate proembryo, wall formation in upper four nuclei. F Later stage, eight embryonal units and radiating cytoplasm. G Two-celled embryonal unit with elongated embryonal suspensor (A-D after Buchholz 1939, E-G after Lehmann-Baerts 1967).

In *Ephedra* the zygote nucleus divides in situ, and the two nuclei move apart. Two more mitoses follow which result in eight nuclei. Each nucleus is surrounded by a densely-staining cytoplasmic sheath which radiates out strands. At the eight-nucleate stage, walls are laid down around each nucleus resulting in eight independent units (Fig. 9E, F).

Further development in both the taxa is more or less similar. Each unit gives rise to a tubular outgrowth. The nucleus divides, a transverse wall is laid down and gives rise to an embryonal cell and a suspensor cell (morphologically *es*). The latter elongates (Fig. 9D, G).

4. *Welwitschia* and *Gnetum*. The zygote, both in *Welwitschia* and *Gnetum*, have a prominent cell wall. In *Gnetum* the early behaviour of the zygote varies in different species. In *G. africanum* the zygote and daughter cells divide and produce a row of cells. Each of these cells elongates and forms a suspensor tube. In *G. gnemon* the zygote (usually in a pair) forms a branched tube and the nucleus migrates into one of the branches. The tubes—suspensor tubes—become septate, and grow irregularly in the endosperm (Fig. 10A-D). Finally, terminal cells become differentiated at the apex of the septate and branched tubes. By further divisions in different planes in the terminal cell, a globular embryo is formed.

In *G. ula* the zygote divides to form two cells. Both the cells elongate and develop into tubular structures. Further division followed by elongation of daughter cells produces a bunch of uninucleate suspensor tubes which grow in all directions in the female gametophyte. Each tube has a prominent cell wall and a nucleus located at the apex. The nucleus divides to form a cell which is at first lenticular, and later becomes pyriform (Fig. 10E, F). By further divisions the pyriform cell gives rise to a globular mass of cells (Fig. 10G, H) which eventually form the embryo.

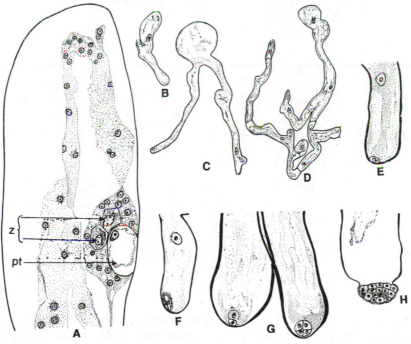

Fig. 10 A-H. **Proembryogeny. A-D *Gnetum gnemon*. A Upper portion female gametophyte shows pollen tube (*pt*) and adjacent pair of zygotes (Z). B-D Stages in the formation of elongated, branched and septate suspensor tubes. E-H *G. ula*. E, F Tip of suspensor tube shows a lenticular (E) or pyriform (F) embryonal cell. G, H Further development of embryonal cell. (A-D After Sanwal 1962, E-H after Vasil 1959).**

In *Welwitschia* the zygote is located in the fertilization bulb (Fig. 11E). The zygote elongates and divides to form a long suspensor cell and a small embryonal cell (Fig. 11F). The latter divides transversely and produces a succession of embryonal suspensors which elongate but remain within the prothallial tube. The young embryonal cell descends in the nucellus while still inside the prothallial tube. After it reaches the female gametophyte, further development occurs to form the embryo.

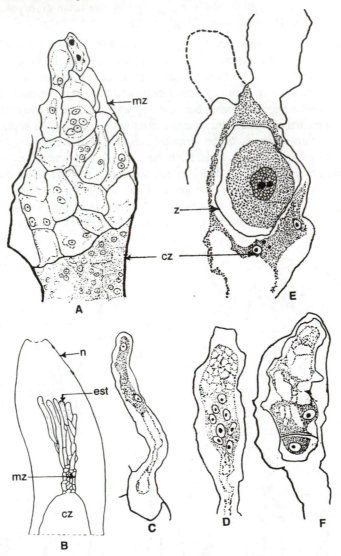

Fig. 11 A-F.   *Welwitschia mirabilis.* **A Longisection upper portion of female gametophyte shows micropylar (*mz*) and chalazal zone (*cz*). B Longisection upper portion of nucellus (*n*) with female gametophyte, shows chalazal zone, micropylar zone and endospermic tube (*est*). C Elongated endospermic tube. D Tip of endospermic tube shows swollen fertilization bulb. E Zygote (Z) in the fertilization bulb with binucleolate nucleus; two haploid nuclei (arrowed) in the residual cytoplasm of the bulb. F Two-celled embryo (after Martens and Waterkeyn 1974).**

Early Embryogeny. This includes the elongation and proliferation of suspensors, and the development of the embryonal mass (Fig. 12A).

Late Embryogeny. This includes further development of the embryonal mass and establishment of root and shoot meristems (Fig. 12B, C).

Mature Embryo. The number of cotyledons in a mature embryo varies. These may be one (*Ceratozamia*), two (most cycads, *Ginkgo* —Fig. 13A, *Cupressus, Podocarpus*, araucarias, taxads, *Ephedra, Welwitschia* and *Gnetum*), three (*Encephalartos*, occasionally *Ginkgo*, Taxodiaceae, taxads and *Gnetum*), four (*Sequoia*) or multiple (Pinaceae Fig. 13B). In *Cedrus* plumular leaves appear after the cotyledons have reached their full size (Fig. 13C).

In *Welwitschia* (Fig. 14D, E) and *Gnetum* (Fig. 14A-C) the embryo has a special structure—the feeder. This is a vascularized lateral protuberance produced by the hypocotyl and is even more prominent than the embryo.

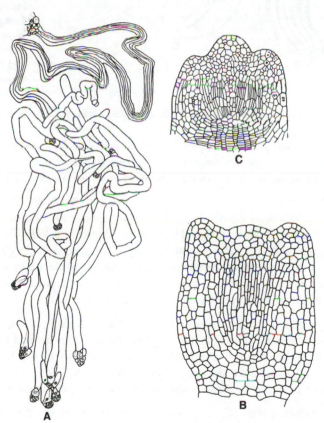

Fig. 12 A-C. Embryogeny. A *Sciadopitys*, elongation and proliferation of suspensor system. B *Podocarpus*, C *Pseudotsuga*. Longisection late embryo shows stele promeristem, procambium in the middle and cotyledonary primordia (A after Buchholz 1931, B after Brownlie 1953, C after Allen 1947).

7. Seed

A mature seed consists of a hard seed-coat, and an 'endosperm' rich in reserve food material. With maturity,

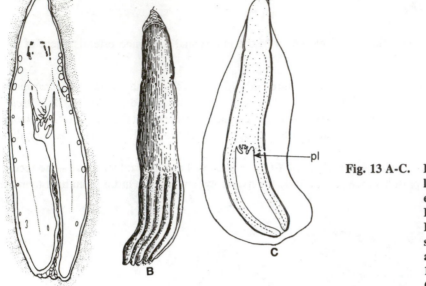

Fig. 13 A-C. Embryo. A *Ginkgo biloba*, longisection, dicotyledonous embryo. B, C *Cedrus*. B Polycotyledonous embryo. C Longisection mature embryo shows plumular leaves (*pl*) (A after Coulter and Chamberlain 1917, B, C after Roy Chowdhury 1961).

the seed becomes dehydrated, the nucellus becomes compressed and it persists as a thin cap-like 'perisperm' over the endosperm.

A mature seed in taxads and podocarps has a bright red aril. The fleshy seeds of cycads, *Ginkgo* and *Gnetum* are red or orange. In *Ginkgo* the outer (orange-coloured) fleshy portion of the mature seed is rich in butyric acid. It emits an odour like rancid butter. The seeds in *Pinus*, *Cedrus*, araucarias, and *Welwitschia* are winged.

Seed-Coat. The seed-coat develops mainly from the chalazal portion of the ovule (cycads, members of Pinaceae, *Cephalotaxus*) or from both chalaza and integument (cupressads, *Gnetum, Ephedra*). In podocarps the epimatium (ovuliferous scale) forms the outer portion of the seed-coat. In *Ephedra* and *Welwitschia*, the outermost layer and in *Gnetum* all the three layers covering the ovule are involved in the formation of the seed-coat. A mature seed-coat comprises three layers—an outer parenchymatous sarcotesta, a middle stony sclerotesta, and an innermost parenchymatous endotesta (Fig. 15).

Reserve Food. During seed maturation deposition of reserve food such as starch, fat and protein takes place in the endosperm. In general several layers of gametophytic cells around the embryo are devoid of any visible reserve products. The deposition of food reserves coincides with the development of the embryo in the seed. In *Ginkgo* and cycads the storage products are formed even in the embryoless seeds. The accumulated food reserve in the endosperm is utilized during seed germination.

Germination. In most gymnosperms (*Cycas*, conifers, *Ephedra*, *Gnetum*) the germination is epigeal. In taxads it is hypogeal.

8. Concluding Remarks

Gymnosperms comprise an important group of plants which have immense economic importance. The life history of several gymnosperms is incompletely and imperfectly understood. Many gaps exist in our present knowledge of seed development. By using techniques such as histo- and cytochemistry, and electron

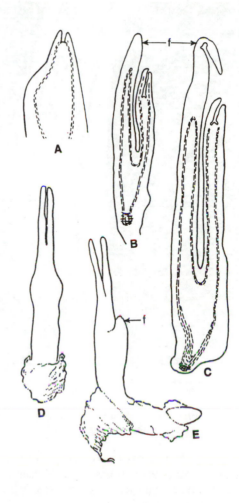

Fig. 14 A-E. A special feature, feeder (*f*) is present in the embryo of *Gnetum* (A-C) and *Welwitschia* (D, E) (A after Vasil 1959, B, C after Sanwal 1962, D, E after Martens and Waterkeyn 1964).

(TEM, SEM), fluorescence and interference microscopy, significant information will emerge which so far has not been reported because of exclusive use of the light microscope. Intensive investigation on any one aspect of the life cycle such as pollen biology, role of pollination drop and pollen tube, and wall/alveoli formation in the female gametophyte in different plants may yield important data.

Better understanding of seed development in this group is urgently called for in view of its economic importance as well as academic interest in this group, since gymnosperms are the last group of the archegoniates.

References

Allen GS (1947) Embryogeny and the development of the apical meristems of *Pseudotsuga*. II Late embryogeny. Am J Bot 34: 73–80

Bino RJ, Dafni A, Meeuse ADJ (1984) Entomophily in the dioecious gymnosperm *Ephedra aphylla* Forsk. (-*E. alte* C.A. Mey.) with some notes on *E. campylopoda* C.A. Mey. 1. Aspects of the entomophilous syndrome. Proc K Ned Akad Wet Ser C 87: 1–13

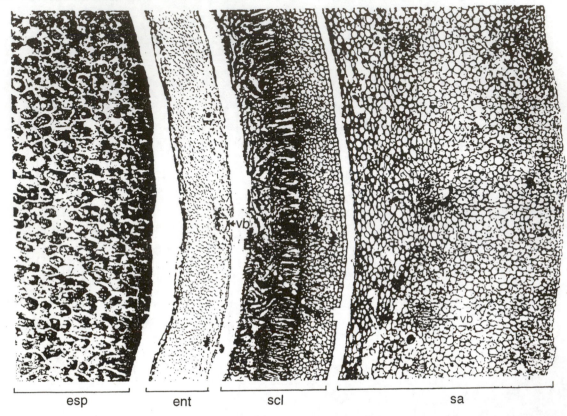

Fig. 15 *Gnetum gnemon*, transection of seed shows endosperm (*esp*), endotesta (*ent*), sclerotesta (*scl*), sarcotesta (*sa*) and vascular bundle (*vb*) (after Rodin and Kapil 1969).

Biswas C, Johri BM (1997) The Gymnosperms. Narosa Publishing House, New Delhi, 494 pp

Bracegirdle B, Miles PH (1973) An Atlas of Plant Structure vol 2. Heinemann Educational Books, London, 104 pp

Brownlie G (1953) Embryogeny of the New Zealand species of the genus *Podocarpus*, section *Eupodocarpus*. Phytomorphology 3: 295–306

Bryan GS (1952) The cellular proembryo of *Zamia* and its cap cells. Am J Bot 39: 433–443

Buchholz JT (1929) the embryogeny of conifers. Proc 4th Intl Congr Plant Sci, New York 1: 359–392

Buchholz JT (1931) The suspensor of *Sciadopitys*. Bot Gaz 92: 243–262

Buchholz JT (1939) The embryogeny of *Sequoia sempervirens* with a comparision of the sequoias. Am J Bot 26: 248–257

Burlingame LL (1915) The morphology of *Araucaria brasiliensis*. 3. Fertilization, the embryo and the seed. Bot Gaz 59: 1–39

Carmichael JS, Friedman WE (1995) Double fertilization in *Gnetum gnemon*: the relationship between the cell cycle and sexual reproduction. Plant Cell 7: 1975–1988

Chamberlain CJ (1935) Gymnosperms: Structure and Evolution. Chicago University Press, Chicago, pp 484

Choi JS, Friedman WE (1991) Development of the pollen tube of *Zamia furfuracea* (Zamiaceae) and its evolutionary implications. Am J Bot 78: 544–560

Coulter JM, Chamberlain CJ (1917) Morphology of gymnosperms, 2nd edn. Chicago University Press, Chicago, pp 466

Dogra PD (1966) Embryogeny of the Taxodiaceae. Phytomorphology 16: 125–141

Eames AJ (1913) the morphology of *Agathis australis*. Ann Bot (Lond) 27: 1–38

Favre-Duchartre M (1984) Homologies and phylogeny. *In:* Johri BM (ed) Embryology of Angiosperms. Springer, Berlin Heidelberg New York, pp 697–734

Friedman WE (1990a) Double fertilization in *Ephedra*, a non-flowering seed plant: its bearing on the origin of angiosperms. Science 247: 951–954

Friedman WE (1990b) Sexual reproduction in *Ephedra nevadensis* (Ephedraceae): further evidence of double fertilization in a non-flowering seed plant. Am J Bot 77: 1582–1598

Friedman WE (1991) Double fertilization in *Ephedra trifurca*, a non-flowering seed plant: the relationship between fertilization events and the cell cycle. Protoplasma 165: 106–120

Friedman WE (1992) Evidence of a pre-angiosperm origin of endosperm: implications for the evolution of flowering plants. Science 255: 336–339

Friedman WE (1994) The evolution of embryogeny in seed plants and the developmental origin and early history of endosperm. Am J Bot 81: 1468–1486

Friedman WE (1995) Organismal duplication, inclusive fitness theory and altruism: understanding the evolution of endosperm and the angiosperm reproductive syndrome. Proc Natl Acad Sci USA 92: 3913–3917

Gangulee HC, Kar AK (1982) College Botany vol 2. New Central Book Agency, Calcutta, pp 1184

Jeffrey EC, Torrey RE (1916) *Ginkgo* and the microsporangial mechanisms of the seed plants. Bot Gaz 62: 281–292

Johri BM (1936) Contribution to the life-history of *Cedrus deodara* Lond. 1. The development of the pollen grains. Proc Indian Acad Sci B 3: 246–257

Johri BM (1992) Haustorial role of pollen tubes. Ann Bot (Lond) 70: 471–475

Kato M, Inoue T, Nagamitsu T (1995) Pollination biology of *Gnetum* (Gnetaceae) in a lowland mixed dipterocarp forest in Sarawak. Am J Bot 82: 862–868

Khan R (1943) Contribution to the morphology of *Ephedra foliata* Boiss 2. Fertilization and embryogeny. Proc Natl Acad Sci India (Allahabad) 13: 357–375

Konar RN, Oberoi YP (1969) Studies on the morphology and embryology of *Podocarpus gracilior* Pilger. Beitr Biol Pflanz 45: 329–376

Kubitzki K (1990) Gnetaceae. *In:* Kranmer KV, Green PS (eds) The families and genera of vascular plants 1. Springer, Berlin Heidelberg New York, pp 378–391

Lehmann-Baerts M (1967) Étude sur les Gnétales-12. Ovule, gamétophyte femelle et embryogenése chez *Ephedra distachya* L. Celluie 67: 53–87

Loze JC (1965) Étude de l'ontogenése de appareil reproducteur femelle de l'if *Taxus baccata*. Rev Cytol Biol Vég 28: 211–256

Lyon HL (1904) The embryogeny of *Ginkgo*. Minn Bot Stud 3: 275–290

Maheshwari P, Biswas C (1970) *Cedrus*. Bot Monograph 5: Council Sci Industr Res (CSIR), New Delhi, pp 115

Maheshwari P, Konar RN (1971) *Pinus*. Bot Monograph 7. Council Sci Industr Res (CSIR), New Delhi, pp 130

Maheshwari P, Vasil V (1961) *Gnetum*. Bot Monograph 1: Council Sci Industr Res (CSIR), New Delhi, pp 142

Martens P (1961) Étude sur les Gnétales 5. Structure et ontogenese du cone et de la fleur males de *Welwitschia mirabilis*. Cellule 62: 7–87

Martens P (1963) Étude sur les Gnétales 6. Recherches sur *Welwitschia mirabilis* 3. L'ovule et le sac embryonnaire. Cellule 63: 309–329

Martens P, Waterkeyn L (1964) Recherches sur *Welwitschia mirabilis* 4. Germination et plantules. Structure, fonctionnement et productions du meristeme caulinaire apical. (Études sur les Gnétales 7). Cellule 65: 5–64

Martens P, Waterkeyn L (1974) Étude sur les Gnétales 13. Recherches sur *Welwitschia mirabilis* 5. Evolution ovulaire et embryogenese. Cellule 70: 163–258

McWilliam JR (1958) The role of the micropyle in the pollination of *Pinus*. Bot Gaz 120: 109–117

Mehra PN (1938) The germination of pollen grains in artificial cultures in *Ephedra foliata* Boiss and *Ephedra gerardiana* Wall. Proc Indian Acad Sci B 8: 218–230

Moussel B (1978) Double fertilization in the genus *Ephedra*. Phytomorphology 28: 336–345

Oliver FW 91982) The natural history of plants, vol 2. A J Reprints Agency, New Delhi

Pant DD (1973) *Cycas* and Cycadales. Central Book Depot, Allahabad (India), pp 255

Rodin RJ, Kapil RN (1969) Comparative anatomy of the seed coats of *Gnetum* and their probable evolution. Am J Bot 56: 420–431

Roy Chowdhury C (1961) The morphology and embryology of *Cedrus deodara* Loud. Phytomorphology 11: 283–304

Sanwal M (1962) Morphology and embryology of *Gnetum gnemon* L. Phytomorphology 12: 243–264

Singh H (1978) Embryology of Gymnosperms. Bornträger Berlin, pp 302

Singh H, Chatterjee J (1963) A contribution to the life history of *Cryptomeria japonica* D. Don. Phytomorphology 13: 429–445

Singh H, Maheshwari (1962) A contribution to the embryology of *Ephedra gerardiana* Wall. Phytomorphology 12: 361–372

Tiagi YD (1966) A contribution to the morphology and vascular anatomy of *Ephedra foliata* Boiss. Proc Natl Acad Sci 36: 417–436

Troup RS (1921) The Silviculture of Indian Trees. vol 3. Oxford University Press, Oxford (UK)

Vasil V (1959) Morphology and embryology of *Gnetum ula* Brongn. Phytomorphology 9: 167–215

Ziegler H (1959) Über die Zusammensetzung des "Bestaubungstropfens" und den Mechanismus seiner Sekretion. Planta 52: 587–599

11

Reproductive Biology of Angiosperms

B.M. Johri, P.S. Srivastava and Nandita Singh

1. Introduction

In flowering plants reproduction may be sexual, parthenogenetic or by adventive embryony (see Johri 1984, Johri et al. 1992).

In the sexual process the pollen grains produce male gametes, and the egg and secondary (fused polars) nucleus are formed in the embryo sac. One male gamete fuses with the egg (syngamy), and the other with the secondary nucleus or the two polars, resulting in triple fusion. These processes (the syngamy and triple fusion) are referred to as 'double fertilization', a unique characteristic of the angiosperms. The zygote (fertilized egg) usually develops into a bipolar structure, the embryo, that is the progenitor of the next generation. The primary endosperm nucleus, usually a product of triple fusion (fertilized secondary nucleus), produces the endosperm (a nutritive tissue).

In parthenogenesis, in a reduced (n) embryo sac, the haploid egg develops directly (without fertilization) into a haploid embryo (haploid parthenogenesis), e.g. *Solanum nigrum*. In 'haploid apogamy', the synergid develops into an embryo (e.g. *Solanum forvum*). In 'generative apospory', an archesporial or sporogenous cell, and in 'somatic apospory' a nucellar cell develops into an unreduced (2n) embryo sac, e.g. *Hypericum perforatum*. The embryo may arise from the egg (diploid parthenogenesis) or some other cell of the embryo sac (diploid apogamy), e.g. *Potentilla argentea*. Sometimes, the embryo (2n) develops from a nucellar or integumentary cell (adventive embryony), e.g. *Citrus microcarpa*.

2. Sexual Reproduction

The stamens (male) and carpels (female) may develop in the same flower (bisexual) as in *Hibiscus rosa-sinensis,* or in separate flowers (unisexual) as in *Ricinus communis*.

3. Microsporangium

The mature anther usually has four pollen sacs, and the half-anther has two pollen sacs as in Malvaceae. Some anthers are polysporangiate, a condition brought about by partitions laid down in pollen sacs as in Viscaceae. At dehiscence, the adjoining pollen sacs become contiguous.

3.1 Anther Wall

The anther wall comprises the epidermis followed by the endothecium, middle layer/s and tapetum (Fig. 1A-G). The epidermal cells may persist in a healthy state, or collapse and appear to be absent as in plants of dry habitats. The endothecial cells, in the majority of angiosperms, usually develop fibrous thickenings (Fig. 1G) which are absent in the region of dehiscence (Fig. 1A). In cleistogamous flowers as in *Colocasia*,

Fig. 1 A-M. Microsporangium, microsporogenesis and male gametophyte. A Transection of a four-lobed anther showing fusion of locules of each side, degenerated tapetum, and endothecium with fibrous thickening. B Archesporium surrounded by developing tapetum, middle layers, undifferentiated endothecium and epidermis. C Pollen mother cells initiating meiosis and prominent tapetal cells; note the cytoplasmic channels; middle layers are getting curshed. D Meiosis in pollen mother cells surrounded by callose and disintegrating middle layers; note the binucleate tapetal cells. E Pollen tetrads with callose wall. F Wall-less microspores, middle layers are totally crushed and tapetal cytoplasm invades the anther locule. G Pollen with vegetative and spindle-shaped generative cell; note the degenerated tapetum and middle layer; shows fibrous thickening of endothecium. H-M Development of male gametophyte. H Uninucleate pollen. I Appearance of vacuole and migration of nucleus to one side. J Division of nucleus; vegetative and generative cell. K, L Pollen with male gametes and vegetative cell. M Germinated pollen grain; male gametes in pollen tube (*en* endothecium, *ta* tapetum) (A; H-M, adapted from P Maheshwari 1950; B-G, Shivanna et al. 1996)

and members of Hydrocharitaceae as in *Vallisneria*, a fibrous endothecium is said to be absent. The middle layers mostly collapse due to expansion of the endothecium and tapetum. The tapetum is of two types. The glandular tapetum breaksdown, while the amoeboid tapetum forms a periplasmoduim. The tapetum nourishes the pollen grains. The tapetum also secretes the sporopollenin, a constituent of the wall of pollen grains. The pollen sacs of the mature anther contain pollen grains which may be two- or three-celled: vegetative cell and generative cell (Fig. 1G, H-J) or two male gametes dervived from generative cell, and the vegetative cell (Fig. 1K-M). The generative cell may sometimes divide on the stigma (after pollination) as in *Holoptelea*, or in the pollen tube in the style as in *Gossypium*.

3.2   Tapetum

The tapetum, the innermost layer of the wall surrounding the sporogenous tissue, plays an important role in pollen development. In most angiosperms the tapetum is of dual origin (Periasamy and Swamy 1966). Towards the outer side the tapetum is derived from the primary parietal layer, whereas on the inner side it develops from the cells of the connective tissue. Periasamy and Kandasamy (1981) report trimorphic tapetum in *Annona squamosa*.

In the beginning the tapetal cells are uninucleate with a marked increase in DNA content. However, due to irregular mitotic divisions they become two- to four-nucleate or at times the number goes up to 16. Very often, due to fusion, the nucleus becomes polyploid. In some marine angiosperms, the nuclear divisions are followed by periclinal divisions in the tapetal cells resulting in a two- to four-layered condition.

In an amoeboid tapetum the inner tangential and radial walls breakdown, the protoplasts enlarge, and their fusion results in a plasmodium around the microspore mother cells/dyads/tetrads/microspores. It plays an important role in subsequent development of microspores. The breakdown of the tapetal cell wall is probably due to the hydrolytic enzymes released by dictyosomes in the tapetal cells. After meiosis II the individual microspores remain surrounded by the tapetal cytoplasm. The enzymes responsible for the degradation of the callose wall are also considered to be derived from the tapetal periplasmodium.

The deposition of callose (β-1-3-glucan) around the pollen mother cells (pmc) initiates at the onset of meiosis and is contributed by dictyosomes. Prior to meiosis the pmc are interconnected through plasmodesmatal connections which are later replaced by cytoplasmic channels. The callose deposition usually starts at the corners of the cells between the plasma membrane and the original wall. By the end of the meiotic prophase the callose walls close up, cutting off the cytoplasmic channels. Subsequently, additional callose walls are formed between the daughter cells of the pmc. In *Pergularia daemia* which has pollinia, there is no callose deposition.

Pacini and Juniper (1983) studied the ultrastructure of the development of the amoeboid tapetum in *Arum italicum*. At the premeiotic phase of microspore mother cells, the tapetal cells are connected to each other by plasmodesmata. During meiosis the plasmodesmatal connections widen and are transformed into cytoplasmic channels. The radial walls of tapetal cells disappear and the tapetal cytoplasm starts to intrude between the meiocytes/microspores to lay the foundation for initial development of the pollen wall.

In secretory or glandular types of tapetum, characteristic of most dicotyledons, the cells remain intact in their original position throughout the microspore development and provide nutrition. Vigorous RNA and protein synthesis occurs in tapetal cells during the premeiotic period which continues during the meiotic process. By the tetrad stage, the cells of the parietal tapetum reach peak metabolic activity with well-developed mitochondria, RER and conspicuous plastids.

The tapetal cells develop pro-orbiscules. The mitochondria produce sporopollenin which is resistant to chemical deposition. The sporopollenin is deposited on the pro-orbiscules. It has been suggested that orbiscules have a role in pollen exine foundation (Banerjee and Barghoorn 1971) and in dispersal (Heslop-Harrison 1968).

The tapetum is physiologically the most active layer of the anther wall. The tapetum synthesizes certain substances which are released into the anther locule during maturation of pollen grains. These include fibro-granular proteins and carotenoid-containing lipid droplets. The fibro-granular proteins enter the exine of the pollen, and are stored in the region of the bacula. They are extra-cellular and capable of recognizing stigma compatibility. The carotenoid-containing lipid granules spread over the exine and help in pollination and are described as "pollen kitt".

4.   Pollen Wall

Pollen wall formation occurs in two phases. In the first phase, the wall material is contributed to by the cytoplasm of the spore alone (gametophytic) which occurs during the tetrad stage. In the second phase, which occurs after the release of the microspores, the wall materials are contributed by the tapetal cells (sporophytic) in addition to the spore cytoplasm (Fig. 2A-F).

The wall of a pollen grain is made up of two layers—the inner pectocellulosic layer, intine, which is degraded during acetolysis, and the outer layer, exine, which is acetolysis-resistant and also resistant to physical and biological degradation (Fig. 2E, F). Kress and Stone (1982), using cytochemical studies, distinguished two layers in the intine. An outer pectic polysaccharide layer and an inner cellulosic layer, termed the exintine and the endintine, respectively.

The exine is composed of sporopollenin, a highly resistant material derived from carotenoids by oxidative polymerization (Broods and Shaw 1971). The exine is differentiated into two layers, the outer sculptured layer, the sexine, and the inner non-sculptured layer, the nexine (Fig. 2E, F).

The sculptured part of the exine is made up of radially oriented rod-like baculae (columella) which may either remain open (pilate grain; Fig. 2F) or may be covered by a roof (Fig. 2E) or tectum (tectate grain). The baculae (in pilate grain) may be enlarged above and stand free, or may be fused together to give a raised wall often arranged in a reticulate pattern. In a tectate grain, the voids between baculae open to the outside through perforations termed micropores. The exine covers the entire pollen grain surface except

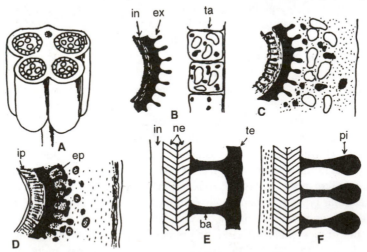

Fig. 2 A-F.   Pollen wall formation. A Transection of four-lobed anther (diagramatic). B Part of pollen wall. C Part of tapetal cytoplasm and pollen wall. D Incorporation of proteins and enzymes in pollen wall that leads to recognition-acceptance-rejection during pollen germination and subsequent growth of pollen tube. E, F Part of pollen wall. E Tectate grain. F Pilate grain (*ba* bacula, *ex* exine, *ep* exine proteins, *in* intine, *ip* intine proteins, *ne* nexine, *pi* pilate, *te* tectate) (adapted from Shivanna and Johri 1985).

germinal apertures. According to the different ornamentation of the exine, it is variously described as foveolate (pitted), scabrate (very fine projections), gemurate (sessile pilae), punctate (minute perforations), etc. The intine proteins are the products of pollen grain cytoplasm (gametophytic) and the exine proteins, on the other hand, originate in cells of the tapetum (sporophytic).

The pollen grains are characterized on the basis of shape, size, thickness and ornamentation of exine, the most important is the form of aperture. There are two basic forms of apertures—porate and colpate. The latter have elongated furrows along the axis. The compound forms are called corporate or pororate whereas the outer region bears a furrow with a pore in the centre. The number of apertures vary from zero to eight and may be distal in position (outwards in a tetrad–anatreme) or distributed on the equatorial plane at equal distance–zonotreme.

At maturity, the pollen grains may remain attached in tetrads as in Orchidaceae (Fig. 3A), *Drymis* and *Drosera*, or dyads (Fig. 3B, Podostemaceae), or all the pollen grains remain attached in 'pollinia' (Fig. 3 Ca, Cb, D; *Acacia, Calotropis, Pergularia*) or massula as in *Peristylis* (Fig. 3E).

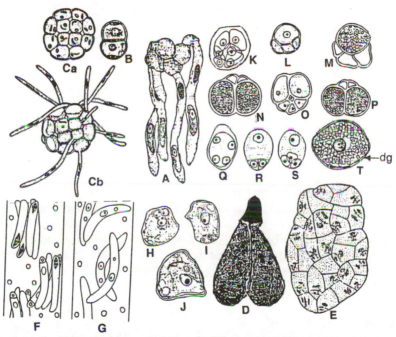

Fig. 3 A-T. **Taxa showing unusual features of pollen. A** *Cymbidium bicolor*, germinated pollen tetrad. **B** *Dicraea stylosa*, double pollen grain with generative and vegetative cell, **Ca** *Acacia dealbata*, pollinium with two-celled pollen grains, **Cb** Germinated pollen grains. **D** *Pergularia daemia*, pair of pollinia. **E** *Peristylis spiralis*, massula of microspores showing mitotic synchrony. **F, G** *Halodule*, highly-elongated pollen grains. **F** Filiform pollen showing first mitotic division. **G** Two-celled pollen. **H-J** *Carex wallichiana*, pollen. **H** One functional and three non-functional nuclei. **I** Degenerated three nuclei and prominent functional nucleus. **J** Two-celled grain with vegetative and generative cell. **K** *Richea sprengelioides*, polyad. **L-O** *Cyathodea dealbata*, development of functional pollen. **L** Astroloma type of pollen. **M** Tetrad with single functional pollen grain. **N** Tetrad with two functional pollen grains. **O** Polyad with two functional pollen grains. **P** *Pentachondra involucrata*, Pentachondra type with two functional pollen grains. **Q-T** *Astroloma humifusum*, Sytphelia type of pollen development. **Q** Pollen tetrad with four microspore nuclei. **R** One nucleus in the upper part and three in the lower. **S** Three nuclei separated from the functional nucleus. **T** Monad with two-celled pollen and remnants of three degenerated (*dg*) microspores. (Adapted from Johri et al. 1992)

In aquatic plants the pollen grains are filiform (Fig. 3F,G), as in *Zostera, Halodule*, etc. The microspores are arranged linearly in a tetrad, and on separation, the microspores elongate and become filamentous, e.g. Hydrocharitaceae. Pollen grains are somewhat triangular in Cyperaceae (Fig. 3H-J) with three microspore nuclei segregated at the broad base. These nuclei may remain free or walls may appear between them, as in *Styphelia*. In Epacridaceae the pollen grains are round and three nuclei are enclosed in the wall of the pollen grain while the fourth occupies a central position (Fig. 3 K-T). Rarely, pollen grains simulate the development of the embryo sac, as in *Hyacinthus orientalis* (Stow 1930, 1934) and *Heuchera micrantha* (Vijayaraghavan and Ratnaparkhi 1977).

5.   Male Gametophyte

The generative cell gets separated from the wall of the pollen grain, and comes to lie in the cytoplasm of the vegetative cell. The generative cell undergoes mitosis to form two sperms (the male gametes) either before, or after the dehiscence of anthers. The vegetative cell continues to grow, accumulate reserve food in the form of starch and fat. It becomes rich in proteins, and its nucleus may undergo endoduplication to become highly polyploid. The pollen grains may be liberated at the three-celled stage. However, if the pollen grains are dispersed at the two-celled stage, the division of the generative cell may occur after the deposition of pollen on the stigma, or in the pollen tube while traversing the style or, rarely, after reaching the embryo sac.

6.   Ovary

The ovary may be uni- or multi-locular, and bears ovules (Fig. 4A) on free-central, axile or parietal placenta. In some taxa, the ovules occur on the inner wall of the ovary as in *Butomopsis* (Johri 1936).

6.1   Ovule

In relation to funicle and micropyle, the mature ovule is considered: (1) orthotropous—the micropyle faces upward and the funicle is in the same line at the base (Fig. 4B); (2) anatropous—the condition is the reverse of 1 (Fig. 4C); (3) campylotropous—the curvature is much less than in 2 (Fig. 4D); (4) hemianatropous— (Fig. 4E) the condition is midway between 1 and 2 but the micropyle is at right angles to the funicle; (5) amphitropous—the ovule and the embryo sac become curved like a horse-shoe (Fig. 4F) and in (6) circinotropous—the ovule first becomes anatropous, and then the curvature continues till the micropyle faces upward (Fig. 4G-I).

6.2   Integuments

The ovule has one (uni-) or two (bitegmic) integuments (Fig. 4A). The nucellus may be massive (crassinucellate) or scanty (tenuinucellate). The micropyle is formed either by the outer or inner, or both integuments, and sometimes, due to unequal growth of the two integuments, the endostome (micropyle formed by inner integument) and exostome (outer integument) are not in alignment and the micropyle is zig-zag. The apical nucellar cell may divide and form a beak which extends beyond the micropyle. In some dicots, the nucellus collapses and the inner epidermis of the integument differentiates into a glandular tissue, the endothelium, which is supposed to transport nutrients to the embryo sac. An endothelium is uncommon in monocots.

7.   Megasporangium

The female archesporium/sporogenous cells differentiate in the hypodermal layer of nucellus. A parietal cell is cut off towards the micropyle and the inner cell functions as the megaspore mother cell. If the parietal cell divides repeatedly, the parienal tissue pushes down the megaspore mother cell. This may be

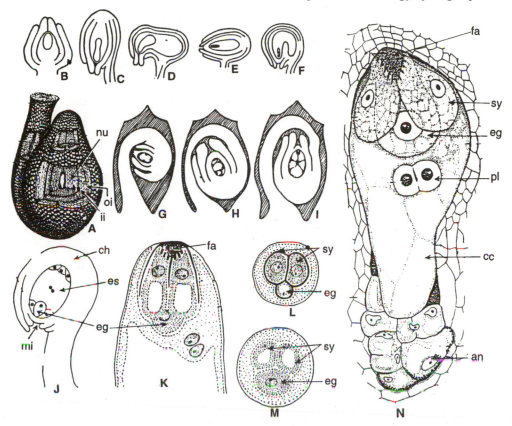

Fig. 4 A-N. **Ovule and female gametophyte. A Bitegmic anatropous ovule with segments exposed to show internal structure. B-F Types of ovule. B Orthotropous or atropous. C Anatropous. D Campylotropous. E Hemi-anatropous. F Amphitropous. G-I Development of circinotropous ovule, striated area represents ovarian cavity. J Bitegmic anatropous ovule with eight-nucleate female gametophyte. K Upper part of embryo sac. L, M Transections through region of egg apparatus. N Embryo sac with egg apparatus, polar nuclei and eight antipodal cells, note filiform apparatus in the synergids. (*an* antipodal cells, *cc* central cell, *ch* chalaza, *eg* egg, *es* embryo sac, *fa* filiform apparatus, *ii* inner integument, *mi* micropyle, *nu* nucellus, *oi* outer integument, *pl* polar nuclei, *sy* synergid). (A adapted from Bouman 1984; B-F; G-I adapted from P Maheshwari 1950; J-M adapted from Jensen 1965; N adapted from Diboll and Larson 1966).**

further accentuated by periclinal divisions in nucellar epidermal cells. Meiosis I in the megaspore mother cell may or may not be followed by a wall. The dyad cells undergo meiosis II and, again, a wall may or may not be formed in each dyad cell. Thus, the terad comprises four megaspore cells, or a dyad of two megaspores, or a coenomegaspore with four nuclei.

8. Female Gametophyte—Embryo Sac Types

The different types of embryo sacs are depicted in Figure 5.

8.1 Monosporic

In the majority (ca. 80%) of angiosperms, the basal megaspore of the tetrad undergoes three mitotic

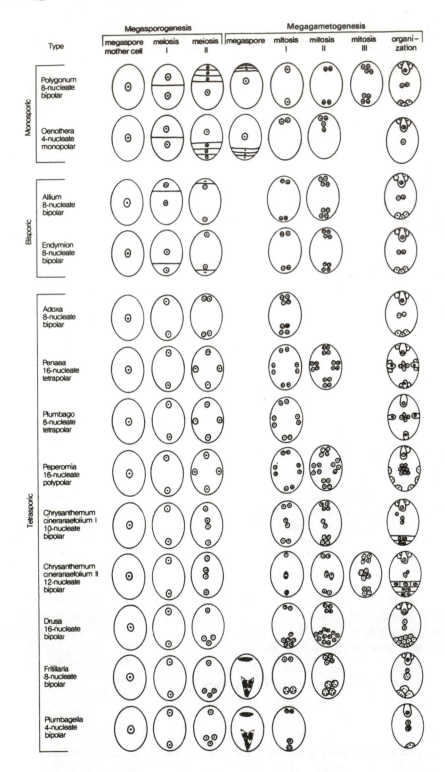

Fig. 5 Development of various types of embryo sacs (after Johri 1984; adapted from Johri et al. 1992).

divisions and the resulting nuclei differentiate into a three-celled egg apparatus at the micropylar pole, two polar nuclei in the central cell, and three antipodal cells at the chalazal pole—this is the eight-nucleate Polygonum type of embryo sac. If the megaspore undergoes only two mitotic divisions, the embryo sac is four-nucleate with a three-celled egg apparatus and one polar nucleus—the Oenothera type.

8.2 Bisporic

When meiosis II is not followed by a wall, the lower dyad cell produces an eight-nucleate embryo sac—the Allium type. Sometimes, the upper dyad cell develops into an eight-nucleate embryo sac—the Endymion type.

8.3 Tetrasporic

In this case a wall is not formed either after meiosis I or II, and the arrangement of four megaspore nuclei determines the type of tetrasporic embryo sac.

A 2 (micropylar) + 2 (chalazal pole) arrangement of megaspore nuclei followed by one mitotic division produces eight nuclei which organize like a Polygonum type of embryo sac, and is called the Adoxa type. A 1 + 1 + 1 + 1 (two polar and two lateral) disposition of megaspore nuclei undergo one mitotic division and, of the eight nuclei, four nuclei migrate to the centre and function as polars, of the remaining four one organizes as egg cell, one an antipodal cell, and one each as lateral cells—Plumbago type. If the four megaspore nuclei undergo two mitotic divisions, the 16 nuclei organize into four groups, the upper three as egg apparatus, the chalazal three as antipodals, and two laterals of three cells each, while four nuclei (one from each group) function as polars—Penaea type. The 16 nuclei may also organize as a two-celled egg apparatus, six antipodal cells, and eight polar nuclei (variations are common)—Peperomia type. A1 + 3 disposition of megaspore nuclei undergo two mitotic divisions and the 16 nuclei organize as a three-celled egg apparatus, two polars, and 11 antipodal cells—Drusa type. 1 + 3 megaspore nuclei may undergo only two mitotic divisions, and during the first division the three chalazal mitotic spindles fuse. The next mitotic division generates four haploid micropylar and four triploid chalazal nuclei. Three haploid nuclei form the egg apparatus, one haploid and one triploid nucleus function as polars, and three triploid nuclei as antipodals—Fritillaria type.

There is also a 1 + 2 + 1 arrangement of megaspore nuclei; while the micropylar and chalazal nuclei divide twice, the two central nuclei do not divide. The embryo sac comprises the three-celled egg apparatus, four antipodal cells, and three polar nuclei (one micropylar and two central). If the two central nuclei fuse and, along with micropylar and chalazal nuclei, undergo two mitotic divisions, the embryo sac comprises three-celled egg apparatus, one haploid and one diploid nucleus function as polars, and the remaining four haploids and three diploids organize as antipodals—Chrysanthemum cinerariaefolium type.

The embryo sac lacks synergids in Plumbago and Plumbagella types, there is only one synergid in the Peperomia type, and two in all the other types.

Sometimes, more than one type of embryo sac occurs in the same taxon. Mono-, bi- and tetra-sporic types occur in *Erigeron* spp. (Harling 1951). In *Delosperma cooperi*, Kapil and Prakash (1966) reported five types of embryo sacs—Polygonum (mono-), Endymion (bi-) and Drusa, Penaea and Adoxa (all three tetrasporic). Many other variations are also known (see Johri et al. 1992).

8.4 Structure of Embryo Sac

The mature embryo sac usually shows three-celled egg apparatus, three antipodal nuclei (Fig. 4J-N) or cells, and two polar nuclei or fused polar nuclei as a secondary nucleus (Masand and Kapil 1966; Johri and Bhatnagar 1973). Synergids are, however, absent in Plumbago and Plumbagella types, and antipodals in the Oenothera type. The ploidy level and number of antipodals also vary. The cells of egg apparatus and the antipodals are usually uninucleate and haploid, whereas the central cell is binucleate or diploid. In the majority of angiosperms, the embryo sac remains confined to the nucellus during its growth and maturation.

The egg apparatus with its egg cell and two synergids remains within the limits of the embryo sac wall except in Asteraceae, Ericaceae and Stylidiaceae where the synergids project into the micropyle. The egg cell shows a distinct polarity with its nucleus embedded in the dense cytoplasm at the lower end, while the upper end is vacuolate (Fig. 4K-N). In the absence of synergids, the egg cell takes over the function of synergids as in *Plumbago*. The wall of the embryo sac in maize is multilayered and pectocellulosic (Chebotaru 1970) and in cotton it is rich in pectic substances. According to Diboll and Larson (1966), the innermost layer of the female gametophyte boundary represents the wall of the functional magaspore and outer layers are the remains of crushed nucellar cells. The wall of the gametophyte in *Stipa* consists of three parts. The innermost is gametophytic, whereas the middle and outer walls represent the degenerated megaspore and nucellar cells, respectively. The walls separating the cells within the embryo sac are traversed by plasmodesmata.

The embryo sac develops micropylar and chalazal extensions (Fig. 6A-C), as in Acanthaceae, Santalaceae, Stylidiaceae and Opilliaceae. In *Exocarpus* (Fig. 6D, E), finger- like outgrowths are produced in the apical and middle region of embryo sac. The apex of the embryo sac extends in the stylar canal, broadens laterally and develops tubular processes which grow downward from the micropylar region of the embryo sac, along the side of the placenta (Johri et al. 1992).

8.5 Synergids

The synergids are elongated cells with the tip prolonged into a 'beak'. The cell wall is thickest in the micropylar region, becomes membraneous adjacent to the polar nuclei, and finally disappears. They show a characteristic structure, the filiform apparatus at the tip. The synergids are strongly polarized with the nucleus occupying the upper narrow region. The majority of organelles are situated between the nucleus and the filiform apparatus in the micropylar region of the cell (Fig. 4N), where mitochondria, endoplasmic reticulum and dictyosomes are especially concentrated.

When present, one or both of the pollen tubes enter through one of the persistent synergids and bursts. When the synergid is absent, the pollen tube directly enters the embryo sac through the egg cell, e.g. *Plumbago*. The egg cell in such embryo sacs has filiform apparatus, a feature otherwise of the synergids. The synergids are considered to provide nutrition and a chemical attractant to the pollen tube (see Vijayaraghavan and Bhat 1983). If this is true, these functions may be presumed to be performed by the egg cell in the absence of one or both synergids. The filiform apparatus and transfer cell-like thickenings in the synergids do reflect its glandular activity.

The synergids are ephemeral but play an important role during fertilization by secreting chemotropic substances that guide the growth of the pollen tube. The synergids are also haustorial in Compositae, Crassulaceae, Santalaceae and Gramineae (Johri et al. 1992).

8.6 Antipodals

The antipodals exhibit maximal variation in number, size, ploidy level, and may be ephemeral or persistent. Normally, there are three antipodal cells but in many plants they undergo secondary multiplication and may become polyploid due to endomitosis, or may become polytenic, and become multinucleate. There are, however, 11 antipodals in Drusa. The number also increases if chalazal nucellar nuclei migrate into the embryo sac as in *Pandanus*, or due to their mitotic divisions as in Gramineae. If the walls are not laid down after divisions, the antipodal cells become multinucleate, e.g. Dipsacaceae (Johri et al. 1992). The antipodal cell may also become polyploid due to endomitosis. Rarely, the antipodal cells enlarge as in *Aconitum napellus*. The tip reaches as far as the egg apparatus (Maheshwari 1950). Even though the multicellular

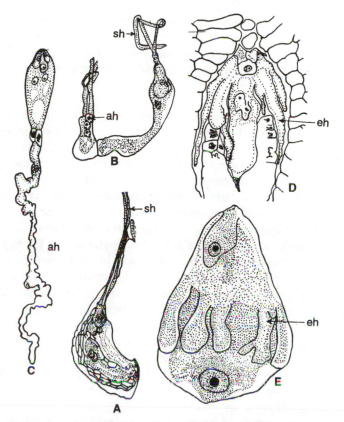

Fig. 6 A-E. **Embryo sac haustoria. A-B *Quinchamalium chilense*. A Embryo sac with synergid haustoria. The haustoria reach up to the base of the style, and the antipodal chamber extends into the funiculus. B Extensive synergid and antipodal haustoria in the mature embryo sac (*WM*). C *Galium mollugo*, mature embryo sac shows highly elongated haustorium of basal antipodal cell. D *Exocarpus menziesii*, downward growth of finger-like haustoria from organized embryo sac. E *E. sparteus*, embryo sac, haustoria develop from middle region of embryo sac. *ah* antipodal haustorium, (*eh* embryo sac haustorium, *sh* synergid haustorium). (A, B after Agarwal 1962, C after Fagerlind 1937, D after Fagerlind 1959, E after Ram 1959; adapted from Johri et al. 1992).**

antipodals may simulate early stages in embryogeny (*Ulmus*), there is no unequivocal report of embryo formation from an antipodal cell (Johri et al. 1992). Some of these cells show a papillate wall projecting into the cytoplasm, as has been noticed in poppy (Jensen 1973) and rice. The antipodals are rich in ascorbic acid, oxidases, starch, lipids and proteins. However, they show a low concentration of RNA and polysaccharides.

The antipodal cells show haustorial behavior in many plants. A unique condition, for example, has been observed in *Quinchamalium chilense* (Johri and Agarwal 1965) where the three nuclei at the chalazal end of the embryo sac are contained in an antipodal chamber which develops into a prominent branched haustorium (Fig. 6A-B).

8.7 Central Cell

The central cell is separated from the egg cell and synergids by the plasma membrane, and from antipodals by a thin wall traversed by plasmodesmata. The outer wall of central cell may show transfer cell-like wall ingrowths concerned with absorption and transport of nutrients. Normally, there are two polar nuclei in the central cell, except in Oenothera, Penaea, Plumbago, Peperomia, and Chrysanthemum cinerareaefolium types. When there are more than two polar nuclei, all of them come together and fuse simultaneously to form a secondary nucleus (polyploid). Sometimes, the two polar nuclei remain distinct until triple fusion.

8.8 Female Gamete (Egg)

The female gamete, the egg, is arranged in a triangular fashion with the adjacent synergids forming the egg apparatus. The egg shares common walls with the two synergids, and the central cell. The wall of the egg is thickest near the micropyle, and becomes thinner towards the chalazal side (see Fig. 4N); the wall is absent at the chalazal end in cotton, maize and *Torenia*, but present all over the egg cell in *Epidendrum*. The egg cell wall is traversed by plasmodesmata on the wall adjoining the two synergids and the central cell. The egg cell becomes highly polarized early in its development. It is marked by the presence of plastids; starch content is variable and is abundant around the nucleus. The organelles become scarce at maturity, indicating poor physiological activity. In *Plumbago* (synergids are absent) finger-like wall projections resembling filiform apparatus of synergids arise at the micropylar end of the egg cell (Cass 1972).

In Loranthaceae and Santalaceae there are no ovules, and in the former the embryo sac grows up to various lengths of style and stigma (Fig. 7A-E). Scanty information is available on the physiological aspects of megasporogenesis. However, sporadic reports indicate the special nature of the megaspore mother cell. Ultrastructural studies show phasmodesmatal connections between the megaspore mother cell and surrounding cells in *Dendrobium* and *Zea*. In addition, deposition of callose in the wall differentiates the functional megaspore mother cell. In *Fuchsia* (monosporic type of embryo sac) for example, during the first meiotic prophase callose appears on the entire wall. The cross walls formed after each meiotic division are also callosic. At the end of the second meiotic division, callose first disappears from the functional micropylar megaspore. The reverse is true in the Polygonum type of embryo sac. Thus, the three nonfunctional megaspores are completely separated by callose-rich walls. In tetrasporic embryo sacs this feature is absent. In *Epipactis* (Polygonum type of embryo sac), interestingly, the callose wall is highly porous, giving the appearance of a sieve plate.

9. Pollination

The transport of pollen to the stigma is brought about by wind (anemophily), water (hydrophily), insects (entomophily), and birds (ornithophily). Most aquatic plants which bear flowers on aerial axes are pollinated by wind or insects. Water pollination may take place either on the surface of water or when the flowers are submerged. The individual pollen grains join to form large floating rafts that adhere to stigma (see Cox and Knox 1986). In entomophily, the insects visit flowers for nectar, and rarely for resin as in *Delicampia* (Euphorbiaceae). Bees which collect resin use it for their nest.

9.1 Reception of Pollen

Stigma. The stigma is either dry or wet. The wet stigma secretes an exudate. Both types of stigmatic exudate contain lipids, proteins, glycoproteins, carbohydrates, amino acids and phenols, and non-specific esterases on the receptive surface. Dry stigma has an extracellular membrane, the pellicle. The pellicle components originate from the epidermal cells of the stigma or its papillae and contain aralsinogalactins.

During the early stages of floral development, the wet stigma is comparable to the dry stigma with a cuticle-pellicle layer. Later, secretions from the cells of stigma accumulate below the cuticle-pellicle layer. Upon its disruption, the exudate spreads on the surface of stigma.

9.2 Style

The style can be 'solid' (Fig. 8A-C) or 'hollow' (Fig. 8D-F). In the solid style a core of transmitting tissue

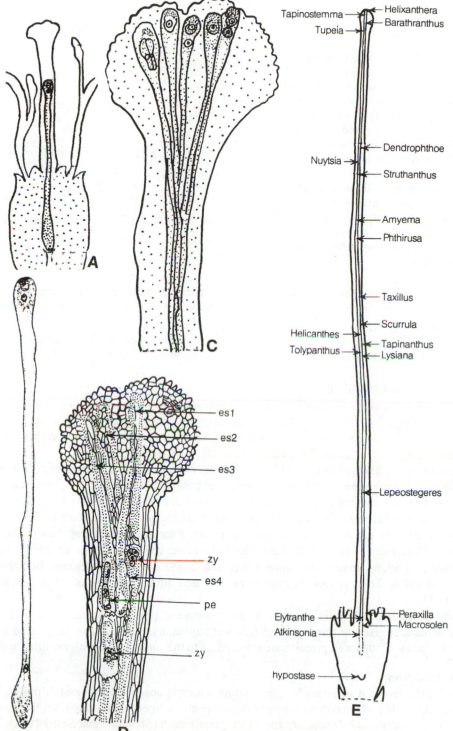

Fig. 7 A-E. Loranthaceae, the tip of embryo sac extends into style and stigma. A, B *Struthanthus vulgaris.* A Flower, note elongation of embryo sac into style. B Eight-nucleate embryo sac from A. C *Helixanthera ligustrina,* L.s. of stigma showing embryo sacs reaching up to the tip of stigma. D *Moquiniella rubra,* part of style and stigma shows tips of four embryo sacs (es₁, es₄), three reach up to the stigma and curve backward so that the embryo sac/proembryo acquires reverse polarity. Two embryo sacs show zygote and one four-celled proembryo. E L.s of carpel. The names of different taxa indicate the height up to which the tip of the embryo sac reaches in the ovary, style and stigma. (*es* embryo sac, *pe* proembryo, *zy* zygote) (after Johri 1984; adapted from Johri et al. 1992).

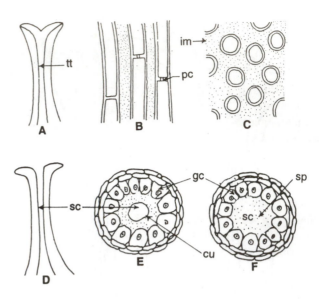

Fig. 8 A-F.   Types of style in angiosperms. A Closed style (L.s), e.g. *Gossypium*. B Transmitting tissue (L.s). C T.s transmitting tissue. D Open (hollow) style, e.g. *Lilium*; stylar canal in L.s. E, F T.s to show glandular cells bordering stylar canal. The secretion product is released into the canal, as cuticle layer is disrupted. (*cu* cuticle, *gc* glandular cells, *im* intercellular matrix, *pc* plasmodesmetal connection, *sc* stylar canal, sp secretion product, *tt* transmitting tissue). (Adapted from Shivanna and Johri 1985).

starts from the stigmatic secretive tissue and traverses the entire length of the style. The transmitting tissue contains plasmodesmata. The intercellular spaces become filled with secretion products through which the pollen tubes grow down the style.

The stylar canal of hollow type runs from the stigma up to the ovarian cavity. The secretions from the cells accumulate under the cuticle of glandular cells that line the stylar canal (Fig. 8E, F).

10.   Pollen-Stigma Interaction

The pollen-pistil interaction plays a significant role in sexual reproduction. The pistil has developed mechanisms to recognize pollen grains, and to permit the growth of tubes through stigma and style. This screening and selection of male partners are essential components of sexual reproduction (Shivanna and Johri 1985). Stigma receptivity is generally maximal soon after anthesis.

Pollination initiates many changes in the pistil and ovary including wilting of the corolla (Gilissen and Hoekstra 1984), followed by ethylene production in the pistil and corolla (Peach et al. 1987), and accumulation of flavonoids in the stigmatic exudate and outer cell layers of stigma (Vögt et al. 1994) which are essential for pollen germination and tube growth. Post-pollination related changes in the pattern of RNA, protein synthesis and several enzyme activities occur in the lower part of the style and ovary, before the arrival of the pollen tubes (Bredemeijer 1982). Pollination also induces degeneration of one of the synergids in many species including cotton (Jensen and Ashton 1981), and delays embryo sac degeneration as in pear (Herrero and Gascon 1987). These responses seem to be mediated through the release of gibberellic acid (Shivanna et al. 1997).

Pollen adhesion is followed by pollen hydration, the moisture is provided by the stigma, driven by osmotic potential difference (Heslop-Harrison 1979). In wet stigma, hydration is generally rapid whereas in the dry type it is gradual. Pollen wall proteins are released on to the surface of the stigma after hydration.

10.1   Pollen Germination

The stigma provides the required substrate for germination which is absent or deficient in pollen. In wet stigma, the role of exudate in pollen germination varies from species to species. In some taxa, like *Amaryllis* and *Crinum*, the exudate seems to be necessary for pollen germination (Shivanna and Sastri 1981), but not in *Petunia* and *Nicotiana*. In dry stigma, the pellicle is involved in pollen recognition and germination

(Knox et al. 1976, Clarke et al. 1979). The stigmatic surface also provides boron and calcium which are generally deficient in pollen but are required for germination.

10.2 Pollen Tube Entry into Stigma

In wet stigma, the cuticle of the stigmatic surface/papillae is already disrupted during secretion of the exudate, and the pollen tube enters the intercellular spaces of the transmitting tissue of the stigma. In dry stigma, the cuticle—which provides a physical barrier for the pollen tube entry—is eroded by the activation of cutinase. In solid-styled systems, the pollen tubes enter the intercellular spaces of the transmitting tissue and grow through the intercellular matrix of the stigma and style. However, in hollow-styled systems, pollen tubes enter the stylar canal and grow on the surface of the canal cells. Pollen tubes also come into contact with the extracellular components secreted by cells of transmitting tissue. The pollen tubes continue their growth further into the ovary along the placenta, drawing water and nutrients from the surrounding tissue (Shivanna and Johri 1985). In two-celled pollen systems, pollen tube growth occurs in two phases, a period of slow growth in which pollen reserves are utilized, and a period of rapid growth during which the tube utilizes nutrients from the style. However, in three-celled systems, there is a short lag phase (period between hydration and germination) and a faster rate of pollen tube growth compared with two-celled pollen systems.

10.3 Pollen Tube Entry into Ovule

The pollen tubes, after entering the ovary, grow along the surface of the placenta towards the ovules and enter the micropyle. Additional structures such as the obturator (a placental outgrowth in the micropylar region of the ovule) are present in some species to regulate the passage. In *Utricularia* the tip of the embryo sac grows through the ovule and comes in direct contact with the placenta and penetrates the placental nutritive tissue; in fact, it becomes buried in the nutritive tissue (Fig. 9A, B), and fertilization is exogamous (Fig. 9C). The pollen tube grows through the micropyle and enters one of the two synergids through filiform apparatus. In many species, such as *Gossypium* and *Linum*, one of the two synergids degenerates before the arrival of the pollen tube (Russell 1992). In such cases, the pollen tube invariably enters the degenerated synergid (Jensen 1973). However, in some species both the synergids remain intact and the pollen tube enters either of the two. The synergid that receives the pollen tube starts degenerating and the other one remains intact for some time.

Studies on the structure of three-celled pollen grains of *Plumbago zeylanica* (Russell 1983, Russell and Cass 1983), *Brassica oleracea* (Dumas et al. 1984), *B. campestris* (McConchie et al. 1987), *Spinacea oleracea* (Wilms 1986) and *Zea mays* (McConchie et al. 1987) reveal that the sperms are markedly dimorphic (Fig. 9D, E). One sperm cell is larger than the other and contains nearly all the mitochondria, while the second sperm contains nearly all the plastids. However, in *Petunia* the sperm cells are not significantly dimorphic (Wagner and Mogensen 1987), and in barley the sperms are isomorphic (Mogensen and Rusche 1985).

The two sperms in three-celled pollen grains are linked by a common cross wall (*Plumbago zeylanica*) or by evagination of the plasma membrane (*Brassica*) and lie within a single enclave of the pollen cytoplasm. One of the sperms is also connected to the vegetative nucleus by a cytoplasmic extension, thus forming a three-celled male-germ-unit (MGU; Fig. 9D, E). The special organization of the three cells in a MGU seems to be fixed and determines which sperm will fuse with the egg and which one will participate in triple fusion. In *Plumbago*, for example, the mitochondria-rich larger sperm, connected to the vegetative nucleus, preferentially fuses with the egg while the smaller sperm without any direct contact with vegetative nucleus participates in triple fusion (Russell 1984, 1989). Although the formation of MGU is recorded in a number of plants, the time of its appearance varies. In species with two-celled pollen, for example, *Rhododendron*,

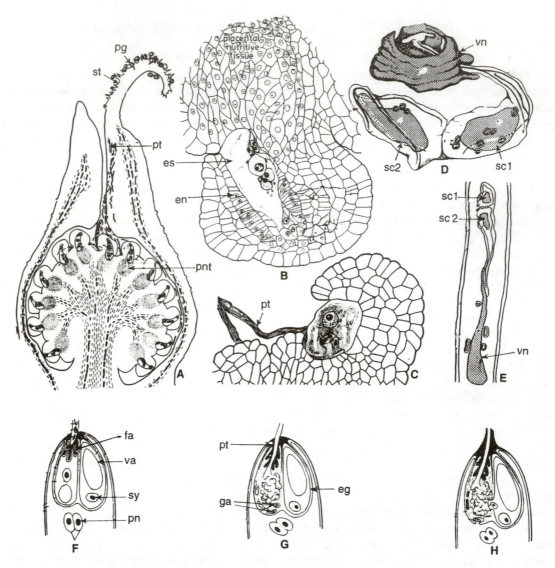

Fig. 9 A-H. Transfer of male gametes to embryo sac during fertilization. A-C *Utricularia flexuosa*. A L.s of carpel to show disposition of pollen tubes, placental nutritive tissue, and embryo sacs. B Portion of A, note the endothelium and tip of embryo sac extending into the placental nutritive tissue. C 'Exogamous' entry of pollen tube into embryo sac. D *Brassica campestris*, three-dimensional (computer assisted) diagram of 'male germ unit' (*mgu*); note the relative position of vegetative nucleus (*vn*) and sperm cells (*sc*). E *Plumbago* sp., MGU during pollen tube growth. F-H Stages in fertilization. F The pollen tube enters (the embryo sac) through the filiform apparatus of the persistent synergid. G The pollen tube passes through the filiform apparatus and releases the gametes. H One of the released male gametes migrates towards the egg and the other towards polar nuclei, or the secondary nucleus and double fertilization occurs. (*eg* egg, *en* endothelium, *es* embryo sac, *fa* filiform apparatus, *ga* gametes, *pg* pollen grain, *pn* polar nuclei, *pnt* placental nutritive tissue, *pt* pollen tube, *sc* sperm cells, *st* stigma, *sy* synergid, *va* vacuole, *vn* vegetative nucleus) (A-C, adapted from Johri et al. 1992; D, E, adapted from Knox et al. 1986; F-H, adapted from Jensen 1973).

Petunia, and *Gossypium*, the MGU is organized in the pollen tube. In some three-celled pollen, species like *Hordeum* and *Spinacea*, the MGU appears only after pollen germination. The organized MGU travels through the pollen tube as one entity but after reaching the ovule, the vegetative nucleus separates from the sperms first, followed by the separation of the two sperms from each other. It is likely that the quality and spatial arrangement of the sperms in the MGU allows their targeted fusion with the egg and the secondary nucleus (Mathys-Rochon and Dumas 1988).

11. Double Fertilization

The pollen tube discharges its contents (two sperm cells, vegetative nucleus and cytoplasm) into the synergid (Fig. 9F) by a pore at the tip or through a subterminal pore (Fig. 9G, H). One of the sperms enters the egg and the other the central cell. In many cases, sperm dimorphism has been reported so that there is preferential fertilization of the egg and central cell by specific sperms (described earlier).

Fertilization takes place through the contact and fusion of plasma membranes which form a bridge through which the sperm nucleus enters (Jensen 1973). Nuclear fusion is also mediated through the fusion of nuclear membranes. In the majority of species, the cytoplasmic organelles of the sperm cells are excluded from fertilization (Hagemann and Schroder 1989).

12. Endosperm

Endosperm, a product of double fertilization (and triple fusion in the majority of cases), develops only in angiosperms. A male gamete fuses with the polar nuclei, or the fusion product of polar nuclei, the secondary nucleus, and forms the primary endosperm nucleus. The division of the primary endosperm nucleus, and repeated divisions of daughter nuclei, lead to the formation of endosperm tissue (Bhatnagar and Sawhney 1981). The endosperm is a nutritive tissue which nurses the developing embryo until it becomes autotrophic, and regulates its pattern of development (Krishnamurthy 1988). The mature endosperm stores carbohydrates, proteins and fatty food reserves.

The developing endosperm derives nutrients from food reserves stored in the nucellus and integuments. The formation of chalazal and micropylar endosperm haustoria help in absorption of nutrients from the surrounding cells. Depending on the mode of development, the endosperm is said to be of the Nuclear, Cellular or Hellobial type.

12.1 Nuclear Type

The primary endosperm nucleus undergoes free-nuclear divisions (Fig. 10A-C). The endosperm remains free-nuclear—throughout as in *Limnanthes* or may become cellular at a later stage (in most cases). Wall formation is mostly centripetal or initiated at the micropylar end and proceeds to the chalaza. However, it commences from the chalazal end and proceeds upward in *Frankenia* (Walia and Kapil 1965) and *Pistacia*. The degree of cellularization varies. Usually the endosperm becomes completely cellular, but in *Phaseolus* cellularization occurs only around the embryo.

12.2 Hellobial Type

This type of endosperm is restricted largely to the monocotyledons. The primary endosperm nucleus moves to the chalazal end of the embryo sac where it divides and forms a large micropylar chamber and a small chalazal chamber (Fig. 10D-G). Free-nuclear divisions occur in the micropylar chamber, cell formation may start at a later stage. In the chalazal chamber, the nucleus either remains undivided or divides only a few times, usually free-nuclear.

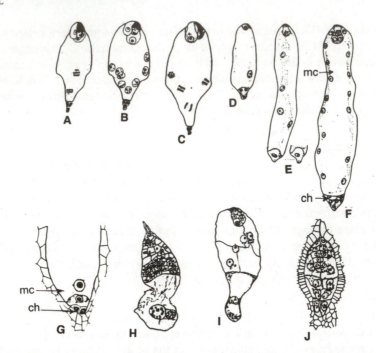

Fig. 10 A-J. Three types of endosperm development. A-C Nuclear type. A Division of nuclei. B Free nuclei along the periphary. C Mitotic divisions. D-G Hellobial type. D First division followed by wall formation. E, F Free-nuclear divisions in the micropylar chamber. G Micropylar and chalazal chamber. H-J Cellular type. H Upper part is cellular and the lower part develops into a haustorium. I Cellular upper part and haustorial lower part. J Entire endosperm is cellular (*mc* micropylar chamber, *ch* chalazal chamber) (adapted from Johri et al. 1992).

12.3 Cellular Type

The division of the primary endosperm nucleus, and the division of daughter nuclei are followed by wall formation (Fig. 10H-J). The walls may be transverse resulting in a linear row of cells as in *Cercidiphyllum* (Swamy and Bailey 1949) or cross-wise as in *Parrotiopsis* (Kapil and Kaul 1972).

12.4 Ruminate Enodosperm

The ruminate condition of endosperm occurs during the maturation of seed so that the surface of the endosperm becomes uneven. The ruminations are caused due to localized activity of the seed-coat or endosperm (see Johri et al. 1992).

12.5 Endosperm Haustoria

The occurrence of haustoria is a common feature of endosperm (Fig. 11A-G). Sympetalous families are characterized by the formation of either micropylar, or chalazal, or both types of endosperm haustoria. In *Crotalaria*, the wall formation is confined to the upper region and the free-nuclear chalazal region often elongates and behaves like a haustorium (Fig. 11A). In Gramineae, the peripheral layer of the cellular endosperm becomes meristematic. This peripheral layer develops into an aleurone layer. The endosperm develops a micropylar haustorium in *Frankenia hirsuta* (Walia and Kapil 1965) and a chalazal haustorium in Cucurbitaceae (Fig. 11B), Leguminosae and Euphorbiaceae. The development of micropylar and chalazal haustoria is noteworthy in Loasaceae, Scrophulariaceae and Orobanchaceae. In *Iodina rhombifolia* (Fig.

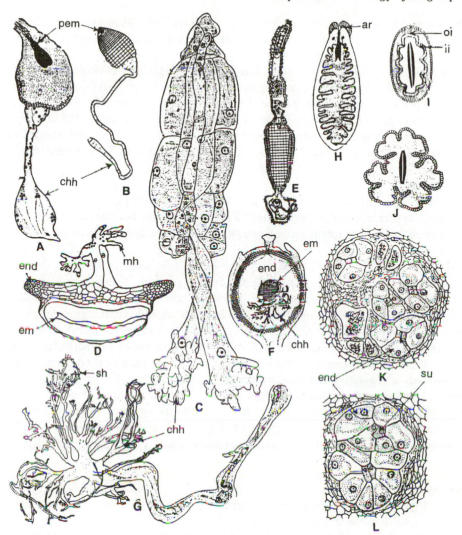

Fig. 11 A-L.   A-G Endosperm haustoria. H-J Ruminate endosperm. K, L Composite endosperm. A *Crotalaria*, multinucleate coenocytic chalazal haustorium with 'balooned' base. The upper region of endosperm is partially cellular. B *Cucumis*, cellular upper region of endosperm and chalazal multinucleate coenocytic tubular haustorium. C *Exocarpus*, elongated proembryo and endosperm haustoria with branched tip. D *Thunbergia*, dicot embryo, endosperm proper, secondary haustoria, and much-branched micropylar haustorium. E *Blumenbachia*, micropylar and chalazal haustoria. The micropylar is cellular and linear and the chalazal is an enlarged cell. F, G *Iodina*. F L.s of fruit to show well-developed endosperm with highly-branched chalazal haustorium. G Much-branched lower end of haustorium and hypertrophied nucleus. H-J Ruminate endosperm. H *Annona squamosa*, L.s of mature seed shows irregular endosperm, and aril at the micropylar end. I *Passiflora calcarata*, T.s of mature seed with outer and inner integument; note the wavy endosperm. J *Coccoloba uvifera*, T.s of seed shows endosperm and exotesta. K, L *Tolypanthus involucratus*, T.s inner part of ovary to show biseriate suspensor of embryo passing through composite endosperm. K Four separate endosperm with suspensor of proembryos passing through each endosperm; note degenerated embryo sacs. L Composite endosperm with biseriate suspensor of two proembryos (*ar* aril, *chh* chalazal haustorium, *em* embryo, *end* endosperm, *ii* inner integument, *mh* micropylar haustorium, *oi* outer integument, *sc* seed coat (exotesta), *sh* secondary haustoria, *su* suspensor) (adapted from Johri et al. 1992).

11F, G), the chalazal haustorium branches extensively at the free end (Bhatnagar and Sabharwal 1968). The endosperm in Acanthaceae (Fig. 11D) is characterized by asymmetrical development of the central cellular endosperm, formation of 'basal apparatus', primary and secondary haustoria and ruminations. Ruminate endosperm also occurs in *Annona* (Fig. 11H), *Passiflora* (Fig. 11I), and *Coccoloba* (Fig. 11J). Because of the lack of definite ovule, the embryo sacs elongate up to various lengths of style and stigma in Loranthaceae. Fertilization and triple fusion takes place in the stylar and stigmatic region. The primary endosperm nucleus migrates to the lower part of the embryo sac in the ovary. With the expansion of the developing endosperm in several embryo sacs, the intervening ovarian tissue is crushed and all the endosperms fuse—resulting in a composite endosperm (Fig. 11K,L). This feature is known only in the Loranthaceae.

13.   Perisperm

The nucellus that is usually consumed by the developing endosperm persists in certain families, such as Amaranthaceae, Capparidaceae, Portulacaceae and Zingiberaceae, and becomes a reservoir of food materials. The persistent nucellus, the perisperm, provides nutrients to the developing embryo (through the endosperm).

14.   Embryo

The zygote is formed by the fusion of the male and female gametes. As a rule the zygote goes through a resting period which is usually short but more prolonged when the endosperm is Nuclear. The zygote, following a predetermined mode of embryogeny, gives rise to an embryo that provides the complete plant.

The zygote is usually spherical or ovoid but, sometimes, pyriform or cylindrical. Its cytoplasm may have primordia of organelles that characterize all plant cells. The zygote shows distinct polarity, the potentialities for further development are restricted to the upper pole which have the so-called germinative function. The lower pole has only a vegetative and nutritive function. There are two distinct phases in the development of the embryo. The first is the proembryo, the second is the embryo proper (Fig. 12,13). The proembryo is characterized by an axial symmetry which is inherited from the egg and maintained until the appearance of the cotyledonary protuberances in dicotyledons (Fig. 12A-I) and the differentiation of the shoot apex in the monocotyledons (Fig. 13A-J). The axial symmetry finally gives rise to the bilateral symmetry of embryo.

14.1 Embryogeny

In a two-celled proembryo, the basal cell (towards the micropylar end) either remains undivided, or undergoes transverse division to form two cells. Based on the plane of division of the apical cell (towards the central cell) and the contribution of the basal and apical cell in the formation of embryo proper, five chief types of embryogeny (Fig. 14) have been recognized (Maheshwari 1950).

1. The apical cell of the two-celled proembryo divides longitudinally.
    (a) The basal cell plays a minor or no role in the subsequent development of the embryo proper—Crucifer or Onagrad type, as in *Capsella*,
    (b) The basal cell and apical cell both contribute to the development of the embryo—Asterad type, as in *Crepis*,
2. The apical cell of the two-celled proembryo divides transversely. The basal cell plays only a minor role, or no role at all, in subsequent development of the embryo proper.
    (a) The basal cell usually forms a suspensor—Solanad type, as in *Solanum*,
    (b) The basal cell does not undergo any further division and the suspensor, if any, is always derived from the apical cell—Caryophyllad type, as in *Myriophyllum*,
3. The basal and apical cells both contribute to the development of the embryo—Chenopodiad type, as in *Myosotis*

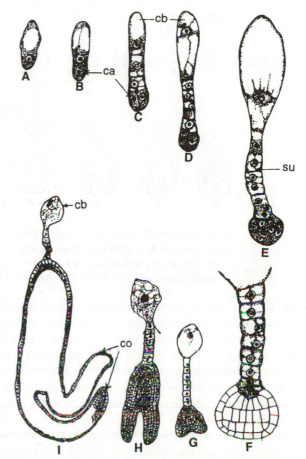

Fig. 12 A-I. Development of dicot embryo, *Capsella*. A Zygote. B Division (transverse) of zygote. C The larger basal cell elongates, and the smaller apical cell divides further. D Note the enlarged basal cell, linearly placed transversely divided cells of the suspensor (products of apical cell), and quadrant. E, F Later stages. G Pre-heart shaped embryo and suspensor. H Heart-shaped embryo, suspensor cells and basal cell with prominent nucleus. I Young dicot embryo (*co* cotyledon, *su* suspensor) (adapted from Maheshwari 1950).

Johansen (1950) recognized a sixth type of embryogeny, where the division of the zygote is vertical—Piperad type as in *Scabiosa*. The division of the zygote may also be obliquely vertical as has been reported in *Triticum*. In *Paeonia*, the zygote undergoes repeated free-nuclear divisions (coenocytic). The nuclei become distributed along the periphery with a central vacuole (Fig. 15A, B). Centripetal wall formation occurs (Fig. 15C) to produce a proembryonal mass. Subsequently, several peripheral meristematic centres develop and produce embryonal primordia (Fig. 15D). Only one of these primordia develops into a dicotyledonous embryo (Fig. 15E).

In a number of taxa belonging to both di- and monocotyledons, the embryo remains small and reduced and lacks the cotyledon/s and apical meristems, and may even lack delineation of tissues when the seed is ready for dispersal. The embryo therefore remains undifferentiated and unorganized. In *Eriocaulon* (Fig. 16 A-F) for example, the proembryo in the seed appears bell-shaped with cotyledonary and epicotylary loci. In *Burmannia*, a saprophyte, the seed carries a quadrant proembryo only (Fig. 16 G-J). The embryos of *Orobanche* remain globular or ovoid with no histogens except the dermatogen (Fig. 16 K-M). *Aeginetia* also sheds seed at the globular stage of the proembryo (Fig. 16 N). The simplest embryo is exhibited by *Monotropa*, with only a two-celled proembryo (Fig. 16 O, P).

14.2 Suspensor

During embryogenesis, the basal cell and its derivatives may develop into a nutritive tissue, the suspensor

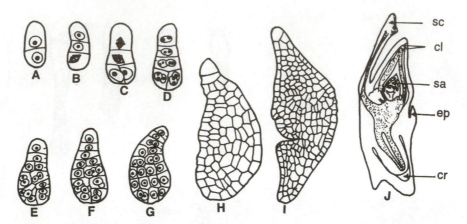

Fig. 13 A-J. Development of monocot embryo, *Triticum*. A Bi-celled proembryo. B-H Further stages of development. I, J Differentiation of monocot embryo. J Organized monocotyledonous embryo with epiblast, scutellum and coleorrhiza (*cl* coleoptile, *cr* coleorrhiza, *ep* epiblast, *sa* shoot apex, *sc* scutellum) (adapted from Maheshwari 1950).

(Fig. 17 A-J), which is an ephemeral structure at the radicular end of the embryo. At initial stages of embryogeny, the suspensor grows much faster than the embryo proper and usually attains its maximal size by the globular or early heart-shaped stage, and in a mature seed only its remnants persist. The suspensor shows much variation in its size and shape (see Fig. 17A-J).

The suspensor anchors the embryo to the embryo sac and pushes it deep into the nutritionally- favourable environment of endosperm. In the Loranthaceae, the egg is fertilized at the tip of the embryo sac in the style or stigma, the suspensor is therefore exceptionally long to bring down the embryo into the endosperm in the ovary.

Suspensor haustoria occur widely and show structural similarities to 'transfer cells', supporting the concept of absorption of nutrients from various ovular and extraovular tissues and translocating this to the embryo.

The embryo in Orchidaceae, Podostemaceae and Trapaceae (where the endosperm is absent) has extensively-developed suspensor haustoria. In Orchidaceae (Fig. 17E, F) the suspensor may be single-celled, and may enlarge to become conical, tubular, sac- or cyst-like, e.g. *Cypripedium*, *Dendrobium*, or develop into a uniseriate filament penetrating the placenta by issuing haustorial branches, e.g. *Habenaria*, *Orphis*, or look like a bunch of grapes, e.g. *Epidendrum*, *Sobralia*, or the suspensor initial may divide by three vertical divisions, and the resultant eight cells cover more than half of the embryo, e.g. *Cottonia*, *Lusia*, *Vanda* and others.

In Podostemaceae, e.g. *Dicraea stylosa*, the basal cell enlarges and contains two hypertrophied nuclei. Subsequently, several haustorial branches develop which grow in between the two integuments (Fig. 17B).

Suspensor haustoria are also common in Crassulaceae, Fumariaceae, Haloragaceae, Leguminosae, and Tropaeolaceae. In Crassulaceae, e.g. *Sedum*, the apical cell of the two-celled proembryo gives rise to the embryo as well as the four-celled suspensor. The basal cell, however, enlarges and grows through the nucellar epidermis, forming a vesicle at the micropyle. The vesicle may grow further and develop tubular branches on either side of the embryo sac. In the Haloragaceae, the suspensor haustoria simulate synergids, e.g. in *Haloragis* (Fig. 17G), the large basal cell divides longitudinally to form two daughter cells which enlarge and occupy the entire space in the micropylar region.

In *Diplotaxis erucoides* and *Eruca sativa* (Fig. 17 I), members of Brassicaceae, the globular embryo has

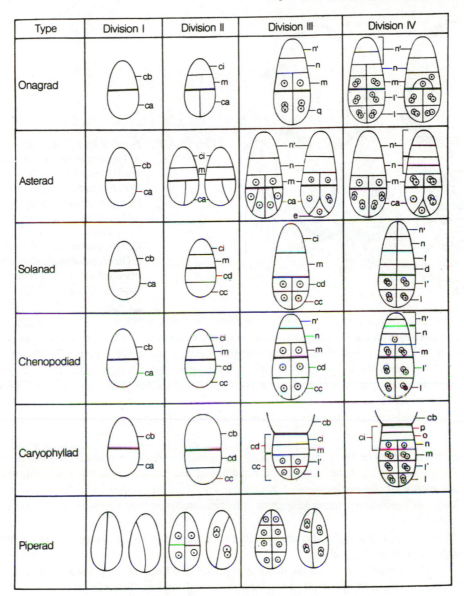

Fig. 14   **Types of embryogeny. Outline diagramms show initial stages in the embryogenic process representing various categories (adapted from Johri et al. 1992).**

a fairly long and tapering basal cell followed by, towards the embryo proper, a fairly large 'bellied' cell and two to five rectangular cells. The suspensor is joined to the embryo through the hypophysis. Some of the modifications in suspensor structure are depicted in Fig. 17A-J.

Detailed studies on the suspensor of *Eruca sativa* and fine structural studies in *Phaseolus* (Yeung and Clutter 1978), *Capsella* (Schulz and Jensen 1969), and *Stellaria* (Newcomb and Fowke 1974) reveal that the nutrient transfer into the embryo may be through the suspensor. In *Capsella bursa-pastoris* the basal cell divides transversely into a large cell towards the micropyle that enlarges further, and a small cell

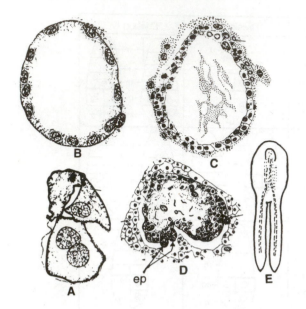

Fig. 15 A-E.   *Paeonia lactiflora*, unusual embryogeny. A Two-nucleate proembryo (there is no wall formation following division of zygote); note the persistent synergid. B Peripheral arrangement of free nuclei and vacuole in the central region. C Wall formation in the proembryo. D Differentiation of embryonal primordia. E Dicotyledonous embryo (*ep* embryonal primordia) (after Yakovlev 1969; adapted from Johri et al. 1992).

toward the apical cell that forms a row of six to ten uniseriate suspensor cells. The enlarged single cell towards the micropyle becomes haustorial, and remains anchored to the micropylar end of the embryo sac. There are also certain ingrowths of the wall that become more prominent with the development of the embryo. The ingrowths are similar to the filiform apparatus of synergids, and the 'transfer cells' associated with short distance transfer of materials in glandular tissues, nectaries, and haustoria of parasitic plants, etc.

The following evidences also support the active involvement of suspensors in the nutrition and development of the embryo:

1. The chromosomes in the suspensor cells of *Eruca* and *Phaseolus* become highly polytenic, especially in *Phaseolus*; highly-lobed nuclei and presence of micronucleoli in the nucleoplasm indicate the active nature of the nucleus.
2. In *Stellaria media* the primary suspensor cells become filled with protein bodies which are absorbed by the embryo.
3. In *Phaseolus coccineus* the suspensor cells contain more RNA and protein, and synthesise them at a higher rate than the embryo.
4. Yeung (1980) reported that in the developing pods of *Phaseolus coccineus*, fed with ^{14}C- sucrose, the radioactivity first appeared in the suspensor and later in the embryo.
5. In suspensor cells of *Ipomoea*, smooth endoplasmic reticulum, a characteristic feature of many glandular cells in plants, has been observed.

In *Phaseolus coccineus*, the suspensor cells show gibberellin-like activity 30-times higher than in the embryo proper. Subsequently, at later stages, the level of gibberellin in suspensor cells drops dramatically and there is a corresponding increase in the embryo.

In vitro studies also indicate a positive role for the suspensor in nutrition of embryos. Older embryos of *Phaseolus coccineus* grew well when cultured with or without suspensors. In the younger embryos, removal of the suspensor significantly reduced the frequency of plantlet formation. The growth-promoting activity of the suspensor was maximal at the early heart-shaped stage of the embryo. Gibberellin substituted for a suspensor most effectively (Cionini et al. 1976; see also Yeung and Sussex 1979).

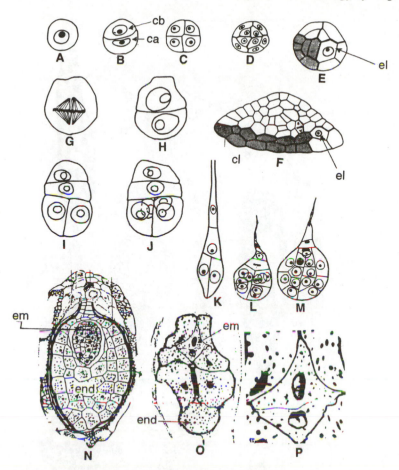

Fig. 16 A-P. Undifferentiated mature embryos. A-F *Eriocaulon xeranthemum*. **A** Zygote. **B** Two-celled proembryo with apical (*ca*) and basal (*cb*) cell. **C** Four-celled proembryo. **D** Later stage. **E** Differentiation of cotyledonary locus (*cl*) and epicotyl-ovary locus (*el*). **F** Some degree of internal differentiation into radicular pole and shoot apex; note the epicotyl locus (*el*) and *cl*. **G-J** *Burmannia pusilla*, Embryogeny leading to quadrant stage in a seed at shedding stage. **G** Division of zygote. **H** Two-celled proembryo. **I** Later stage. **J** Quadrant proembryo at seed shedding. **K-M** *Orobanche cernua*. **K** Vertical division in embryonal tier. **L** Globular proembryo. **M** Proembryo at seed-shedding. **N** *Aeginetia indica*, globular proembryo (in L.s of seed) in mature seed. **O, P** *Monotropa uniflava*. **O** Mature seed (L.s). **O** Two-celled proembryo surrounded by endosperm cells. **P** Two-celled proembryo from **O** (*ca* apical cell, *cb* basal cell, *cl* cotyledonary locus, *el* epicotyledonary locus, *em* embryo, *en* endosperm) (adapted from Johri et al. 1992).

Natesh and Rau (1984) have given an excellent account of the origin, development and differentiation of embryos.

15. Seed

In angiosperms, most seeds have an embryo. The basic structure of the seed is that of the ovule. As the ovule matures into a seed, the integuments undergo conspicuous changes. Mostly there is reduction in

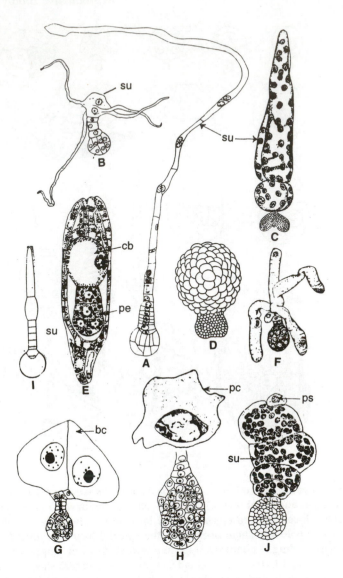

Fig. 17 A-J. Variations in suspensor structure. A *Pedicularis*, highly elongated uniseriate multicellular suspensor at globular stage of proembryo. B *Dicrarea*, the basal cell develops tubular extensions. C *Orobus*, vesicular multinucleate suspensor. D *Cytissus*, highly 'globose' multicelled suspensor. E *Dendrobium*, L.S. seed with embryo and highly enlarged vacuolate suspensor (basal cell). F *Cymbidium*, embryo with tubular suspensor haustoria. G *Haloragis*, hypertrophied daughter cells of basal cell, globular proembryo. H *Machaerocarpus*, globular proembryo with large basal cell. I *Diplotaxis*, uniseriate multi-celled suspensor at globular proembryo. J *Cuscuta*, multinucleate vesicular suspensor cells at globular proembryo; note the persistent synergid. (*cb* basal cell, *pe* proembryo, *ps* persistent synergid, *su* suspensor) (adapted from Natesh and Rau 1984).

thickness and disorganization but, sometimes, additional layers may be formed. The funiculus usually abscises leaving a scar, called the hilum.

15.1 Seed Coat

The integument/s of the ovule form the seed coat or testa of the seed. In bitegmic ovules the seed coat may be derived from both the integuments, or the inner or outer integument.

15.2 Mature Seed

The outer integument becomes differentiated into an outer epidermis, a pigmented zone of four or five layers, and a zone of two or three colourless layers. The inner integument is distinguishable into a palisade layer, inner pigmented zone of fifteen to twenty layers, and the fringe layer.

In cotton, which has a bitegmic ovule, both the integuments contribute to the seed coat (Ramchandani et al. 1966; Joshi et al. 1967). The outer integument consists of four to eight layers, distinguishable into three zones: (1) outer epidermis, (2) outer pigmented zone of two to five layers of cells filled with tannin and starch, and (3) inner epidermis. With the maturation of the seed, there is considerable enlargement of cells. The nucellus is absorbed and the endosperm is consumed by the developing embryo. Fibers are distributed all over the surface of the seed in cotton which arise from the epidermal cells of the outer integument.

In Cucurbitaceae also the ovule is bitegmic but only the outer integument takes part in the development of the seed coat, and the inner integument degenerates. The cells of the outer integument divide periclinally and, at maturity, the seed coat can be differentiated into five zones: (1) seed epidermis which is single-layered and consists of cells with rod-like thickenings on the radial wall, (2) hypodermis is one- to many-layered with uniform thin or thick-walled cells, (3) main sclerenchymatous layer of brachy-, osteo- or branched-sclerides, (4) aerenchyma derived from hypodermis comprising stellate cells that show prominent air spaces, and (5) chlorenchyma of thin-walled cells formed from the remaining layers of integument.

The Acanthaceae have unitegmic ovules. The massive integument is consumed by the developing endosperm, and only the epidermis persists in the mature seed (see Wadhi 1970). The seed epidermis develops various types of hairs or thickenings (Bhatnagar and Puri 1970). In *Elytraria* (Johri and Singh 1959), *Andrographis* (Mohan Ram and Masand 1962) and *Haplanthus* (Phatak and Ambegaokar 1961), the mature seeds are naked and the outer layers of endosperm take up the functions of the seed coat.

The chief function of seed coat is protection against attack by microorganisms or insects, mechanical injury, and desiccation, and to help in dispersal.

15.3 Endosperm in the Seed

The developing endosperm, in early stages of embryo development, provides nutrition to the embryo. The endosperm may be consumed before the seed is ripe but, in some cases, persists in the ripened seed. Seeds with endosperm are described as albuminous as in maize and castor bean, and without endosperm as exalbuminous as in cucurbits. However, *Pyrus* has scanty endosperm, although it is described as an exalbuminous seed. In angiosperms, albuminous seeds are more common in the monocotyledons than in dicotyledons. When there is no endosperm and the cotyledons are poorly-developed, food may be stored in other parts of the embryo, especially the hypocotyl.

16. Apomixis

The formation of the sporophyte from the gametophyte without the sexual process signifies apomixis. It relates to the replacement of alternation of a reduced gametophyte and an unreduced sporophyte by an unreduced gametophyte and sporophyte (summarized in Fig. 18A-E). It occurs in relatively few families, but is common in Compositae and Graminae. Apomixis can be classified into two types: recurrent (in non-reduced embryo sacs), and non-recurrent (in reduced embryo sacs). The latter relates to development of haploid embryos without the actual fusion of gametes. In a reduced embryo sac, the egg can develop into an embryo without pollination stimulus and hence there is no fertilizaton (Fig. 18A). Syngamy is shown in Fig. 18B.

16.1   Recurrent Type

Because of irregular meiosis, euspory is replaced by aneuspory (diplospory). When the spore mother cell functions directly as the embryo sac initial, it is gonial apospory, and when one or more of the somatic cells of the nucellus or chalaza act as the embryo sac initial, it is somatic apospory.

16.2   Aneuspory (Diplospory)

Aneuspory includes taxa in which, during spore formation, the meiotic process is irregular. Depending upon the extent of irregularity in meiosis, aneuspory is further recognized in four types.

16.2.1   Datura Type

Two unreduced nuclei undergo two mitotic divisions to produce an eight-nucleate embryo sac as in *Datura*.

16.2.2   Taraxacum Type

The first meiotic division ends in a restitution nucleus, meiosis II produces unreduced cells, and ultimately an eight-nucleate embryo sac is formed. The egg directly produces the embryo as in *Taraxacum vulgare*.

16.2.3   Ixeris Type

It is like *Taraxacum*. A single binucleate gynospore (megaspore) undergoes two mitotic divisions leading to the formation of an eight-nucleate embryo sac as in *Ixeris dentata*.

16.2.4   Allium Type

A premeiotic-endomitotic doubling makes meiotic prophase to start with a double chromosome number, and an unreduced embryo sac is formed as in *Allium nutans*.

16.3   Gonial Apospory

The megaspore mother cell does not undergo meiosis. It enlarges, a small vacuole appears above and below the nucleus—a one-nucleate embryo sac. Three successive mitotic divisions result in an eight-nucleate embryo sac as in *Antennaria alpina*.

In *Brachycome*, the polar nuclei function directly as the first two endosperm nuclei.

16.4   Somatic Apospory

Meiosis is apparently normal but megaspores are not concerned in the embryo sac formation. Either during, or soon after meiosis, one or more sporophytic cells (nucellar or chalazal) enlarge and invade the nucellus destroying and replacing the megaspore. After three successive nuclear divisions, each such cell differentiates into an aposporic 2n embryo sac with normal organization. Thus, the gametophytic generation is completely eliminated, e.g. *Artemisia nitida*.

17.   Unreduced Embryo Sacs

The fate of a nucleus in the embryo sac depends upon its position. Many irregularities in the disposition of nuclei in early polarization have been recorded. The mature embryo sac shows a normal organization with two synergids, three antipodals, an egg, and proendospermic (central) cell.

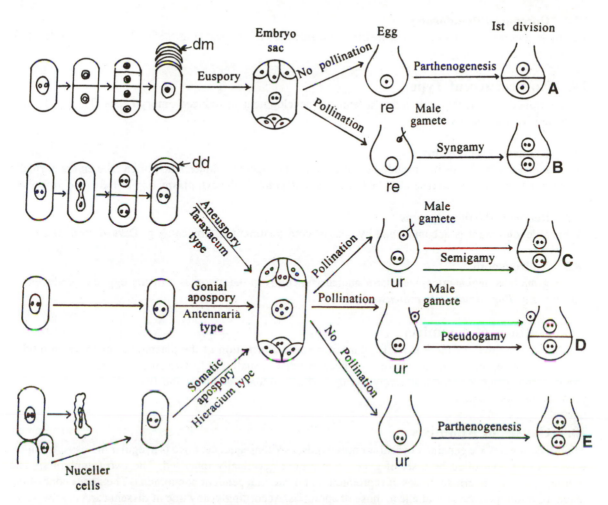

Fig. 18 A-E. **Apomixis in angiosperms. A** Parthenogenesis resulting into a haploid embryo; there is no pollination and fertilization. **B** Normal sexual process involving pollination and fertilization result in a diploid embryo. **C** Pollination results in semigamy, promote the development of the embryo from unreduced egg. **D** Development of diploid embryo from unreduced egg. **E** Diploid embryo developed parthenogenetically without the pollination stimulus (*dd* degenerated dyad, *dm* degenerated megaspore, *re* reduced, *ur* unreduced) (adapted from Battaglia 1963).

17.1 Embryogenesis in Unreduced Embryo Sacs

Embryo development in unreduced embryo sacs can take place as a consequence of pollination. Those taxa associated with pollination are called 'eugamous', 'semigamous' or 'pseudogamous', and those taxa which develop without the stimulus of pollination are 'parthenogenic'.

17.2 Eugamy

Normal fertilization of the apomictic egg takes place to produce the zygote, e.g. *Hypericum perforatum*.

17.3 Semigamy

The male gamete penetrates into the egg but does not fuse with the egg nucleus (Fig. 18C). Both the nuclei divide independently but the division of the male nucleus stops early, e.g. *Rudbeckia speciosa*.

17.4   Unreduced Pseudogamy

The male gamete degenerates either inside or outside the embryo sac (Fig. 18D). Thus, the egg develops without fertilization, e.g. *Zephyranthes texana*.

18.   Non-recurrent Type

As mentioned earlier, this type refers to embryo development in reduced embryo sacs. The mechanism identified for such cases are:

18.1   Reduced Pseudogamy

The reduced egg develops in a pseudogamous, and parthenogenic manner. The pollen tube enters normally but the male gamete fails to fuse with the egg and disintegrates in the cytoplasm, e.g. *Triticum monococcum*.

18.2   Reduced Parthenogenesis

Embryo development is accomplished by heat or cold treatment to flowers, e.g. *Datura stramonium*.

18.3   Androgenesis

The egg nucleus degenerates, the sperm nucleus functions in the cytoplasm of the egg and produces the embryo, e.g. *Poa alpina* × *P. pratensis*,

19.   Anthers of Apomicts

Meiosis is abnormal, or there is total failure of meiosis. Formation of the plasmodial microspore mother cell has been reported, polyads have also been noticed. Usually, only two microspores of a tétrad are normal. The generative nucleus rarely divides. Thus, pollen remains at the two-celled stage and male gametes are not formed.

20.   Causes of Apomixis

Apomictic species are generally hybrids or polyploids; as a consequence, there is irregular meiosis. Apomixis appears to be controlled by a set of genes, this trait is genetically inherited. The two (amphimixis and apomixis) are not alternate modes of reproduction, but are independent phenomena. The genes controlling sexual reproduction are non-allelic to those of apomixis. Accordingly, any line of dissent carrying the genes for apomixis will produce both types; apomicts as well as sexually-reproducing plants.

It has been proposed that apomixis is governed by recessive genes. The three genes (AABBCC) determine the breeding behaviour. In the homozygous condition *a* forms unreduced eggs, *b* prevents fertilization, and *c* promotes egg development without fertilization. Thus, aaBBCC will have unreduced egg but cannot develop without fertilization, AAbbCC produces reduced egg but embryo development can not take place because fertilization is prevented, and AABBcc will show normal sexual behaviour because the gene C has no effect in the presence of A and B.

As a consequence of apomixis, genetic variability in such species is frozen since they have the same genotypes as their parents. However, facultative apomicts have an advantage as they have retained both kinds of reproduction.

21.   Parthenogenesis

The diploid egg in the embryo sac—during diplospory and apospory—develops into an embryo without fertilization, thus maintaining the sporophytic level of chromosomes (Fig. 18E). This process of embryo development from an unfertilized egg is called 'parthenogenesis'. The stimulus to form embryos may be pollination-dependent. For example, in grasses pseudogamy operates that involves pollination stimulus, while in apomictic taxa of Compositae and Rubiaceae no such stimulus is required.

Pseudogamy has been credited with: (1) supply of a male nucleus for endosperm development, (2) activation of the growth of ovules and ovaries, and (3) stimulation of parthenogenesis.

Pollination is reported to initiate the development of adventive embryos of *Citrus*. Likewise, parthenogenetic development of embryos proceeds in apomictic grasses, but normal embryos result only when endosperm is also formed.

Parthenogenesis can be separated into 'reduced' and 'unreduced' eggs, and accordingly the developed embryo can be haploid or diploid.

22. Conclusions

Recently there has been a surge of information on ultrastructural details of the embryo sac and pollen, including the male gametes. In addition, understanding of physiological and biochemical changes that trigger the shift from one stage to another is also in progress. Certain problems remain to be explored and understood to unravel the necessity of specific cells during reproduction in angiosperms.

The factors that trigger the onset and regulation of meiosis remain to be solved. Does the synthesis of the callose wall around the pollen mother cells (PMC) and megaspores indicate the isolation of the gametic cell? The disposition of nuclei in the coenomegaspore is another area that requires further investigation. What are the compelling factors and what is the significance of the arrangement of nuclei in Adoxa, Plumbago, Penaea, Drusa, Plumbagella and Fritillaria types of embryo sacs? The mechanism of relative movement of the two male gametes remains a mystery. The factors that govern the mechanism of recognition of three types of endosperm need to be worked out. Also, in taxa where both the zygote and primary endosperm nuclei are diploid, what morphogenic mechanism operates that selectively induces the zygote to form a bipolar structure but not the primary endosperm nucleus. If one synergid and the antipodals are redundant structures which degenerate later, what is their role in the embryo sac? Even the dehiscence of seeds when the embryos are only at globular stage remains unexplained. Nucellar polyembryony and apomixis are other areas which require in-depth study. The recent success with the identification of an apomictic gene in *Arabidopsis*, and isolation of egg and sperm cells of cereals open new vistas in angiosperm embryology research.

Acknowledgements

We thank Dr (Ms) Nisha Bharti Verma, Ms Alka Narula, Mr Sanjeev Kumar, Ms Sapna Malik, and Ms Garima Khare for their help during the preparation of the manuscript. Expert computer typing and formatting was completed by Mr Satish Chand.

Glossary

Adventive embryony. Embryos arise directly from the diploid sporophytic cells such as nucellus or integuments. The zygotic embryo either degenerates or competes with the apomictic-embryos.

Amphimixis. The normal sexual cycle involving meiosis during gametogenesis and fertilization in which the two haploid gametes of opposite sex fuse re-establishing the diploid sporophytic generation.

Amphitropous. The curvature of the ovule also affects the nucellus that becomes horse-shoe-shaped.

Anatropous. The micropyle of ovules lies close to the funiculus due to unilateral growth of the ovule.

Androgenesis. Development of plants (sporophytes) from male gametophytic cells (anthers).

Antipodal. Haploid cells present at the chalazal end of the embryo sac, usually three in number but can vary. May become polyploid and/or multinucleate.

Apomixis. A type of asexual reproduction in place of sexual reproduction in plants. It does not involve meiosis and syngamy.

Apospory. Development of an unreduced embryo sac from a somatic cell of the nucellus. The diploid egg develops parthenogenetically into an embryo. The normal haploid embryo sac either degenerates or may form asexual embryo after fertilization.

Bitegmic ovule. Ovules with two integuments.

Campylotropous. Ovules in which the curvature is less than that in the anatropous.

Central cell. The largest cell of the female gametophyte (embryo sac) that carries egg, synergid/s, and polar nucleus/nuclei and (may be) antipodals. The central cell is flanked by the integument on either side with the micropylar opening towards the egg-side and chalaza at the antipodal end. It is also the seat of embryo and endosperm development.

Circinotropous. A special type of ovule found in Cactaceae. The nucellar protuberance is in line with the axis and due to unilateral growth, the ovule first represents anatropous condition but as the curvature continues the micropyle again faces upwards in a fully-developed ovule.

Cleistogamous. Flowers that do not open yet the pollination occurs.

Colpate. Pollen grains carrying aperture which are slit-like.

Crassinucellate. The sporogenous cell becomes embedded in the massive nucellus.

Dimorphic. Anthers in which tapetal cells are dual in origin, one from the parietal cells and the other from cells of the connective.

Diplospory. A form of apomixis in which the gametophytes are formed from unreduced megaspores.

Double fertilization. The process in flowering plants when one male gamete fuses with the egg nucleus, the other with the fusion product of the polar nucleus/nuclei.

Endosperm. The storage tissue in the seeds of most angiosperms, derived from the fusion of one male gamete with two female polar nuclei. The endosporm is usually a compact triploid tissue, lacking intercellular spaces. Seeds containing an endosperm at maturity are termed albuminous, whereas those lacking endosperm are called exalbuminous.

Endostome. The part of the micropyle formed by the inner integument is endostome (see also exostome).

Exostome. In bitegmic ovules when both the integuments are involved to form the micropyle, the passage formed by the outer integument is exostome.

Filiform apparatus. A mass of finger-like projections of the wall of synergid into the cell cytoplasm at the micropylar end. Each projection of the apparatus has a core of tightly packed microfibrils enclosed by a non-fibril sheath. They are rich in polysaccharides.

Generative apospory. Sporophyte development from cell/s of a gametophyte without gametic fusion.

Hemianatropous. Half-anatropus, the ovule with the funiculus at right angles to the nucellus and the integuments.

Hilum. The cleft on a seed formerly attached to the funicle or placenta

Parthenogenesis. The development of an egg cell into an embryo without fertilization.

Pellicle. A hair-like structure present at stigmatic surface plays a very important role in incompatibility control. Pellicle may be rigid or flexible.

Perisperm. A nutritive tissue, derived from the nucellus, found in the seeds of certain plants in which the endosperm does not completely replace the nucellus.

Pseudogamy. A form of apomixis in which a diploid embryo is formed without fertilization though a stimulus from the male gamete is required and thus pollination is necessary.

Ruminate endosperm. Endosperm that is irregularly grooved or ridged, and so, appears chewed.

Somatic apospory. Development of sporophyte from the tissue of a gametophyte without the fusion of nuclei. Thus, it has the same chromosome number as the gametophyte.

Suspensor. The Structure that differentiates from the proembryo by mitosis and anchors the embryos to the maternal tissue. It also conducts nutrients to the embryos.

Synergids. The haploid cell/s in contact with egg at the micropylar end of the embryo sac that do not participate in the fertilization process. Together with the egg cell they constitute the egg apparatus.

Tapetum. The food-rich layer of cells that surrounds the spore mother cells in vascular plants. In some plants it breaksdown to form a fluid termed the periplasmodium, which is absorbed by the developing microspores. In others the tapetum remains intact until shortly before anther dehiscence and secretes substances into the loculi.

References

Agarwal S (1962) Embryology of *Quinchamalium chilense* Lam. *In*: Symp Plant Embryology. Council Sci Indus Res (CSIR), New Delhi, pp 162–169

Banerjee VC, Barghoorn ES (1971) The tapetal membranes in grasses and Übisch body control of mature exine pattern. *In*: Heslop-Harrison J (ed) Pollen Development and Physiology. Butterworth, London, pp 126–127

Battaglia E (1963) Apomixis. *In*: Maheshwari P (ed) Recent Advances in the Embryology of Angiosperms. Intl Soc Plant Morphologists, Univ Delhi, India, pp 221–264

Bhatnagar SP, Puri S (1970) Morphology and embryology of *Justicia betonica* Linn. Oesterr Bot Z 118: 55–71

Bhatnagar SP, Sabharwal G (1968) Morphology and embryology of *Iodina rhombifolia* Hook. & Arn. Beitr Biol Pflanzen 45: 465–479

Bhatnagar SP, Sawhney V (1981) Endosperm: its morphology, ultrastructure and histochemistry. Intl Rev Cytol 73: 55–98

Bouman F (1984) The ovule. *In*: Johri BM (ed) Embryology of angiosperms. Springer, Berlin Heidelberg New York, pp 123–157

Bredemeijer GMM (1982) Mechanism of peroxidase isozyme induction in pollinated *Nicotiana alata* styles. Theor Appl Genet 62: 305–309

Broods J, Shaw G (1971) Recent development in the chemistry, biochemistry, geochemistry and post-tetrad ontogeny of sporopollenin derived from pollen and spore exines. *In*: Heslop-Harrison J (ed) Pollen Development and Physiology. Butterworth, London, pp 99–114

Cass DD (1972) Occurrence and development of a filiform appartus in the egg of *Plumbago capensis*. Am J Bot 59: 279–283

Chebotaru AA (1970) Maize embryology: ultrastructure of embryo sac before and at the moment of fertilization. 7th congress Intl Micr electr 3: 443–444

Cionini PG, Benici H, Alpi A, D'Amato F (1976) Suspensor, gibberellin, and in vitro development of *Phaseolus coccineus* embryos. Planta 131: 115–117

Clarke AE, Considine JA, Ward R, Knox RB (1977) Mechanism of pollination in *Gladiolus*: roles of the stigma and pollen tube guide. Ann Bot (Lond) 41: 15–20

Clarke AE, Gleeson P, Harrison S, Knox RB (1979) Pollen-stigma interactions: identification and characterization of surface components with recognition potential. Proc Natl Acad Sci USA 76: 3358–3362

Cox PK, Knox RB (1986) Pollination postulates and two-dimensional pollination. *In*: Williams EG, Knox RB, Irvine D (eds) Pollination 1986. School Bot, University of Melbourne, Melbourne, pp 48–57

Diboll AG, Larson DA (1966) An electron microscopic study of the mature megagametophyte in *Zea mays*. Am J Bot 53: 391–402

Dumas C, Knox RB, McConchie CA, Russell SD (1984) Emerging physiological concepts in fertilization. What's new. Plant physiol 75: 168–174

Fagerlind F (1937) Embryologische, zytologische und bestäubungs-experimentelle Studien in der Familie Rubiaceae löst Bemerkungen über einige Polyploidität problem. Acta Hortic Bergeani 11: 195–470

Fagerlind F (1959) Development and structure of the flower and gametophytes in the genus *Exocarpus*. Sven Bot Tidskr 53: 147–282

Gilissen LJW, Hoekstra FA (1984) Pollination induced corolla wilting in *Petunia hybrida*. Rapid transfer through the style of a wilting-induced substance. Plant Physiol 75: 496–498

Grundwag M (1976) Embryology and fruit development in four species of *Pistacia vera* (Anacardiaceae). Phytomorphology 19: 225–235

Hagemann R, Schroder MB (1989) The cytological basis of the plastid inheritance in angiosperms. Protoplasma 153: 57–64

Harling G (1951) Embryological studies in the Compositae 3. Asteraeeae. Acta Hortic Bergiani 16: 73–120

Herrero M, Gascon M (1987) Prolongation of embryo sac viability in pear (*Pyrus communis*) following pollination or treatment with gibberellic acid. Ann Bot (Lond) 60: 287–293

Heslop-Harrison J (1968) Pollen-wall development. Science 161: 230–237

Heslop-Harrison J (1979) An interpretation of the hydrodynamics of pollen. Am J Bot 66: 737–743

Jensen WA (1965) The ultrastructure and histochemistry of the synergids of cotton. Am J Bot 52: 238–256

Jensen WA (1973) Fertilization in flowering plants. Bioscience 23: 21–27

Jensen WA, Ashton MF (1981) Synergid-pollen tube interaction in cotton. *In:* III Intl Bot Congr. Sydney, Australia, pp 61

Johansen DA (1950) Plant Embryology. Chronica Botanica. Waltham, Mass, USA

Johri BM (1936) The life history of *Butomopsis lanceolata* Kunth. Proc Indian Acad Sci B 4: 139–162

Johri BM (ed) (1984) Embryology of angiosperms. Springer, Berlin Heidelberg New York, pp 830

Johri BM, Agarwal S (1965) Morphological and embryological studies in the family Santalaceae. 8. *Quinchamalium chilense* Lam. Phytomorphology 15: 360–372

Johri BM, Bhatnagar SP (1973) Some histochemical and ultrastructural aspects of the female gametophyte and fertilization in angiosperms. Caryologia 25 (Suppl): 9–25

Johri BM, Singh H (1959) The morphology, embryology and systematic position of *Elytraria acaulis* (Linn.) Lindau. Bot Not 112: 227–251

Johri BM, Ambegaokar KB, Srivastava PS (1992) Comparative Embryology of Angiosperms. Springer, Berlin Heidelberg New York, pp 1221

Joshi PC, Wadhwani AM, Johri BM (1967) Morphological and embryological studies of *Gossypium* L. Proc Natl Inst Sci India B 33: 37–93

Kapil RN, Kaul U (1972) Embryologically little-known taxon: *Parrotiopsis jacquemontiana*. Phytomorphology 22: 334–345

Kapil RN, Prakash N (1966) Co-existence of mono-, bi-, and tetra-sporic embryo sacs in *Delosperma cooperi* Hook. f (Aizoaceae). Beitr Biol Pflanz 42: 381–392

Knox RB, Clarke AE, Harrison S, Smith P, Marchalonis JJ (1976) Cell recognition in plants: determinants of the stigma surface and their pollen interactions. Proc Natl Acad Sci USA 73: 2788–2792

Knox RB, Williams EG, Dumas C (1986) Pollen, pistil and reproductive function in crop plants. *In*: Janick J (ed) Plant Breeding Reviews, vol 4, A VI. Wesport (Conn), pp 9–79

Kress WJ, Stone DE (1982) Nature of the sporoderms in monocotyledons with special reference to pollen grains of *Canna* and *Heliconia*. Grana 21: 129–148

Krishnamurthy KV (1988) Endosperm controls symmetry changes in the developing embryos of angiosperms. Proc Indian Acad Sci Plant Sci 98: 257–259

Maheshwari P (1950) An Introduction to the Embryology of Angiosperms. Tata-McGaw-Hill Publ Co Ltd, Bombay New Delhi, pp 453

Masand P, Kapil RN (1966) Nutrition of the embryo sac and embryo. A morphological approach. Phytomorphology 16: 158–175

Mathys-Rochon E, Dumas C (1988) The male germ unit and prospects for biotechnology. *In*: Mulcahy DL, Mulcahy GB, Ottaviano E (eds) Biotechnology and Ecology of Pollen. Springer, Berlin Heidelberg New York, pp 51–61

McConchie CA, Hough T, Knox RB (1987) Ultrastructural analysis of the sperm cells of maize, *Zea mays*. Protoplasma 139: 9–19

Mogensen HL, Rusche ML (1985) Quantitative analysis of barley sperm: occurrence and mechanism of cytoplasm and organelle reduction and the question of sperm dimorphism. Protoplasma 128: 1–13

Mohan Ram HY, Masand P (1962) Endosperm and seed development in *Andrographis echioides* Ness. Curr Sci 31: 7–8

Natesh S, Rau MA (1984) The embryo. *In*: Johri BM (ed) Embryology of Angiosperms. Springer, Berlin Heidelberg New York, pp 377–443

Newcomb W, Fowke LC (1974) *Stellaria media* embryology: the development and ultrastructure of the suspensor. Can J Bot 52: 60

Pacini E, Juniper BE (1983) The ultrastructure of the formation and development of the amoeboid tapetum in *Arum italicum* Miller. Protoplasma 117: 116–129

Peach JC, Latche A, Larriiguadiere C, Reid MS (1987) Control of early ethylene synthesis in pollinated *Petunia* flowers. Plant Physiol Biochem 25: 431–437

Periasamy K, Kandasamy MK (1981) Development of the anther of *Annona squamosa* L. Ann Bot (Lond) 48: 885–893

Periasamy K, Swamy BGL (1966) Morphology of the anther tapetum in angiosperms. Curr Sci 35: 427–430

Phatak VG, Ambegaokar KB (1961) Embryological studies in Acanthaceae 4. Development of embryo sac and seed formation in *Andrographis echioides* Nees. J Indian Bot Soc 40: 525–534

Ramchandani S, Joshi PC, Pundir NS (1966) Seed development in *Gossypium* Linn. Indian Cotton J 20: 97–106

Ram M (1959) Morphological and embryological studies in the Santalaceae. 2. *Exocarpus* with a discussion on its systematic position. Phytomorphology 9: 4–19

Russell SD (1983) Fertilization in *Plumbago zeylanica*. Gametic fusion and fate of the male cytoplasm. Am J Bot 70: 416–434

Russell SD (1984) Ultrastructure of the sperm of *Plumbago zeylanica*. Quantitative cytology and three dimentional organization. Planta 162: 385–395

Russell SD (1989) Preferential fertilization in the synergid-lacking angiosperm *Plumbago zeylanica*. Phytomorphology 39: 1–20

Russell SD (1992) Double fertilization. Intl Rev Cytol 140: 1349–1359

Russell SD, Cass DD (1983) Unequal distribution of plastids and mitochondria during sperm cell formation in *Plumbago zeylanica*. *In*: Mulcahy DL, Ottaviano E (eds) Pollen: Biology and Implications for Plant Breeding. Elsevier, Amsterdam, pp 135–140

Schulz P, Jensen WA (1969) *Capsella* embryogenesis: the suspensor and the basal cell. Protoplasma 67: 139–163

Shivanna KR, Johri BM (1985) The Angiosperm Pollen: Structure and Function. Wiley Eastern, New Delhi

Shivanna KR, Sastri DC (1981) Stigma-surface esterases and stigma receptivity in some taxa characterized by wet stigma. Ann Bot (Lond) 47: 53–64

Shivanna KR, Cresti M, Ciampolini F (1997) Pollen development and pollen-pistil interaction. *In*: Shivanna KR, Sawhney VK (eds) Pollen Biotechnology for Crop Production and Improvement. Cambridge University Press, pp 15–39

Stow I (1930) Experimental studies on the formation of the embryo sac-like giant pollen grains in the anther of *Hyacinthus orientalis*. Cytologia 1: 417–439

Stow I (1934) On the female tendencies of the embryo sac-like giant pollen grains of *Hyacinthus orientalis*. Cytologia 5: 88–108

Swamy BGL, Bailey IW (1949) The morphology and relationships of *Cercidiphyllum*. J Arnold Arbor 30: 197–210

Vijayaraghavan MR, Bhat V (1983) Synergids before and after fertilization. Phytomorphology 33: 74–84

Vijayaraghavan MR, Ratnaparkhi S (1977) Pollen embryo sacs in *Heuchera micrantha* Dougl. Caryologia 30: 105–119

Vögt T, Pollak P, Tarlyn N, Taylor LP (1994) Pollination or wound-induced kaempferol accumulation in *Petunia* stigmas enhances seed production. Plant Cell 6: 11–23

Wadhi M (1970) Acanthaceae. *In*: Symp Comparative Embryology of Angiosperms. Indian Natl Sci Acad Bull 41: 267–272

Wagner VT, Mogensen HL (1987) The male sperm unit in the pollen and pollen tubes of *Petunia hybrida*: ultrastructural, quantitative and three-dimensional features. Protoplasma 143: 93–100

Walia K, Kapil RN (1965) Embryology of *Frankenia* Linn. with some comments on the systematic position of the Frankeniaceae. Bot Not 118: 412–429

Wilms HJ (1986) Dimorphic sperm cells in the pollen grain of *Spinacia*. *In*: Cresti RD (ed) Biology of Reproduction and Cell Motility in Plants and Animals. Siena University, Siena, pp 193–198

Yakovlev M (1965) Embryogenesis and some problems of phylogenesis. Rev Cytol Biol Vég 32: 325–330

Yeung EC (1980) Embryogeny of *Phaseolus*. The role of the suspensor. Z Pflanzenphysiol 96: 17–28

Yeung EC, Clutter ME (1978) Embryogeny of *Phaseolus coccineus*: growth and microanatomy. Protoplasma 94: 19–40

Yeung EC, Sussex IM (1979) Embryogeny of *Phaseolus coccineus*: the suspensor and the growth of the embryo-proper in vitro. Z Pflanzenphysiol 91: 423–433

12

Cytology and Genetics of Reproduction

Sumitra Sen

1. Introduction

In the evolutionary strategy of the plant kingdom, the reproductive patterns are geared towards two distinct objectives: mass propagation, and generation of variability. Undoubtedly, the production of replicas is the principal objective, but the capacity of adaptation to changing environments is dependent on the variability that can be generated in an individual. Such variability is normally achieved in sexual reproduction through recombination Along with sexual reproduction, which brings about variability, reproduction also takes place through asexual means to produce normal prototypes.

The reproductive pattern in different groups of plants is intimately connected with alternations of generations. In the lowermost groups, termed thallophytes, the predominant generation is the thallus, representing a haploid number of chromosomes. This is characteristic of both algae and fungi, the two groups of thallophytes. However, within the algal groups, there has been a gradual evolution from most simple to complex forms, the plant body in most cases representing the haploid generation with a single set of chromosomes (Krishnamurthy 1994).

However, the most primitive group of algae which show a prokaryotic structure are the Cyanophyceae with blue-green pigments, where the cellular structure is highly undifferentiated. The central body in Cyanophyceae represents the nucleus without a nuclear membrane and does not have well-differentiated chromosome structure. The central body itself is of a different consistency from the rest of the cellular plasma and forms a distinct entity within the cell. It can be differentiated from the surrounding cytoplasmic matter. However, the group has a specialized heterocyst nucleus connected with reproduction. Genetic studies involve physiological properties including nitrogen fixation. Since, like Cyanophages, the group is categorized differently it has been excluded from the following discussion.

2. Algae

Of the algal group, Chlorophyceae is most predominant. In Chloro-phyceae, in the primitive forms, the reproduction is mostly asexual with mitotic cell division. Sexuality developed later, possibly from flagellated zoospores with the gradual advent of isogamy, anisogamy and oogamy. The typical haploid number is predominantly gametophytic. It alternates with diploid sporophyte represented in majority of genera by a single cell, the zygote. Meiotic division with reductional separation of chromosomes takes place in the zygote itself, except in certain genera like *Ulva*, where a sporophytic generation with a diploid number alternates with that of gametophyte. Here the reductional separation of chromosomes takes place during the formation of zoospores from the diploid thallus.

Chlorophyceae

In Chlorophyceae, Volvacales has been studied extensively in relation to their chromosomal features. The chromosomes are mostly small with very little differentiation of centromeric structure. The haploid chromosomes so far reported in *Chlamydomonas* are mostly 8 or 16 with a few exceptions. Sixteen linkage groups have also been established (Levine and Ebersold 1960, Abbas and Godward 1965, Hastings et al. 1965). Intraspecific cytotypes have been recorded in a few taxa such as *Astrophomene* sp. with n = 4–8; *Gonium* n = 16–18 and *Volvox* sp.with n = 7 chromosomes. An interesting case was reported in *Eudorina* where *E. elegans* and *E. illinoisensis* (Goldstein 1964) are both dioecious and heterothallic with n = 14 chromosomes. Hybrid forms have been reported with 2n = 24, 28 chromosomes indicating possibly autogamous types, where the two nuclei undergo fusion in self-fertilization. This genus is characterised by polyploidy as well, which might have affected the sexuality (Godward 1966).

Volvocales. In Volvocales, clear cytological study on zygotic meiosis has been carried out in species of *Chlamydomonus* (Buffaloe 1958, Maguire 1976, Treimer and Brown 1976, Storms and Hastings 1977), *Gonium* (Starr 1955) and *Volvox* (Starr 1975). Even a synaptonemal complex has been recorded and 16 bivalents have been observed in species of *Chlamydomonus*. The origin of 16 chromosomes has been suggested to have arisen through endoreduplication.

Chlorococcales. In Chlorococcales, all the species are uninucleate except *Hydrodictyon* which shows a coenocytic condition. The chromosome number in this group ranges from n = 4–80 indicating both aneuploidy and euploidy. The nuclear divisions are very rapid during autospore and zoospore formation. Meiotic analysis has been carried out in the germinating zygotes of *Hydrodictyon*. Discrepant reports exist with regard to the nature of the thallus and time of meiosis in *Hydrodictyon* (n = 18). However, later studies indicate that haploidy is maintained in the thallus and meiosis takes place at the zygotic level.

Ulotrichales. In Ulotrichales, the haploid number ranges from 3–46 and 4–56 in Chaetophorales (Sarma 1983) which is normally present in the thallus. Here the plant body represents a typical haploid stage, but there are a few genera such as *Ulva* and *Enteromorpha* where a diploid constitution has been noted even in the plant body, which indicates an alternation of haploid and diploid phases. Extensive chromosomal races have been recorded in several genera indicating variability of chromosomal races at the haploid level. A feature of special interest in reproduction is that different cytotypes may look identical in morphology, but differ in their physiological characteristics, as in species of *Cephaleuros* (Jose and Chowdary 1977). Evidently, the origin of aneuploids through vegetative reproduction is one of the common features in the Ulotrichales-Chaetophorales complex. In the evolution of reproductive mechanisms, it has been suggested (Abbas and Godward 1964) that the disappearance of sexuality has been associated with the development of chromosomal races—cytotype.

Cladophorales. The group Cladophorales is characterised by multinucleate cells and both haploid and diploid thalli have been recorded (Sinha 1968, Shyam 1980). The chromosome number generally ranges from 2n = 12 to 124 with basic chromosome number n = 6 (Wik-Sjostedt 1970). The reduction division occurs in the sporangia of diploid sporophyte before zoospore formation, where the life cycle is haplodiplontic. The two generations may be iso- or heteromorphic. In other cases, such as in *Cladophora glomerata* where the gametophyte is diploid, reduction division is noted prior to gamete formation. Interspecific polyploidy has been recorded as also hybridization at the interspecific level. However, the gametes produced from triploid and hexaploids are normally non-viable (Godward 1966).

Oedogoniales. In Oedogoniales, the chromosome number ranges from n = 9–46. The filaments may be monoecious or dioecious. In dioecious species, there is no difference in chromosome constitution, indicating that sex in *Oedogonium* is not associated with a specific sex chromosome (Srivastava and Sarma 1979).

Both aneuploidy and hybridization play a distinct role in evolution. Evidently, such changes occur before the onset of the gametophytic generation of the thallus during reproduction.

Conjugales. The reproductive system in Conjugales is very distinct. The sexual reproduction is through conjugation of aplanagametes, both in unicellular forms, as in *Desmids* and filamentous forms as in *Spirogyra*. In this group, as in *Spirogyra*, a polycentric or diffuse centromere have been located (Godward 1966). The survival of irradiated fragments as chromosomes in reproduction is cited as evidence of their polycentric nature. The chromosome number in *Spirogyra* varies from n = 2–92 (Sarma 1983) whereas in *Desmodium* the lowest number is n = 8 chromosomes. The very high chromosome number and their survival through vegetative or sexual reproduction may be due to fragmentation of chromosomes. Polyploidy (vide Sarma 1983) is common both in *Spirogyra* and *Desmids* and even the aneuploids survive through vegetative reproduction. The origin of aneuploids, as in *Micrasterias* sp., has been suggested through agmatoploidy.

In *Desmids* it has been claimed that the change of chromosome number affects the mating viability (Goldstein 1964), in the sense that fusion of gametes with unequal numbers leads to non-viability of the zygotes. Viability is ensured through fusion of the gamete with an equal number of chromosomes (Brandham 1965). Zygotic meiosis has been recorded in *Spirogyra* (Godward 1961) with polycentric chromosomes. It differs from that of higher plants, in that the first meiotic division is equatorial rather than reductional.

In the reproductive mechanism in species of *Pleurotaenium* and *Cosmarium botrytis* (Brandham and Godward 1995, Ling and Tyler 1976), the zygospore (at the time of germination) may stay even at diplotene or diakinetic stage, though the ultimate filaments become haploid. This group presents a remarkable reproductive mechanism through development of the vegetative thallus, the non-motile gametes which may or may not have equal numbers of chromosomes and formation of a conjugation tube. Despite the equality of gametes and the formation of a fusion zygote, the viability of the latter in reproduction is controlled to a great extent by true gametic equality.

Siphonales and Siphonocladales. In the groups Siphonales and Siphonocladales, both multinuclear and uninuclear thalli are present. The Siphonaceous forms are either septate or non-septate and are multinucleate throughout, whereas members of the Dasycladaceae remain uninucleate throughout the vegetative phase of the life cycle, except in the genus *Cympolia*. An interesting feature noted in this group is the gradual increase in the size of the nucleus through replicated endomitosis, as in *Acetabularia* and *Batophora*. The dissociation of the primary nucleus is comparable to that of nuclear fragmentation. It is remarkable that the secondary nuclei divide mitotically with a uniform chromosome number. The mechanism through which the regularity is maintained through dissociation is still not knwon. The tendency of vegetative reproduction by fragmentation or a sexual process by diploid zoospores is also common in this group. It has been claimed that there are two types of thalli, (i) haploid—where meiosis occurs during the germination of zygote as in *Protosiphon botryoides*, and (ii) diploid—where meiosis occurs before gametogenesis. The zygote develops into the diploid form in species of *Valonia*, indicating a monogenetic cycle. There are different types of mechanism, as in the digenetic cycle, where the diploid thallus produces zoospores as in *Derbasia*. In this genus, during meiosis three spores abort and one remains functional. The thallus may be haploid and produce gametes as in *Halicystis*.

In *Acetabularia* (Koop 1979) the reduction division sets in during the division of the primary nucleus, though so far no crossing over or synaptonemal complexes have been recorded.

Throughout, in this group, intraspecific polyploidy has been recorded and the races are adapted to different geographical regions. Certain species are characterised by very long chromosomes, such as *Valonia*, whereas in others, e.g. *Bryopsis*, the chromosomes are very short.

Vegetative reproduction is principally through fragmentation of the thallus as in Siphonales or formation of propagules as in *Codium*. Fragmentation is normally associated with nuclear division. It finally leads to the formation of zoospores in the diploid thallus of digenetic species. There is the formation of zoospore in the diploid thalli of digenetic species and formation of gametes in monogenetic species.

In Dasycladaceae, the resting cysts or specialized organ is a special method of reproduction. Such cysts are diploid which, on perennation, lead to the formation of zoospores. The zoospores may develop algal features or may lead to the formation of gametes of two kinds, which again undergo fusion. The resulting zygote leads to the development of plants resembling the parents. While gametes are formed in the cysts, the last mitosis is followed by meiosis leading to haploid nuclei. The latter divide further to constitute the gametes. Thus, the life cycle is monogenetic continuing either through an asexual zoospore or by fragmentation, direct germination of cysts, or through meiosis and fusion of gametes.

Extensive endomitosis has been recorded with 2n = 16 chromosomes in species of *Batophora*. In the genus *Acetabularia*, during fusion, in addition to isogamy, slight anisogamy has been recorded as in *A. mediterranea*.

The group Siphonales (including Dasycladaceae) is remarkable since the diploid thallus may remain diploid throughout, can reproduce asexually through fragmentation or by diploid zoospores, with complete suppression of meiosis.

Charophyta. Alled to the Chlorophyceae is Charophyceae, where not only is the body structure highly differentiated, but reproduction is quite complex. In addition to vegetative reproduction, sexual reproduction involves the formation of globule and nucule, homologous to antheridium and archegonium. Most of the dividing nuclei have been studied from spermatogonial filaments from antheridia, and a few from vegetative thalli. The chromosome number is very variable, ranging n = 7–17 in *Chara*. A comparison of the karyotype of the spermatogonial filaments and vegetative cells of the famele plants shows a similar chromosome number to indicate that the vegetative cell is strictly haplontic. There are conflicting reports with regard to the sex chromosome mechanism in reproduction. In species of *Nitella*, a heterotypic sex chromosome has been claimed, whereas in the rest of taxa there is no clear sex chromosome mechanism with n = 6 chromosomes (Kanahori 1971). Even then there is a record of numerical difference of males and females in a species of *Nitellopsis*, with n = 14 in males and n = 15 in females. However, confirmatory reports are yet to be obtained for sex chromosome mechanism. In the life history of Charophytes, there are two opposing views with regard to the time of reduction division: zygotic meiosis, or before the formation of gametes (Tuttle 1924, 1926) though reduction division could not be recorded in this group. As Charophytes are haplontic, zygotic meiosis seems to be the rule. The distinct chromosmal races at an intraspecific level is also on record (Abbas and Godward 1963).

In Bacillariophyceae, representing the diatoms, though studied by different authors, no clear chromosome counts are reported. A sex chromosome mechanism has not been observed though phenotypic differentiation of the sexes is distinct (Subramanyan 1946). Sexual reproduction is oogamous in centric diatoms, and isogamous in pennate diatoms. The latter, though isogamous, behaviourally are anisogamous, the one active and the other passive. Sexual fusion results in an auxospore and the zygote is often surrounded by a membraneous perigonium which increases after fusion. The auxospores germinate and give rise to a diploid vegetative phase by mitotic division. The reduction division takes place, in the spermatogonia or oogonia, and leads to a large number of sperms from several microspores, or sperm mother cells are produced by mitosis. During spermatogenesis too, a few male nuclei may persist, whereas in oogonia one or two eggs may ultimately survive. There may be autogamous reproduction, that is the haploid nucleus from a single cell may fuse to form auxospore, or auxospores are formed from diploid vegetative cell without meiosis.

In Euglenoids, reproduction is principally vegetative. The chromosome division has been termed 'reduplication' which is linear, resulting in two chromatids with the apparent existence of a centromere. No clear equatorial plate has been recorded, and the nuclear envelope persists through mitosis. The daughter chromosome movement is considered to be autonomous, with the absence of a normal spindle, and reproduction is through cell cleavage (Godward 1966). Counts of chromosomes ranging from n = 14–95 (Shashikala and Sarma 1980) suggests the existence of polyploid strains. The meiotic configuration has been reported in some genera. Autogamy involving fusion of nuclei which are the product of one cell (Krichenbauer 1937) has been claimed but the similarity of chromosome number in this vegetative filament and the spermatogenous filament suggests the existence of zygotic meiosis. This family contains both dioecious and monoecious taxa, which are important criteria for specific identification as well. In the diversification of taxa, both sexual and asexual means of reproduction and polyploidy have played a very distinct role, including even the intraspecific existence of duplication of chromosome numbers. Thus, the reproductive mechanism of *Chara*, though complex, is not necessarily associated with the complexity of the sex chromosomal mechanism.

In Xanthophyceae, represented by *Vaucheria* and related genera, chromosome studies have been carried out using both vegetative and antheroidal filaments. Similarities in these further suggest the haplontic nature of the thallus. Karyokinesis has been observed to be intranuclear, also confirmed through ultrastructural study (Guerlesquin 1972). The occasional existence of a viable cell with a reduced chromosome number, suggests their origin from high polyploid nuclei. Euglenoid flagelletes are sexual clones with high chromosome numbers, and endopolyploid constitutions, but with no distinct spindle or centromere in chromosomes. Despite the occurrence of meiosis, sexuality is lacking.

The Dinoflagellates represent a diverse group of organisms which are considered as intermediate between pro- and eukaryotes. This group of organisms shows a distinct pattern of chromosomes in mitosis, differing from both pro- and eukaryotes. The dinokaryotes have chromosomes in a condensed state throughout the cell cycle, but lacking a visible centromere. No metaphase and anaphase arrangements, absence of a conventional spindle, and the presence of a nuclear envelope during mitosis characterize the group. The chromosomes contain DNA but not the histone protein as known in eukaryotes. Two types of nuclei have been recorded, the dinokaryotic (Zingmark 1970) with condensed chromosomes at interphase, and noctikaryotic where the chromosomes are dispersed. In the genus *Noctiluca*, the dinokaryotic nucleus is present in the gametes, whereas the vegetative cells are noctikaryotic, indicating that the two stages represent phases of differentiation.

The Dinoflagellates are generally haploid in nature and meiosis occurs in the zygote. In the zygote, a resting cyst is normally formed. Nuclear cyclosis has been reported (von Stosch 1973) marking the beginning of meiosis with rotation of nuclear content and an increase in nuclear volume. This is suggested to be related to synapsis. The group, in terms of their nuclear nature, behaviour and reproduction represent an unusual assemblage.

Phaeophyceae

In Phaeophyceae the thalli represent two distinct types: haplobiontic with a single free-living diploid phase, or diplobiontic with alternating free-living gametophytic and sporophytic phases. They reproduce through sexual and asexual means. The plant body is diploid and dioecism has been recorded. The reproduction of the individual occurs through gametophyte and sporophyte. In this group of algae, clear alternation and generation has been recorded, though the gametophyte is smaller in size and of shorter duration. The two groups Fucales and Laminariales have been extensively studied and present interesting data about both chromosomes and nuclear division.

Fucales. Mitosis has been studied mostly from the paraphyses of the conceptacle in species of *Fucus*. In the antheridium, the division is synchronous and reductional separation of chromosomes occurs in the first division of meiosis. The second mitotic division is successive leading to 64 nuclei. The oogonium, on the other hand, shows clear meiosis with the formation of tetrads. The ultimate result is the formation of 8 nuclei, followed by a period of rest. The 2n number is 64 in some species of *Fucus* (Evens 1962), and there is no difference in chromosome size and number in male and female haploid cells. However, a dark-staining body has been recorded in oogonia of several species. So far there is no evidence to confirm such bodies as sex chromosomes. There is a rapid growth of the oogonium in the final eight-nucleate stage.

Laminariales. In the Laminariales, there is a distinct alternation of generations with distinct female and male gametophytes which can be obtained from sporophyte culture. A distinct sex chromosome mechanism has been established in this group, especially in *Saccorhiza polyschides*. There is a very large chromosome which is present in the female gametophyte. The analogy of such differentiated chromosomes with female sex has been made with *Sphaerocarpus donnellii* of the bryophytes. In the young sporophyte, the Y chromosome could not be pin-pointed, but during meiosis in the sporangium, a small chromosome was observed to pair with the longest chromosome. Obviously, the products of meiosis lead to two kinds of nuclei: one containing Y, giving rise to male gametophytes, and the other containing X to give rise to female gametophytes. Because of the very small size of the chromosome of the male gametophyte, the sex chromosome could not be fully analysed. There is clear evidence for the absence of a very large chromosome involved which is present in the sporophyte and female nucleus. Sex segregation takes place, therefore, in the first division of meiosis leading to the formation of nuclei with X and Y chromosomes. The chromosome number of Laminariales ranges from 2n = 16–62 though the sex chromosomal mechanism has not been worked out in all the species. This group of algae has not only attained a specialized nature of reproduction in alternating generations, but also has chromosomal heteromorphicity in sexual reproduction.

Rhodophyceae

The red algae have a specialized breeding system, which is both sexual and asexual. There is an alternation of haploid and diploid phases and the whole group is characterised by non-motile male elements. Fertilization results in the development of carposporophyte which may be partially or completely parasitic on female gametophytes. The diploid carposposporophyte (in turn) leads to the development of carpospores which give rise to diploid tetrasporic phases. Meiosis occurs in the tetrasporophyte resulting in tetraspores which yield a predominant haploid gametophytic generation. It may be monoecious or dioecious (Van der Meer and Todd 1980). The characteristic feature in the preceding cycle is the amplification of the zygote in the gametophytic cell leading to the carposporophyte, and the independent tetrasporophyte, both having a diploid chromosome constitution. In fact, there is triphasic alternation of generation with a succession of haploid sexual, diploid tetrasporic stages as in *Polysiphonia*, where the carposporic and tetrasporic phases are morphologically identical. In the breeding system a single diploid spore may be directly dispersed after a single mitotic division of the zygote as in Rhodocheateae, or one of the diploid cells directly undergoes meiosis and the haploid spores are dispersed to give rise to a gametophyte as in *Rhodophysema*. On the other hand, the zygote may undergo many mitotic divisions and the resulting diploid phase may be mitosporangial, borne on a haploid generation, or directly meiosporangial, which may be very small, borne on a haploid gerneration with the liberation of several haploid spores as in *Liagora tetrasporifera*. In another genus, *Palmaria*, the multicellular meiosporangial generation over-grows the diminutive haploid phases, producing many haploid sperms. However, the tetrasporphyte in the majority of red algae is free-living and meiosis occurs in the tetrasporangium.

The triphasic generations may again involve the addition of asexual cycles with the spores developing from both the gametophytic and sporophytic generations, without the formation of gametes or without reduction division. There is further complication of the breeding system in the case of dioecious species where the male and female gametophytes' containing the spermatogonium and carpogonium are identical in appearance. In tetrasporophytes responsible for reduction division, there may be monosporangia which lead to diploid spores for vegetative reproduction, as in *Coliospora*. Sexual dimorphism has been reported in the male and female gametophytes.

There have been varying reports of the occurrence of meiosis in *Porphyra tenera*, such as a fertilized carpogonium, a concosporangium, or during concospore generation (Tseng and Chang 1955 1989). In North Atlantic species no alternating haploid and diploid species have been recorded. The chromosome number in the group ranges from $n = 3$ in *Porphyra* (Cole 1990) to $n = 72$ in *Polydes* (Rao 1956). Polyploidy is rather common. Endopolyploidy arising out of polyteny, fusion of multinucleate cells, fusion in axillary cells and the apical cell, and anlargement are also reported (Goff and Coleman 1986). Chromosome studies so far carried out do not reveal clear evidence of a sex chromosome except in *Wrangelia argus* (Rao 1971) where XX_2, Y_1Y_2 heterochromatin knob chromosomes have been recorded. Telocentric chromosomes with clear bands as well as translocation heterozygotes are also on record. The presence of clear haploid and diploid phases also make facultitative asexual reproduction possible.

3. Fungi

In fungi, the reproductive mechanism is normally associated with the formation of spores, both sexual and asexual. Asexual reproduction may involve a haploid nucleus, or a dikaryotic cell with two nuclei side by side, both haploid. Fragmentation of the vegetative body, splitting of cell into two daughter cells by constrictions, formation of buds or special spores are the features of asexual reproduction with same chromosome number. However, there are cases where such asexual spores may ultimately behave as male gametes, and participate in sexual reproduction leading to a diploid nucleus. The spores may often be flagellate or non-flagellate, the former are termed zoospores. The sporangia may be formed in a normal mycelium, or after meiosis (meiospores), and the nucleus remains haploid. Thus, the sporangium of *Mucor*, a phycomycete, is produced by the mitotic division of haploid nuclei, whereas the ascus in ascomycete (homologous to sporangium) results after meiosis. The reproductive life cycle in fungi may involve three distinct phases: (1) the haploid monokaryotic phase, (2) the diploid dikaryotic phase preceded by plasmogamy, two nuclei lying side by side, and (3) diploid diplophase arising out of fusion or karyogamy. In order to maintain the haploid stage in the thallus, the reduction division follows immediately after karyogamy, so that the spores become haploid. Both monoecism—bisexuality, or dioecism—unisexuality, have been recorded, and the gametangia may be isogametic or heterogametic, the former being indistinguishable, and the latter being distinguishable. In several fungi, clear formation of gametes has not been recorded, and in such a method of reproduction, the genetic term 'compatibility' is used to denote genetic difference.

Genetics of Parasexuality

In addition ot the normal method of reproduction, where plasmogamy, karyogamy and meiosis occur in succession, there are fungi which do not follow the conventional methods of these events at specified period, the process is termed 'parasexuality'. A parasexual cycle may involve fusion of two hyphal nuclei of the same mycelium, resulting in a homozygous diploid. Similarly, through mutation of a hyphal nucleus followed by plasmogamy and fusion, a heterozygous diploid nucleus is formed. During mitotic division, a diploid nuclear recombination may occur, otherwise termed 'mitotic recombination'. Finally, as noted in *Aspergillus nidulans*, diploid mycelium may give rise to haploid conidia in the somatic cell, through somatic reduction. Such parasexual cycle is common in Deuteromycetes and Basidiomycetes which have a normal sexual cycle as well. Except for the time sequence, parasexuality and sexuality do not show any quantitative differences. The terms compatibility and incompatibility, are used to denote the status of the spores in relation to their potential to undergo union and fusion. This genetic differentiation of reproductive bodies is necessary when the two parents do not show any visible sexual difference. Genetic compatibility in reproduction may involve homothallism and heterothallism, and indicate that the thallus is sexually self-fertile or self-sterile.

Further complexity in the genetics of sexual reproduction arises from the fact that the genetic make-up may lead to bipolar heterothallism or tetrapolar heterothallism, leading to two or four mating types. It has been well established that the compatibility in bipolar heterothallism is controlled by a pair of factors 'A' 'a' located in identical loci in homologous chromosomes. Similarly, two pairs of factors, Aa and Bb, are involved in tetrapolar heterothallism.

Range of Sexuality and Genetic Characteristics

In the sexuality of fungi, there is a range of both haploid and diploid cycles in their genetic constitution. The haploid cycle may originate directly after nuclear fusion followed by meiosis, the zygotic diploid phase is of short duration. Such types are rather common in many Phycomycetes and in a few Ascomycetes.

In Phycomycetes, there is almost no interval between plasmogamy and karyogamy, and on germination the zygote gives rise to the haploid mycelium. In Ascomycetes, there is a distinct interval between dikaryotic structure and diplodization, though the interval between the two is variable. The ascogenous hyphae in Ascomycetes may represent only the dikaryotic phase, whereas in Basidiomycetes, this growth phase may be very much prolonged. In both Ascomycetes and Basidiomycetes, there are three phases: (1) haplophase, (2) dikaryophase, and (3) diplophase, but there is a fundamental difference between these two groups. In Ascomycetes, dikaryophase is dependent on haplophase, whereas in Basidiomycetes, it is completely independent. The specialized cell of the dikaryotic hypha of a haploid mycelium gives rise by reduction division to asci and an ascospore, whereas in Basidiomycetes, the basidia develop from secondary dikaryotic mycelium. In Ascomycetes, moreover, the dikaryotic stage arises out of the development of gametangia, spermatia or somatogamy resembling to some extent the sexuality of Phycomycetes. In Basidiomycetes, on the other hand, the dikaryotic stage develops through hyphal fusion or fusion of spore and hyphae, that is spermatization and somatogamy. Distinct sexual organs have not been recorded, indicating complete degeneration of sexuality. But, generation of bipolar or tetrapolar spores indicate complexity.

In Ustilaginalles of Basidiomycetes, the dikaryotic condition with two haploid nuclei persists almost throughout the life cycle, and the haplophase originating after nuclear fusion and post-meiotic spore formation is short-lived. the other growth phase, the haploid phase (as in Oomycetes) is restricted only to the gamete, and the entire thallus represents the diploid phase as in higher organisms. Finally, there are certain fungi as in aquatic Oomycetes where both haploid and diploid phases alternate regularly—a state which is rather rare in fungi. The above-mentioned patterns of nuclear cycles are indices of wide genetic diversity in fungi, associated with the reproductive cycle.

Chromosome Study in Relation to Reproduction

Extensive studies have been carried out on the cytology of fungi (see AK Koul and Pushpa Koul 1994, Staben 1995), but not necessarily on the mechanics of reproduction.

In Zygomycetes and Oomycetes, the mitotic division during reproduction may or may not involve spindle formation, the latter is comparable to amitosis. Chromosome number ranges from n = 3 to high polyploids in Oomycetes. During sexual reproduction, meiosis may take place in the zygote, or in the gametangia. The division of the oospore nucleus has been considered as reductional, leading to multinucleate mature oospores (Thirumalachar et al. 1949) whereas gametangial meiosis has been recorded by others (Tyagi et al. 1982).

In Basidiomycetes, the range of chromosome number is between n = 2 and 30 and both aneuploids and polyploids are on record. Such wide variability has been considered responsible for their growth and reproduction in a wide range of hosts. Lack of sexual organs and even somatogamy or heterokaryosis through vegetative nuclear division, is also on record (see AK Koul and Pushpa Koul 1994).

In Basidiomycetes, chromosome studies have been carried out in several genera including rusts, smuts and mushrooms. The base number ranges from n = 3–30, and both aneuploids and polyploids have been reported. In Uredinales, nuclear fusion occurs in teliospore (in many genera) and, finally, the nucleus divides in the basidium to give rise to haploid basidiospores through meiosis. In several rust fungi, a four-celled basidium develops two-nucleate basidiospores, whereas quadrinucleate basidiospores with haploid nuclei in two-called basidia are also on record (see AK Koul and Pushpa Koul 1994). There are contradictory reports of nuclear fusion in hyphae and the development of diploid mycelium as recorded in certain genera (Rajendran 1967; Chinnapa and Sreenivasan 1968).

The smuts represent a clear haplodiplonic life cycle in which the promycelium and sporophytic mycelium are distinct. During meiosis in species of *Ustilago*, the first division in germinating chlamydospores is claimed as equational, and the second as reductional (Singh and Pavgi 1975).

In Deuteromycetes, the cytological studies indicate six nuclear mycelial cells from germinating conidia which ultimately differentiate into uninucleate cells (AK Koul and Pushpa Koul 1994). The binucleate and uninucleate conditions of the mycelium are associated with the host-range species.

Special Genetic Features in Ascomycetes and Basidiomycetes

The special genetic characteristic in Ascomycetes have been extensively studied in yeast *Saccharomyces*, and *Neurospora* (Egel et al. 1990, Herskowitz et al. 1992). Yeast, unlike filamentous fungi, do not have a clear differentiation of sexes. The plasmogamy is normally followed by karyogamy but, in certain species, such as in *S. cerevisiae*, diploids can proliferate indefinitely by budding. Meiosis followed by haploidy in this species can be induced by specific nutritional conditions. Haploids can also continue to proliferate by budding before mating. The genetics of mating is controlled by certain specific types. The haploid 'a' cell expresses α-pheromone. The diploid state is signalled by the presence of both mating type alleles in the single nucleus. In different species of *Saccharomyces*, different mating types are recorded, such as four mating type genes in *S. pombe* (Kelly et al. 1988). The physiological response between haploidy and diploidy in yeast is clearly expressed, and regulating factors of meiosis and recombination have been worked out.

In *Neurospora crassa*, on the other hand, the vegetative form is a multinucleate myclium. The sexual spore, the macro-conidia and micro-conidia can also serve as male mating types in sexual reproduction. The trichogynes, the female hypha specialised for mating, respond to the pheromones secreted by the opposite mating type, and orient their growth towards the latter (Bistis 1983). After hyphal fusion within the ascogonium between the opposites, dikaryotic nuclei undergo synchronous mitotic division and develop ascogenous hyphae. Fusion occurs in the penultimate cell followed immediately by meiosis and formation of haploid ascospores (Raju 1980). Several maturing types have been studied so far (Glass and Kuldau 1992). Both pseudohomothallic and heterothallic forms have been observed in *N. crassa*. Moreover, there are four heterothallic species of *Neurospora* which can be distinguished by their ability to form fertile off-spring in interspecific crosses. In pseudohomothallism, a single ascospore may be derived from generally 'a' and 'A' nuclei in the ascus and is, therefore, self-fertile. The special features of reproduction in *Neurospora* are the ascogenous hyphae in a special fruiting structure, whereas yeast asci do not have any fruiting body. In yeast, switching of the mating type is a very special phenomenon. If there are two mating types, one is stable and the other is switchable. The exact molecular mechanism of this event and genetic control of the phenomenon is not fully known.

An unusual case has been noted in *Pyronema confluens* where, in the ascogenous hyphae, the diploid nuclei undergo a second fusion resulting in a tetraploid state. In ascus, the first meiotic division is succeeded by a mitosis which in turn is succeeded by a second reduction division, resulting in the haploid chromosome number in spores. Because of the second meiotic division, the term 'brachymeiosis' is applied. However, there are alternative descriptions which involve brachymeiosis as well (Staben 1995).

In Basidiomycetes, the reproductive mechanism and its genetics differ, to some extent in homo- (rusts and smuts) and heterobasidiomycetes (mushrooms), the former have septate and the latter unicellular basidia.

Normally in *Ustilago*, mating involves the formation of a conjugation tube which may be preceded by a thin thread, which connects mating partners (Day 1976), both are haploid. The haploid dikaryotic filamentous growth can only occur if the mating partners have both *a* and *b* loci. The dikaryons with two haploid nuclei continue to propagate and differentiate in the host plant. They do not form clamp connections as in mushrooms, and can form diploid brandspores by fusion; within the plant tissue. These diploid brandspores again germinate to give rise to mycelium and basidium, reduction division in the basidium leads to haploid basidiospores. These spores form sporidia and can propagate by mitotic division and budding. In *Ustilago* (Gow 1995), there are at least two distinct loci controlling mating. The *a* locus has two *a*1 and *a*2 alleles, whereas the *b* locus is multiallelic. For example, in the *a* locus idiomorphic DNA has been located encoding two genes, a precursor and a receptor for pheromone. The dikaryotic state is maintained by interaction of pheromone receptor pairs. Similarly, for the *b* locus as well, idiomorphic DNA has been located with many forms (Schutz et al. 1990, Gillissen et al. 1992).

The mycelium of *Schizophyllum commune* and *Coprinus cinereus* of homobasidiomycetes is monokaryotic and contains uninucleate haploid cells. Each haploid nucleus of a homokaryon mycelium has two incompatible factors A and B. If two homokaryons with unlike factors come near each other, they mate. After fusion, the hyphae become dikaryotic. the dikaryotic hyphae divide synchronously.

The mating types in this group are determined by four loci As, Ab, aB, Bb. The A and B loci are on different chromosomes. The requirement for mating requires difference at least in one A and one B loci. The genetic complexity of mating and polarity is very distinct.

Comparative Pattern of Genetics of Sexual Reproduction in Thallophytes

In Thallophytes, reproduction through sexual means has not necessarily been associated with genetic complexity. There has been an evolution from isogamy-anisogamy to oogamy, monoecism to dioecism, and homothallism to heterothallism. But, genetically, the haploid and diploid chromosome number has the predominant feature without sex chromosomal mechanism associated with sex, except in certain cases.

However, in algal forms, despite the absence of genetic complexity, the organ differentiation associated with reproduction is quite evident, as exemplified in *Chara* as well as in some brown and red alga. Even with such complexity in male and female reproductive organs, and development of clear haplophases, leading to the evolution of higher forms, there has not been a high degree of genetic complexity.

On the other hand, in fungi, specially in yeasts as well as in Basidiomycetes, genetic differentiation of + and – sex, the donor and the recipient is marked. The two mating types do not show much morphological difference unlike in algae, but genetic polarity exhibits a high level of complexity. In addition to existence of a number of alleles, the Basidiomycetes show clear quadropolarity in the genetics of spores, and mating depends to a high degree on specificity of alleles. Thus, the two groups, algae and fungi, represent certain basic contradictions in their genetics of reproductive systems. In algae, the organ differentiation with heterosexuality has reached a high level of differentiation without a concomitant increase in genetic complexity. In fungi, on the other hand, the genetics of mating (homologous to sex) is highly complex, but without differentiation of sex organs. The fungi, with such complexity in genetics of mating, in a saprophytic or parasitic existence, have thus evolved without contributing to any other eukaryota. Algae, on the other hand, with simple genetic systems along with evolution of heterosexuality and organ differentiation have helped the evolution of higher eukaryotes.

4.   Bryophyta

In Bryophyta, the delay in meiosis after the formation of the diploid zygote and the intervention of a diploid sporophyte is the most characteristic feature. The two principal groups in Bryophyta sensu lato, Hepaticae and Musci, differ markedly in haploid gametophyte and diploid sporophyte.

Genetics of Reproduction and Reproductive Organs in Liverworts (Hepaticae)

In Sphaerocarpales the haploid gametophytes are dioecious and are usually formed in groups of four. Two males and two females develop on the dorsal surface of the thallus. The male plant is smaller than the female. The mature diploid sporophyte shows a globular capsule with a foot attached by a short neck of seta. The spore tetrads usually remain adhering together and develop into four gametophytes. Two males and two females thus grow in a cluster, and facilitate reproduction.

In Marchantiales, the genus *Riccia* is mostly monoecious or homothallic, but some species such as *R. discolor* are dioecious or heterothallic. The fertilized egg becomes the oospore or zygote and it increases in size, completely filling up the ventor. The zygote, forms the sporophyte by mitotic divisions. The amphithecial layer forms the jacket of the sporophyte, while the endothecium forms the sporogenous tissue, the archesporium. The archesporial cells divide further to form two types of cells, the diploid spore mother cells or sporocyte, and the diploid nurse cells with watery vacuolate cytoplasm. Ultimately, the nurse cells and the jacket layer (both diploid) distintegrate, followed by the disintegration of the haploid inner cell layers of the ventor wall—all these form a nutritive viscous fluid within which spore mother cells remain suspended and undergo meiosis. The liberation of the spore takes place later and the spores may germinate even while attached to the mother plant. The ventor wall is a gametophytic tissue of the previous generation. The spore wall comprises three layers: exosporium, mesosporium and endosporium.

In the entire life cycle of *Riccia*, the thallus is the gametophyte starting from the spore onwards and only the zygote is the sporophytic generation. for the development of an undifferentiated embryo originating from the zygote, the protective tissue is of supreme importance. This keeps the embryo responsible for the development of spores and reproduction. The archegonial ventor remains embedded within the archegonium which contains a haploid chromosome, along with the rest of the gametophytic tissue. Moreover, the nourishment to the developing spore mother cells is provided by the nurse cells which are diploid. Biciliate sperm with haploid nuclei are characteristic of the group. In the comparatively more highly evolved genus *Marchantia*, the antheridium and archegonium develop on the antheridiophore and archegoniophore. All these have the typical haploid chromosome number. Out of reproductive mechanisms, vegetative reproduction is rather common, with the formation of gemma cups (in *Marchantia*), which represent typical haploid nuclei. The development of a diploid sporophyte follows progressive sterillization of potential sporogenous tissue, adding complexity to the sporophytic structure.

Dioecism and Sex Chromosomes in Liverworts

In Bryophyta, the chromosome number (in hepatics and mosses) ranges from n = 6–38, or more. Polyploidy is more common in the mosses than in hepatics.

The dioecism in liverworts is considered to be associated with heterochromatic chromosomes forming the longest and smallest types, referred to as 'H' and 'h' chromosomes, respectively (Tatuno 1941; Yano 1957; Khanna 1971; Smith 1983). However, there are conflicting reports about the nature of these chromosomes, their relationship to one another, and their role in sex determination (Chattopadhyay and Sharma 1991). The best example of a sex-determining chromosome in reproduction is in *Sphaerocarpos donnellii* (Smith 1983). In the sex chromosomal mechanism, multiple sex chromosomes have been reported. It has been further suggested that the X chromosomes bear genes for femaleness, as plants with damaged X chromosomes may also be male, even in the absence of a Y chromosome (Knapp 1936). Such plants produce non-motile sperms, indicating further that the gene for motility of the sperm is present on the Y chromosome. The location of a gene for femaleness in X is supposed to be near the centromere (Lorbeer 1941). Regarding the size of X and Y and their heteropycnosity, there are contradictory opinions and 'sex-associated-chromosome' is preferred rather than the 'sex-chromosome' (Smith 1978). Such heteromorphicity of the chromosomes has been recorded in several liveworts, including species of *Pellia* and *Plagiochila*. Their heterochromatic nature has been recorded through banding techniques as well (Smith 1983).

In addition to sex-associated-chromosomes, which may be large or small, aneuploidy and polyploidy have also been recorded in liverworts and Anthocerotales, though it is rare in the latter. Intraspecific polyploidy and structural alterations are also on record, suggesting their role in the propagation of individuals.

Accessory Chromosomes and Sex

Accessory chromosomes, often termed as 'm' chromosomes have also been observed in Anthocerotales. Such accessories may be absent from the antheridia but present in thallus and sporophytic tissue as in *Phaeoceros laevis* ssp. *carolinianus* (Proskauer 1957). Another dioecious form of this species does not contain accessories, whereas monoecious forms contain the accessory chromosomes. The relationship between the sex and necessary chromosomes is not completely clear, though the role of accessories in increasing genetic variability in inbred genotypes is clear. The small chromosomes have been recorded in species of *Sphagnum* as well.

Chromosome Number and Hybridity in Mosses

In mosses too, the chromosome study reveals certain unusual cytogenetic features. In general, polyploidy is prevalent in mosses. The existence of polyploidy ranging between $n = 7–15$ and its multiple has been recorded in different genera (Smith 1978, 1983). One of the reasons for the high numbers often noted in the capsules of mosses may be the exposed nature of the capsule, which is subjected to environmental changes. The extent to which such an assumption is valid is yet to be confirmed. However, the existence of cytotypes within a species, as noted in *Pseudobryum cinclidioides* (Lowry 1948; Ono 1970; Wigh 1972), may indicate the viability of structural and numerical changes in reproduction of the individuals. The structural changes of chromosomes as evidenced in meiotic irregularities in non-hybrid sporophytes, suggest the influence of structural heterozygosity in evolution. However, the meiotic irregularity as noted through asynopsis, bridges, and fragments may be due to genetic imbalance (Smith 1983). The difficulty of securing interspecific viable hybrids in the majority of mosses indicates that the isolating mechanism leading to speciation is controlled by complex genetic behaviour. However, meiotic irregularity in mosses does not necessarily indicate a gross structural hybridity, as evidenced by their presence in populations where spore formation is perfectly normal, as in *Atrichum undulatum* (see Smith 1983). In general, polyploidy in mosses has played an important role in evolution at every taxonomic level, ranging from families to varieties. It is likely that in the mosses rather than the liverworts, the polyploids have been more favoured in reproduction because of inherent genetic characteristics for adaptation with high gene dosage.

Polyploidy and Sex Differentiation

Polyploidy in mosses has also been associated with the evolution of monoecism in reproductive features. Like *Sphaerocarpus*, morphology of different sex chromosomes has been studied in several species of *Bryum* (Kumar and Verma 1980). However, the occurrence of both homo- and heteromorphic types of sex bivalents in different populations of *B. argenteum* may suggest heteromorphicity and homomorphicity are derivatives from each other and occur in different stages. In some species of Polytrichaceae and Mniaceae, sex chromosomes have often been observed to be identical with 'm' chromosomes, the two designated as X and Y differ from each other in length and shape (Ono et al. 1977).

The relationship of polyploidy with the reproductive mechanism has been claimed in certain genera. The dioecious species of *Fissidens* have n = 5 or 6 chromosomes whereas the monoecious species have n = 10 or 12 chromosomes in the complement. The majority of polyploid species of mosses are monoecious while related haploid species are dioecious. It has been claimed that polyploid tendency for alternation from dioecism to monoecism has been responsible for promoting self-fertilization (Kumar 1983).

Diversity in Sex Mechanisms in Reproduction

Different types of sex chromosomes associated with sexual reproduction have been observed in bryophytes. There may be a single 'H' chromosome in a monoecious gametophyte as in the species of *Haplozia* and *Nowellia* (Tatuno 1941). In species of *Madotheca* and *Brachythecium*, the 'H' chromosome does not differentiate but the gametophyte is dioecious. There may be cryptic sex differentiation at the genic level. Another step is possibly represented by differentiation of the 'H' chromosome into X and Y, in female and male gametophytes, respectively, as in species of *Philonotis* and *Leucodon* (Yano 1957; Khanna 1960). However, in the genetic mechanism of reproduction in mosses, the autosomes seem to have played a marginal role in the determination of sex.

5. Pteridophyta

Genetics of Reproduction in Lycopsida

In Pteridophytes, the reproductive system is provided with several unusual genetic features. Despite the typical alternation of generation with free-living diploid sporophyte and independent much-reduced haploid gametophyte, the genetics of sexual cycles show wide diversity. This diversity includes homospory and heterospory in diploid sporophytes, and homothallism and heterothallism in haploid gametophytes, comparable to monoecism and dioecism in higher plants. Homosporous ferns are rather common and heterospory developed independently in Lycopsida and Pteropsida, without disturbing the nuclear cycle. The diploid sporophytes of *Lycopodium* are homosporous, giving rise to haploid spores and prothalli which contain both antheridia and archegonia. As such, the uniformity of spores in diploid sporophytes, also transcends the uniformity in the genetic constitution of the haploid gametophyte. The differentiation of sex organs in the same gametophyte is thus an ontogenetic pattern and depends on repression and derepression of genes in the prothallus, at certain phases of development.

The onset of heterospory in this group is marked in species of *Selaginella* and *Isoetes*. The differentiation of diploid megasporangium and microsporangium marks the very beginning of genetic expression of sex differentiation through repression and derepression of genes at the sporophytic level. Genetically, sexuality is associated with differential development. In *Ampelopteris* it has been recorded (Loyal et al. 1984) that the fast-growing gametophytes are archegoniate, the slow-growing ones produce antheridia, whereas nearly 20% of the population features the hermaphrodite condition. In *Selaginella*, such induction of heterospory originates at an early stage where micro- and megasporophylls are well differentiated. The process of sex differentiation, through gene expression, is an otherwise uniform genomic structure and is initiated at an early stage. But, in species of *Isoetes*, such differential gene expression occurs at a very late stage (Sharma 1994a) where, often, the micro- and megasporophyll can not be distinguished before the formation of two types of spores, large and small.

The gametophytes in both the genera have distinct sexes which are haploid. Both in *Selaginella* and *Isoetes*, within the prothallus the gametophyte (endosperm) is haploid and remains at a free-nuclear state for a long time. The nutritional role of this haploid tissue for the diploid embryo is characteristic.

Of the heterosporous forms, a very unique case has been recorded in *Isoetes coramandelina*. Several populations of this species bear only megasporangia and are present as both diploid and triploid sporophytes (Abraham and Ninan 1958; Ninan 1958; Pant and Srivastava 1965). During meiosis, the chromosomes are univalents in several cells and often a ring of four has been noted, suggesting structural heterozygosity. In several cells with complete asynapsis, there is regular anaphase spearation with 2n number at the pole (Verma 1961). No evidence has been recorded of reductional separation. Such megaspore development is parthenogenetic and diploid sporophytes are normally formed even in the absence of microspores.

Compared to *Isoetes*, in *Equisetum*, despite the presence of asynapsis in several populations of hybrid origin, parthenogenesis or agamospory has not been recorded at all (Walker 1984). Evidently, selective development of agamospory implying development of sporophytes from megaspores in absence of microsporangia, as in *Isoetes* and *Selaginella*, is a strategy for maintaining genetic system for reproduction of heterosporous pteridophyte.

Some species of *Selaginella* have a base of nine chromosomes (Kuriachen 1963) whereas in species of *Isoetes* the numbers range from 10 to 56, with a high degree of polyploidy. No sex chromosomal mechanism could be clearly established in the determination of sex. Evidently, duplication of chromosomes has played an important role in the diversification of species of *Isoetes*.

Genetics of Reproduction in Sphenopsida

In Sphenopsida, some species of *Equisetum* represent a remarkable combination of homospory with homo- and heterothallism. Here the spores are all identical and haploid, like other Lycopsida but, on germination, they give rise to male and female or bisexual prothalli, both haploid. However, despite morphological similarity of the haploid spores, it was claimed long ago that the spores show cytoplasmic sexuality, that is, males and females have different internal constitution and yield different chemical reactions indicating incipient heterospory. In some species of *Equisetum*, in the diploid strobilus, the nutritive tissue is diploid which provides nourishment to the developing spore mother cells and, ultimately, to the development of haploid spores.

The prothalli may develop antheridia or archegonia, or both, suggesting apparent haploid homothallism or heterothallism. However, dioecism in gametophytes has been claimed to be under the influences of nutritional requirements in the haploid system. There can be reversal of sexuality through artificial manipulation of nutrients in the evironment (Mohan Ram and Chatterjee 1970). Such reversal of sex as well as the presence of a homothallic prothallus, clearly indicates the presence of both male and female genetic factors for sex in each spore. The development of male and female organs is entirely dependent on gene expression in terms of ontogenetic pattern, working in self-equilibrium with the environment. However, from such a behaviour, *Equisetum* possibly represents an intermediate stage in the genetic system of reproduction towards the development of sexually-differentiated distinct plants.

Genetics of Reproduction in Pteropsida

Amongst the ferns, self-fertilization is very common in the homosporus type, resulting in the rapid formation of tetraploids after the production of unreduced spores. It has been claimed that, as homosporous forms are capable of self-fertilization, there is a tendency towards homozygosity and a consequent loss of genetic variability. It is quite likely, that multiplication of chromosome sets, alterations and, to some extent, pairing between them, may increase genetic variability (Klekowski 1970, 1973, 1976). It is, however, claimed that self-fertilization has also been responsible, to a great extent, for the survival of interspecific hybrids, so common in several fern genera, as in *Asplenium* and *Adiantum* (Walker 1984).

Another fact is of special importance in ferns, and especially in homosporous ferns, the chromosome number is very high, indicating clear polyploidy. Despite such duplication in chromosome number in the wild polyploids, no multivalents are generally formed. This is a clear indication of the fact that the genetic system is operating for the suppression of multivalent formation, thus leading to normal reproduction. As a large number of polyploids have been recorded in ferns, and as hybrids have also been observed on a large scale, it is quite likely that the genetic mechanism for multivalent suppression, as well as apomictic reproduction where necessary, has been favoured in selection. Ferns thus represent an unique example of a very high degree of polyploidy, large scale hybridization and almost obligate apomixis, all genetically controlled in a large number of genera (Walker 1984).

In Pteropsida, or true heterosporous ferns, diploid sporophytes occur only in Hydropteridineae or water ferns. Sporocarps, in diploid sporophytes, contain both micro- and megasporangia. As compared to *Selaginella*, here the gene expression differentiating male and female spores is much delayed until after the formation of the sporocarp.

Genetics of Apogamy and Apospory

Apogamy and apospory are two rather unusual features of reproduction, and consequently there is a shift from the normal genetic constitution of gamete and spore-bearing structures. The term apogamous sporophytes implies the development of the sporophyte from a gametophyte without syngamy, that is combination of male and female gametes. Such sporophytes may result from any cell of the gametophytes, i.e. the vegetative cell or even a sexual cell. The term parthenogenesis is applied in such cases, and, as a consequence, such sporophytes contain the haploid number and there is often formation of restitution nucleus through spindle failure resulting in a diploid cell before meiosis. As a result, the reductional separation restores the haploid number of chromosomes in the spores and, consequently, the haploid gametophytes.

Apospory, on the other hand, is known to produce a diploid gametophyte directly on the sporophytic leaf. They can also be artificially induced. Such diploid gametophytes may lead to the (1) formation of tetraploid sporophytes through fertilization, or (2) a triploid sporophyte by crossing with a normal haploid gamete. But despite the occurrence of such abnormal behaviour, there has always been a tendency in the system to often revert to normal genetic mechanisms. Such an event occurs through a compensating mechanism by an aposporous gametophyte forming an apogamous diploid sporophyte and an apogamous sporophyte producing a aposporous gametophyte. Such reversal of the genetic mechanism, including the absence of fertilization, is rather common in Polypodiaceae. In addition, it is also common in *Osmunda*, *Marselia*, *Selaginella* and *Hymenophyllum*.

Polyploidy and Apomixis

It has also been observed that obligate apogamy and apospory in reproduction is rather common in polyploids (Walker 1984). Possibly, in evolution, the genetic system underlying reproduction in true ferns involving polyploidy is a significant factor. In a polyploid, a diploid gametophyte and a tetraploid sporophyte are rather common. Such a genomic system permits the development of an apogamous sporophyte without going down below the basic diploid number of chromosomes. It is to be regarded as an evolutionary strategy for maintaining a genetic equilibrium in reproduction.

As far as genomic constitution is concerned, except in *Selaginella*, polyploidy is rather common both in ferns and fern-allies, being more frequent in the latter. In species of *Lycopodium*, chromosome numbers have been recorded as high as more than 400, and in *Ophioglossum* more than 1200 have been recorded so far. Intraspecific polyploidy are also on record. The existence of a large number of polyploid and aneuploid cytotypes in Pteridaceae, Aspidianceae and others owe their existence to apomixis—a strategy in the genetic system providing opportunities for variable genotypes to survive through selection.

Apomixis as Reproductive Strategy in Evolution

In ferns, wherever it occurs, agamosporous reproduction is obligate (Walker 1984), and is inherited as a dominant character even though it involves a number of steps in which separate genetic control is necessary. Evidently, this arises out of correlated interactions of genetic features. Possibly, as suggested by Walker (1984), such correlation of separately-controlled features might have been acquired through hybridization from different sources, at different steps. It has also been noticed in ferns that genetic control of mixis or apomixis does not affect the germination of the spores. Evidently, spore germination has an inbuilt genetic system for their viability, irrespective of the nature of sexuality.

Reproduction through vegetative means, without any alternation of generation, is a feature in several species of ferns. The examples are: perpetuation of plants by root bud (Evans 1968) in *Polypodium disporsum* in stress conditions, or by means of gemmae (Farrar and Wagner 1968; Wagner and Wagner 1966), or only as prothalli such as in *Vittaria* and *Trichomanes*. However, except for such examples involving environmental stress, vegetative apomixis is very common, specially in the hybrids (Walker 1966).

Such vegetative reproduction has been responsible for the survival of the triploid *Blechnum* or *Gleichenia* hybrid (Walker 1984). In fact, even the club moss *Lycopodium selago*, which is a high polyploid with ca. 264 chromosomes, shows very irregular meiosis but reproduce through bulbs (Manton 1950). Even hybrids in several *Equisetum* species have been recorded.

In Pteridophytes, and specially in ferns, vegetative apomixis has led to the survival of sterile hybrids. This strategy has been adopted in a number of cases to tide them over unfavourable situations until fertility is restored through mutation (Lovis 1977), controlling pairing or allopolyploidy (Walker 1966). Simultaneously, it is also likely that vegetative apomixis has been responsible to a great extent for the survival of variable genotypes needed for colonisation in new areas. The importance of vegetative apomixis as a genetic strategy for reproduction of polyploids, hybrids and the new genotypes of ferns is very distinct.

6.   Gymnosperms

In gymnosperms, a diploid sporophyte is the predominant generation and the haploid gametophyte is entirely dependant on the sporophyte for its growth. Of the different groups of gymnosperms, mainly represented by Cycadales, Coniferales and Gnetales, unisexuality and dioecism are common in Cycads (Abraham and Mathew 1962) and *Ephedra*, whereas the rest of the species are characterised by monoecious diploid sporophytes. The diploid sporophytic character shows wide variation both in the vegetative and to some extent in the reproductive parts. But the genetic uniformity of diploidy is clearly maintained in general.

As far as the male haploid gametophyte is concerned, the pollen grain is characterised mostly by having three types of nuclei: a prothallial nucleus, a tube nucleus, and a generative nucleus. This comprises the entire nuclear set up of haploid gametophytic generation. The number of haploid prothallial cells may vary from 1 in *Cycas* to 32 in *Araucaria*, and the pollen grain may be shed at the two-called and four-celled stage. In the two-celled stage, the two nuclei are the generative nucleus and the tube nucleus, both haploid, and are shed before the formation of a prothallial nucleus. The nature of the tube also varies, without any effect on the genetic system. Before microsporogenesis, the nutrition to the developing microsporangia is provided by the diploid sporophyte itself, where the strobilus is an inherent part. The development of the haploid female gametophyte at the initial stage, is dependent to a great extent both in monoecious and dioecious forms on the diploid nutritive tissue of the sporophyte, which forms the nucellus.

With the formation of the megaspore after meiosis, the development of the female gametophyte is preceded initially by free-nuclear division of the cell surrounding the archegonia. These haploid cells later constitute the endosperm, which becomes cellular. Following fertilization, the development of proembryo may have free diploid nuclear division, before the formation of suspensor and the embryo proper, all diploid.

Cycadales

A special feature in Cycads is the reproductive cycle which occurs in different seasons of the year. As it is located in tropical regions of the globe, it is genetically adapted not to have a winter rest as in temperate forms. The reproductive cycle requires an unusually long period for completion (see Bhatnagar and Moitra 1996), there being a long interval between pollination and initiation of the ovule. The fertilized embryo also requires several months to mature. Such characteristics in *Cycas* occur in almost all species and are a clear indication of the genetic control of a reproductive system well suited to habit and environment.

Other Gymnosperms

In general in Coniferales, the reproductive cycle is shorter than the Cycads. Being temperate in distribution, they undergo a period of rest in winter. An intervening rest period is a specific genetic adaptation for conifers in the reproductive system. Variation in the normal cycle is also noticed. Diploid pollen grains (Mehra 1946) in *Gnetum* and an occasional polyploid nucleus in the gametophyte have been recorded in a number of cases (Vasil 1959). Similarly, in *Pinus patula*, during male meiosis, the chromosomes may extrude out of the spindle and affect the morphology of pollen (Sharma 1994b). Genera like *Taxus* and *Cupressus* show clear polyembryony, all diploid tissue, mostly arising out of cleavage of the sporophytic diploid embryo. In the species of *Gnetum*, reproduction is also associated with polyembryony originating from the secondary suspensor. Such additional methods of reproduction occurring consistently in certain taxa clearly show genetic control (see Maheshwari and Vasil 1961).

In *Ephedra*, in the formation of the diploid seed, the maternal haploid gametophytic tissue, which is often free-nuclear (as many as 512 nuclei at the beginning; Sharma 1994b), with abundant nutrition, plays a very significant part. In the development of the proembryo, the initial free-nuclear division may vary in a number of species, especially in those of *Taxodium* and *Podocarpus*. In species of *Gnetum*, on the other hand, there is no free-nuclear division of the zygote as in *Welwitschia* and *Sequoia*.

In *Ephedra* a feature of special importance is the occurrence of double fertilization (Friedman 1992) of the two male nuclei, one fuses with egg whereas the other, with ventral canal nucleus. The production of second fertilization also ends to diploid supernumerary embryo (Friedman 1994). This provides some nourishment to the developing embryo comparable to that of endosperm in angiosperms. Such a behaviour provides clues as to the origin of double fertilization and its genetic control in angiosperms.

In *Cupressus* and *Thuja*, the formation of female gametophytes may involve a large number of free-nuclei and development of several archegonia.

As far as chromosomes are concerned, the number is comparatively less: n = 7, 11 in Cycadales and Gnetales as compared to higher numbers often recorded in Cupressaceae, especially in *Juniperus wallichiana* (2n = 32). In general, polyploidy is rare among the conifers (Khoshoo 1959).

Heteromorphicity in sex chromosomes has been recorded in Cycads (2n = 20 + XY; Lee 1954, and *Ephedra foliata* (2n = 12 + XY; Mehra and Khitha 1981). In all three cases, nucleolar chromosomes are involved in heteromorphicity and no clear heteropycnosity has been recorded (Chattopadhyay and Sharma 1991).

A feature worth recording in the development of the reproductive mechanism of the gymnosperms is the nature of the endosperm responsible for the development of the diploid seed. In cycads, the endosperm is a gametophytic tissue with haploid chromosome number, compared with angiosperms, where double fertilization leads to a triploid set of chromosomes in endosperm.

7. Angiosperms

The angiosperms with an enormous number of species represent a vast spectrum of genetic diversity. The genetic strategies of reproduction can be well-understood if species origin is considered in its totality.

Genetics in Relation to Speciation

The genetic basis of origin of species is well known. In a population of individuals belonging to a species, aberrations range from gene mutation to a variety of chromosome abnormalities, providing the basis of variability. The majority of mutations and chromosomally-aberrant cells are invariably eliminated in natural selection. However, several mutants or aberrant gametes may remain viable. If chance permits, they participate in fertilization. Such processes occur at random in different individuals, and may lead to the origin of a variety of recombinants in nature. Of such a wide array of recombinants and mutants, certain heterozygous combinations ultimately become best-suited to the environment. These well-adapted individuals with distinct morphological and physiological characteristics ultimately survive with the gradual exclusion of others in a specific environmental setup. Ultimately, this population maintains an equilibrium with its well-adapted individuals, show phenotypic characteristics, and differ from the parent species. Such a state, often termed 'incipient' species, is at the risk of losing its identity, through gradual crossing with nearby populations. These phenotypes, distinct from the original species, gradually undergo genic alternations, and ensure incompatibility with allied individuals. The incompatibility or sterility barrier, arising out of genic or chromosomal alternations, gives them a distinct identity with the status of a species. This genic isolation may be mainfested at different levels, including seasonal, physiological, morphological or ecological. Thus, the species originates through a reproductive strategy which ensures variability through heterozygosity, mutation and the incompatibility barrier. The entire genetic strategy for reproduction in angiosperms is thus based on these two crucial factors: (1) maintenance of identity through a compatibility barrier, and (2) generation of variability through mutation and heterozygosity. These two genetic issues form the underlying theme of reproduction in angiosperms.

Alternation of Generations

In the evolution of alternation of generations, angiosperms represent the present climax where the plant body is the diploid sporophyte and the haploid gametophytic generation is reduced to a spore and a few nuclei. The sporophytic generation, that is the diploid plants, produce special reproductive structures like flowers in which genetic recombinations and chromosomal reductions occur during meiosis. The diploid meiotic cell produces haploid microspores and female megaspores. The two male and female spores have very limited development, leading to micro- and megagametophyte, the latter embedded in diploid maternal tissue. The male gametophyte ultimately delivers two haploid sperm cells into the female gametophyte, one fuses with the haploid egg to form the zygote which generates the embryo. The other haploid sperm cell usually fuses with the secondary nucleus in the central cell of the female gametophyte. This nucleus has three sets of chromosomes, one from the male parent and two from the female. The double fertilization leads to the development of triploid endosperm, with several nutritive functions, crucial to the development ot the embryo.

In this general pattern of development, nutrition for the developing microspore is provided by the tapetal cell. This is diploid but may often become multinucleate due to nuclear divisions. The endomitotic pattern of nuclear division including even polyploidy is also on record (Maheswari Devi et al. 1994). The tapetal cells which follow degeneration often become a periplasmodium associated with nutritional demand to the meiotic cell (Johri et al. 1992).

Microsporogenesis, which leads to a normal tetrad of four haploid cells, may through further mitotic division, lead to a polysporous state (Maheswari Devi *et al* 1994). Variations are also on record of a single pollen grain with many haploid nuclei. That the tapetum plays an important role in the maintenance of a normal reproductive state is borne out by the fact that premature degeneration may lead to pollen abortion and male sterility.

As far as female gametes are concerned, the integumentary tapetum or endothelium is the repository of food material for the developing embryo sac and the embryo. Though diploid in constitution, they often become polyploid by fusion. The diploid nucellus supplies nutrition to the diploid megaspore mother cell. The embryo sac arises out of meiosis and may be two-nucleate, four-nucleate or eight nucleate, all haploid in constitution. A typical female gametophyte has eight haploid nuclei, two form synergids along with the egg at the upper pole of the embryo sac, two nuclei form upper and lower polars (n + n), and the remaining three nuclei form antipodals at the lower pole—all with haploid chromosome constitutions. The nutrition to the developing embryo after fertilization is provided by the endosperm which is generally triploid, following double fertilization (fusion of diploid nucleus with haploid sperm).

Reproductive Mechanism

In the reproductive strategy of angiosperms, the genetic mechanism for cross fertilization is diverse. It is manifested through organ diversity, like protandry, protogyny and heterostyly, non-correlation between anthesis and stigma receptivity in the same genotype, as well as the heteromorphic relationship between pollen-stigma-interaction at the individual level leading to self-incompatibility (Haring et al. 1990). The genetic control of cross fertilization is aimed at bringing about diversity and recombination at the intraspecific level.

The reproductive mechanism has thus been geared to achieve propagation of genetically stable individuals and generation of diversity to meet the requirements of the environment. Stability is secured through fidelity of replication and regularity of meiotic and mitotic division. In order to achieve diversity, hybridization as well as mutation, both genic and chromosomal, are resorted to in nature. The chromosomal variations range from spontaneous alterations in structure to variations in number. They arise out of deletions, fragmentation and translocation on the one hand and non-disjunction and restitution on the other. The origin of trisomy and tetrasomy, with duplications of one or two chromosomes and polyploidy with entire genome duplication are rather common in the plant kingdom. Structural alterations may lead to structural homozygosity for alterations or heterozygosity for alterations, the best example of balanced structural heterozygosity being provided by *Oenothera lamarckiana*. The importance of structural changes in the reproductive strategy of plants in evolution could not be properly assessed in the past. Developments, of modern techniques of chromosome analysis permits the resolution of even minute differences between chromosomes. These have clearly indicated the universal role of structural changes of chromosomes in plant evolution and breeding strategy. In all families of flowering plants so far studied, structural changes, minute or gross, surviving through reproduction, have affected evolution. In the enrichment of genetic diversity in plant species, the role of structural changes viable through reproduction is significant.

Influence of Polyploidy on Reproduction

One of the important aspects of chromosome behaviour affecting the genetics of reproduction is polyploidy. The direct duplication of chromosomes results in autotetraploidy. The autotetraploids often show non-viability as a consequence of aberrant meiosis. The formation of multivalents and high meiotic irregularities often lead to genetic sterility. As such, the autotetraploid, though having a duplicated gene dosage, does not stand selection. This is the reason why natural autotetraploids are very rare, such as *Allium tuberosum* and species of *Geum. A. tuberosum* is a very good example where despite clear multivalent formation, there is regular meiotic behaviour and formation of genes, which control crossing over and subsequently bivalent formation, as in the 5B genome of wheat. In other cases, such as in *Commelina obliqua*, despite very high chromosome numbers, there is regular bivalent formation and seed-setting.

Leaving aside such rare cases, autopolyploids generally undergo mutations and lead to genotypic diversity at the intra-specific level and ultimately to the formation of stable individuals. Such individuals which arise out of combination of diverse genotypes are not true autopolyploids, but more precisely allopolyploids. The majority of species in the plant kingdom which show high chromosome numbers have undergone heterozygosity at a certain level of reproduction to produce reproductively-viable individuals adapted to the environment. In general, the species with high chromosome numbers, as recorded in different families of flowering plants, have originated out of a lower base at the initial stage, followed by duplication, mutation and heterozygosity at a later period.

The effect of polyploidy on reproduction is also evident as it ensures fertility in hybrids. Following normal hybridization, the hybrids are often sterile due to non-homology of the two genotypes. The polyploidization of two parents followed by hybridization, or hybridization followed by chromosome doubling, lead to the production of amphidiploids. The reproductive fertility and seed-setting are due to normal bivalent formation. Most of the crop species such as *Brassica juncea, Oryza sativa and Triticum aestivum* are amphidiploid. For interspecific hybrids, therefore, polyploidy often becomes a genetic prerequisite to ensure fertility.

Polyploidy ensures interspecific hybridization between parents as well, which normally remain isolated through seasonal isolation. The increased gene dosage may enhance the blooming seasons, simultaneously flowering with the other species where the flowering in the latter is rather later. Through induced flowering at the same season, the seasonal barrier between genotypes is removed, ensuring intergenotype hybridization.

The positive effect of polyploidy is also manifested in reciprocal interspecific crosses. In majority of the species, crosses between a diploid and a tetraploid are successful, provided the tetraploid is used as the female parent. On the reciprocal side, with the diploid as female, the chances of hybrid formation are comparatively meagre. This fact may indicate that a polyploid chromosome constitution remains in a balanced state with its own cytoplasmic setup. In other words, the latter is in equilibrium with its increased chromosome number. As the female parent only provides cytoplasmic backup, the diploid female is unable to accommodate the increased chromosome constitution of the male parent in its cytoplasm. On the other hand, the polyploid female already adapted to an increased chromosome constitution, is well-suited to accommodate fewer chromosomes, and remains viable in reproduction. However, it is true that the entire process of compatibility of reproduction is a gene-controlled process, but polyploidy undoubtedly affects their expression.

In addition to other effects of polyploidy on the genetics of reproduction, the profound influence on sex chromosomes needs special mention. Polyploids occur in dioecious species with heteromorphic chromosomes as in species of *Melandrium* (Riley 1957). Polyploidization and consequent irregular meiotic behaviour leads to an imbalance in the sex chromosomal mechanism. There may be a tetraploid with 2X 2Y, 3X Y, and X 3Y. Such combinations greatly affect the formation of normal progeny and lead to abnormal differentiation of sex. Such an imbalance is considered to indicate that the reproduction and manifestation of sex depend on a distinct balance between autosomes and sex chromosomes, and also as between the two sex chromosomes.

Male Sterility

A genetic factor associated with reproduction with a profound influence on hybridization is male sterility. Male sterility of certain genotypes is manifested in pollen abortion and non-viability of pollen. It has been recorded in several crops, such as *Brassica*, that a factor for male sterility involves abortion and non-viability of pollen and is controlled at the cytoplasmic level mostly by the mitochondrial system. The mitochondrial genome of male sterile and male fertile lines often differ in certain sequences. The genetic characteristics involving mitochondrial genome-induced sterility is another genetic strategy to ensure cross fertilization. The induction of male sterile mutants through artificial means, is thus often resorted to secure hybridization between different genotyupes.

Cytoplasmic Influence on Reproduction

The influence of cytoplasm on compatibility of hybridization is also exhibited in certain reciprocal crosses. In several agricultural and horticultural crops such as in species of *Oenothera*, the hybridization is successful only if certain genotypes are used as females. This brings out clearly the fact that the cytoplasm of a particular genotype can carry with it the hybrid nucleus, whereas the cytoplasm of the other genotype is unable to do so. As the nuclear constitution remains identical, irrespective of the way the crossing is being made, the control of individual cytoplasm in the formation of the hybrid is evident. Such cytoplasmic control is often exerted through cytoplasmic genes located in its organelles, specially in the mitochondria, as in case of male sterility.

Male sterility also arises in a hybrid combination where the cytoplasm belongs to one parent and the nucleus to another. Such genome substitution can be achieved by repeated back-crossing of the hybrid with the male parent, as has been extensively done in Triticineae. Thus, maternal inheritance is a clear example of the control of genetic makeup of the cytoplasm, in addition to nuclear genes, on the origin of fertile hybrids.

Impact of Apomixis

The other genetic mechanism which has a distinct influence on viability of polyploid hybrids and mutants is apomixis. Lately, this behaviour has aroused considerable interest amongst evolutionists (Calzada et al. 1996). Apomixis, as the term implies, is the production of individuals, without undergoing mixis or fertilization. Leaving aside vegetative apomicts, where vegetative cells are involved, apomicts may originate out of unreduced gametes or restitution nuclei, where the chromosome constitution remains the same as the diploid parent. On the other hand, the haploid egg may develop into a haploid individual without undergoing fertilization. Lately, there have been studies on this line (Koltunow 1993) by different workers. Such parthenogenetic development is often noted (parthenocarpy) in some of the horticultural crops.

Apomixis, though a deviation from the normal method of reproduction, has played a very important role in stabilization of aberrant genotypes as in species of *Citrus* and *Potentilla*. Polyploids and hybrids in nature often lead to sterility in view of high meiotic irregularity and consequent formation of aberrant gametes. Such genotypes, as in case of polyploids, where multivalent formation causes absserrations, face extinction because of non-viability. In order to tide them over the intial stage of non-viability until the onset of viability through gradual mutation, several generations may be required. Apomixis, which does not require viable gametes for production of individuals, aids in the stabilization of polyploids and hybrids for a certain period. Once, through gradual mutation in evolution, such genotypes acquire normal meiotic behaviour and seed-setting capacity, apomixis may not be required any more for reproduction. As such, it has been presumed that apomixis has played a very significant role in aiding restorration of fertility, and thus stabilization of species in evolution.

On the other hand, the genetically negative role of apomixis in evolution can not be ignored. Apomicts, by definition, have lost the capacity for recombination. Genetic recombination, as is well known, not only generates variability but confers adaptive advantage to the species. With changing environmental setup, the mutations and recombinations-inherent features of evolution, serve as sources through which the species adapt to change. The selective advantages are ensured by recombinations. Apomicts, on the other hand, permit the survival of the hybrids and mutants under only a constant environment. They have a selective value as long as the environmental conditions remain unchanged. With the onset of any environmental imbalance, the apomicts, which are incapable of sexual reproduction, are at a disadvantage. The inability to produce new recombinants which may survive in the changed situation, is a serious limitation. Their homogeneity devoid of any diversity is a negative factor for their survival in evolution. Thus, apomicts genetically have both positive and negative features. In nature, in order to eliminate the negative effect of apomixis, certain flowering plants have the capacity to revert back to sexual reproduction under conditions of stress. This type of behaviour is otherwise known as partial apomixis. Such type of partial apomicts have been noted in Gramineae, compared with obligatory apomicts as in several species of Rosaceae. Such partial apomicts, because of their genetic flexibility, enjoy a dual advantage of apomixis for stability and sexual recombination for generating variability.

Amongst the angiosperms, the property of partial apomixis has been one of the factors which contributes to grasses standing as a most successful family in the group. Grasses, as is well known, have a large number of species with wide distribution under diverse climatic conditions of the world. The underlying features which are responsible for the success of the family are: (1) a large number of cytotypes adapted to different environmental conditions, (2) innumerable interspecific hybrids, (3) diversity of genotypes, and most important of all, (4) their unique genetic character, partial apomixis.

The family is regarded as a very good example where genetically two methods of reproduction have contributed to successful evolution.

Reproduction in Asexual Species

The importance of the genetics of reproduction in asexual species has been analysed in detail. In monocotyledons, several genera, especially those belonging to the orders Liliales, Arales and Amaryllidales, sensu lato, reproduce principally through asexual means. The method of propagation is principally vegetative, especially through the bulb, rhizome, tuber and corm which serve the purposes of perennation as well. In the majority of these, flowers are formed but seed-setting has been scarcely noticed. Evidently, these categories of plants have become adapted to vegetative reproduction and survive selection pressure, despite the ability to generate variability through sexual means. Even in the absence of sexual reproduction, these genera continually generate new genotypes through vegetative reproduction (Sharma 1983). The majority of bud sports in members of Liliaceae and Amaryllidaceae arise through such reproduction. It was observed initially in *Caladium bicolor* (Araceeae) and later in other asexual species, that somatic tissues often represent a chromosome mosaic in which both subdiploid and hyperdiploid chromosome numbers do occur. Such mosaicism not only involves numerical variations but structural variations as well. Later on, it was recorded that such variant nuclei often enter into the growing shoots, participate in the formation of daughter shoots during development, and ultimately on detachment, develop into new genotypes. Further evidence of this method of speciation through vegetative reproduction and chromosome mosaicism has been recorded through analysis of allied geno types. Intraspecific cytotypes which vary in chromosome number and structure are rather common in these genera. However, in spite of the variations in the somatic tissue, which are regular features of these species, the general diploid number in a genotype occurs in the highest frequency of cells. This complement is considered as to be a normal diploid complement of the genotype conerned. Thus, in such species, origin of new genotypes is facilitated through vegetative reproduction without the act of fertilization of apomicts. Thus, in angiosperms, apomixis plays a very important role in stabilization of variants in sexual forms and origin of new genotypes in asexual species.

Propagation In Vitro

The inherent genetic characteristic of the totipotency of plant cells is a unique example of genetic control of propagation. Every plant cell is capable of generating into a whole plant—a feature termed as totipotency. The natural totipotency is best exemplified in species like *Bryophyllum calycinum*, where even a leaf cell can generate into a whole plant. Regeneration of the entire plant from any organ in vivo is not uncommon in the plant system.

This genetic property of totipotency has been taken advantage of in artificial micropropagation where a large number of individuals can be regenerated from a small mass of tissue. Totipotency permits reproduction of individuals without undergoing fertilization and is a special genetic feature which can control reproduction in the plant kingdom.

During in vitro growth of plant tissues, when differentiated cells become dedifferentiated and form callus, often an instability in chromosome complement is observed. Such cultured cells often show variations in chromosome number and structure categorised under the term 'somaclonal variations'. The somaclonal variations normally stand less chance of selection in competition with the normal complements in regeneration and propagation of individuals. But in a low percentage of cases such variations lead to the origin and propagation of new genotypes. In other words, somaclonal variations are responsible for the enrichment of genetic diversity in tissue culture. Genetically, the property of totipotency—the basis of tissue culture— leads to the production of a genetically homogeneous individual on the one hand, and variable genotypes on the other. This property of totipotency has also been taken advantage of for securing somatic hybridization of widely different incompatible genotypes. Such somatic fusion in culture, facilitated by the fusion of naked diploid protoplasts, were initially worked out in Solanaceous species such as in species of *Nicotiana* by Evans et al. (1980) and *Petunia* by Power et al. (1976). Somatic fusion which leads to the production of tetraploid genotypes has also been obtained in a few other species of plants (Bhojwani and Razdan 1992). In fact, the regeneration and propagation after the fusion of two somatic cells in culture, has opened a new avenue for distant hybridization.

Somatic hybridization may also lead to the production of hybrid cells, with cytoplasm contributed by both parents and the nucleus only of the recipient. Such cells and their regenerants are termed as "cybrids'. They may originate where one of the parents may be an enucleated cytoplasm, or where the chromosomes of one partner are gradually eliminated in successive divisions, after fusion. Such cybrids are often male sterile in their behaviour. The male sterility of the cybrids is an indication of the necessity of nuclear cytoplasmic balance, and the importance of cytoplasm in the propagation of fertile individuals.

Chromosome Dosage and Sexuality

In the angiosperms, flowers may be hermaphrodite or unisexual. Unisexual flowers (male and female) may be present on the same plant (monoecious) or on different plants (dioecious). In spite of unisexuality being prevalent in several families of flowering plants, including Euphorbiaceae, the heteromorphic sex chromosomal mechanism is rather rare. A well-developed sex chromosome mechanism is known in species of *Melandrium*, and *Coccinia indica*, both members of dioecious Cucurbitaceae. Lately, in species of Menispermaceae, such heteromorphicity has been claimed (Chattopadhyay and Sharma 1991). Between the two extremes of hermaphroditism and dioecism, there are several species, such as in Anacardiaceae, where male, female and bisexual flowers occur on the same plant. There are also species where male and female flowers occur in succession, as rarely in *Carica papaya*, and in some species of Araceeae.

In the reproduction of plants, genetic factors responsible for the determination of sex evidently play an important role. In species where such well-defined sex chromosomes are present, as in *Silene dioica*, the genetic segments responsible for sex have been worked out (Charlesworth 1991). It has also been presumed that such heteromorphicism and differentiation of sex have originated from bisexuality, where chromosomes with both male and female factors are present. The genetic evolution of dioecism from monoecism has been demonstrated.

It is interesting that in angiosperms, despite the presence of a large number of species, strict dioecism is proportionately low (Chattopadhyay and Sharma 1991). Moreover, a sex chromosomal mechanism is very rare. Further, the presence of male, female and bisexual flowers on the same individual, such as in *Mangifera indica*, may indicate that strict dioecism and complete sex differentiation in angiosperms is a recent event and it has not met with much success in the reproductive system in this group of plants. Moreover, in the majority of the unisexual species, sex chromosomes have not been observed and the sex is determined by the entire set of autosomes and ontogenetic pattern.

In spite of the presence of a large number of hermaphrodite species, generation of genetic diversity in reproduction is ensured through self-incompatibility of genes, which may occur in various ways. It appears that in angiosperms, the presence of both male- and female-determining genes in the same system have been favoured in selection, compared with strict unisexuality. This feature, coupled with the capacity for totipotency, are unique genetic characteristics which have provided flexibility in their behaviour. These genetic features are responsible for their high selective value against environmental stresses, and achieve success in evolution.

8.   Concluding Remarks

Reproduction in the plant kingdom has been designed towards successful replication and generation of variability. Analyses of the reproductive strategies from thallophytes to phanerogams indicate different degrees of difference between the groups to meet the above objectives.

In the algal system, at least in the primitive groups, the effort of reproduction has been geared principally towards replication and generation of variability only through gene mutation. Such genetic behaviour is prevalent in almost all members of blue-green algae—the Cyanophyceae. With the gradual evolution of complexity in structure, the genetics of reproduction has undergone gradual alterations. There has been reproduction through fusion of identical, non-identical and completely different male and female gametes, evolving genetically, in different groups. Such complexity in reproduction arose out of the need for enrichment of genetic variability through sexual means. Thus, in the algal system, different mechanisms of reproduction both sexual and asexual are evident, and the strategy for genetic recombination has undergone significant changes in different groups. Extremely complex genetic mechanisms are known in brown and red algae, whereas in dinoflagellates, the chromosome structure in even its chemical make-up lies at a very primitive state. Simplicity of chromosome structure such as a diffuse centromere, occurs even in Conjugales where reproduction has attained a high level of genetic complexity through conjugation. Such a wide variety of reproductive mechanisms and diversity of types appear to have been prevalent in the aquatic environment. Even in the nature of the reproductive organs, starting from only male and female filaments with no differentiation of reproductive organs, as in *Spirogyra*, there has been a gradual increase in complexity as in the globule and nuclule stages of *Chara*, or the conceptacle structure in *Fucus*. One of the essential features of the aquatic environment is the abundance of nutrition and compratively less stress. This has enabled the diverse forms of algae to flourish in the rich environment. The phenotypes range from free-floating to deeply embedded types. In other words, the selection pressure for survival have been at a lower level, permitting a variety of genotypes to survive, compared with the stress faced by terrestrial forms.

The other group in Thallophyta, the fungi, has been genetically adapted to restrictive reproductive mechanisms, entirely dependent on the saprophytic and parasitic existence. Being dependant for their survival on the environment, the latter puts a strong premium on the development of complex reproductive forms, except in certain genera. In general, in fungi, the genetics of reproduction involves principally hyphal fusion, excepting in a few genera such as in Ascomycetes where phenotypic complexity is associated with their sexually different genetic status. However, the genetic diversity to cope with limitations posed by a parasitic and saprophytic existence, is met with, in complex genetic mechanisms represented by bipolarity and quadripolarity. Such genetic complexity in the sexuality of fungi is not associated with complexity of reproductive organs. In this respect, their difference from algal forms is evident, as in the latter, the reproductive organs have attained a very high level of differentiation. Of the Thallophytes, the algae indicates a potential for evolution whereas the fungi represent an almost blind line.

The next groups higher in the evolutionary series are the liverworts and mosses. the reproductive features are genetically advanced from those of the thallophytes. This advanced nature is also associated with the evolution of land habit. It is known that modern-day land plants originated from aquatic ancestors inhabiting the tidal zone, where reproduction through ciliate sperms could be accomplished only during high tide. Out of those forms, some of the genotypes established themselves on land where sexual reproduction could be achieved only during rains. Because of this, the plant had to develop genetically different methods for propagation of individuals, through progressive stabilization of potentially sporogenous tissues and the production of spores in the diploid body—termed the sporophyte. The prolongation of mitotic division of the diploid zygote developing into a sporophytic structure, culminating in meiosis and the production of haploid spores, has been the genetic strategy for reproduction in bryophytes. The haploid spores give rise to a gametophyte, which is highly differentiated, and the diploid sporophyte is dependant on the gametophyte. The differentiation of reproductive organs is genetically maintained in the gametophytic structure to bring out recombination and genetic variability. Even in some genera such as *Spherocarpus*, a sex chromosome has been recorded. But the bryophytes, though amphibian in habit, are also adapted to certain restricted environments which are comparatively moist, shaded and where reproduction depends on the presence of water. The plant body, the gametophyte, though independent, does not have features to take advantage of wide climatic ranges, and so mostly thrives in certain seasons, except in drought-resistant genera. The saprophyte with its capacity to produce a large number of genotypes is also genetically handicapped by its dependence on the gametophyte. The principal genetic tendency in this group of plants is to achieve independent sporophytic status, as indicated in the meristematic zone of *Anthoceros* and related genera.

In vascular cryptogams, starting from pteridophytes, the reproductive strategy has undergone heavy genetic alterations. The diploid sporophyte, with the principal objective of producing spores, became the dominant generation. However, its very origin depended on the fusion of gametes borne on haploid gametophytes, which are also independent. The genetic feature in pteridophyta with independence of haploid and well-developed diploid generation is unique. But the gametophytic generation is much reduced and requires favourable conditions for germination, growth and development, essential for the production of gametes and fertilization. The gradual dominance of the diploid sporophytic generation adapted to broader environments, enabled the plants to survive successfully in terrestrial situations. Fertilization through multiciliate sperms, responsible for the origin of the sporophyte by sexual reproduction, still remained dependent on the availability of water. In view of the limitations imposed on fertilization, both the sporophyte and gametophyte—diploid and haploid, respectively, took recourse to apogamy and apospory. Synapsis is avoided. This genetic strategy for reproduction was necessary to cope with unfavourable situations.

In ferns, the development of the sporophyte has also been associated with wide duplication of chromosome numbers. Such duplication of chromosomes might have been one of the genetic strategies for survival. However, in this group of plants, it is not yet known which essential gene sequences are present in the entire chromosome complement. It is likely that species with very high chromosome numbers might have an unusual number of accessories and repeated sequences. Compared to the simple nature of the plant body, such as in *Ophioglossum vulgare*, the chromosome number is as high as 2n = 1260. It is unlikely that all the chromosomes contain unique essential sequences. Highly repeated sequences, if present, might have been a genetic adaptation to their reproduction and survival. However, such an unusually high chromosome number is clear evidence of advanced status, originating possibly out of polyploidy, if not fragmentation of chromosomes. Such advanced genetic status with apparently simple morphological features in *Ophioglossum* may be an index of regressive evolution in this group.

With the advent of seed plants, the gymnosperms, the remarkable feature of reproduction has been the absolute dependence of the haploid generation on the diploid sporophyte. The same plant bears both the generations. The gametophyte is reduced to spore and rudimentary male and female tissues. The need for protection of the embryo arising out of fertilization in the sporophyte, led to the origin of a typical seed-like structure. Sexual reproduction no doubt meets the need for recombination and generates variability. But the sexual gametes originate out of spores borne on the sporophyte. The group of plants, except the Gnetales, are all arborescent and as such restricted to certain specific climatic conditions. The range of distribution of both the Cycadophytes and Coniferophytes is rather limited. Vegetative reproduction is not frequent. Its limited reproductive capacity and restricted geographical distribution has made this group of plants, except the Gnetales, almost a blind line in evolution. However, in the origin of an arborescent habit, the Benettatalean-Magnoliaceous-stock might have been productive.

Despite dioecism, the group does not have a well-developed sex chromosome mechanism except in a few genera. The genetically-adapted arborescent habit of the plant and limited distribution, do not permit a high frequency of chromosomal variants. From the point of view of chromosome evolution, gymnosperms represent a very unusual group, where comparatively primitive genera like that of Cycads have asymmetrical karyotype, and symmetry is noted in the advanced forms. Several authors have advocated such an origin through Robertsonian fusion as in case of animal systems. The extent to which such debated chromosome evolution in gymnosperms is correlated with genetically adapted reproductive mechanisms is not fully known.

In angiosperms, the flowering plants, the reproductive system presents a broad spectrum. Both sexual and asexual reproduction in the group are associated with distinct genetic features. They include polyploidy, aneuploidy and structural variations of chromosomes, in addition to mutations. These influence the origin of several variants and a wide diversity of genotypes. The enormous biodiversity in the plant kindgom in general and angiosperms in particular, owes its origin to remarkable reproductive mechanisms, controlled at the genetic level. The main objective of reproduction had been the generation of variability through different means along with the stability of genotypes. Those objectives have been met by genetic control of the reproductive systems, including production of spores, gametes, fertilization and post-fertilization development. Moreover, a special strategy for the origin of variants has been the production of bud sports or mutants from the vegetative tissue. This is prevalent in species which reproduce through asexual means as in several monocotyledons. The compatibility and incompatibility at different taxonomic levels, has been under strict genetic control with numerous modifications both in male and female floral parts. The stabilization of genotypes, initially with aberrant behaviour due to an increase in chromosome number, aberration in structure or hybridization in several taxa, has been facilitated by apomixis. This is a genetic strategy for adaptation under fixed environmental conditions. The desired compatibility is maintained through the control of male and female sterility, both at the nuclear and cytoplasmic levels. The angiosperms thus represent a climax in the evolutionary series, where remarkable genetic adaptation for reproduction has led to high diversity of genotypes. Thus, the distribution of this group of plants inhabiting all the climatic regions and different ecological nitches, are principally attributable to the complex mechanism of reproduction, controlled at the genetic level.

The above discussion clearly indicates that in all groups of plants, the principal aim of reproduction has been propagation of their own replicas and adaptation to environmental upheavals. The second objective has been met through normal recombination as well as mutation, both chromosomal and genic. From algae to higher organisms, different degrees of organic complexity have evolved to secure variability of genotypes. In angiosperms, the wide spectrum of organic complexity to meet the needs of reproduction has led not only to efficient mechanisms of propagation but also to wide genetic diversity. Thus different steps in evolution, especially involving advances in reproductive mechanisms, are well represented in various groups of plants ranging from algae to angiosperms.

Glossary

Accessory chromosome. Extra chromosomes in the normal chromosome complement without marked effect on phenotype.

Agmatoploidy. Increase in chromosome number by fragmentation as in *Spirogyra*.

Alleles. Alternative forms of a gene which produce different effects for the same inherited feature.

Allopolyploidy. A polyploid having two or more distinct genomes.

Amphidiploid. A tetraploid hybrid with both the genome duplicated.

Anaphase. Stage of cell division following metaphase showing movement of chromosomes to two different poles in the spindle.

Aneuploidy. Cell having chromosome number, not-multiples of basic number (x).

Anisogamy. Sexual reproduction through non identical gametes.

Apomixis. Reproduction without fertilization.

Asymmetrical karyotype. Chromosome complement showing abrupt size difference.

Brachymeiosis. Meiosis equational, followed by reductional separation of chromosomes.

Chromosomal races. Individuals within a species showing different chromosome complements.

Cybrid. A somatic cell hybrid where cytoplasm of two parents have undergone fusion but with one genome.

Cytotypes. Individuals within a species differing in cytological features.

Deletion. Removal of a portion of a chromosome.

Diploid. Having a double set of chromosomes: twice the haploid number.

Diplotene. Post crossover stage during prophase I of meiosis.

Diakinesis. Stage in prophase I of meiosis showing scattered bivalents and terminalized chiasma prior to metaphase.

Endomitosis. Chromosome replication not followed by nuclear or cell division often leading to endoployploidy.

Endopolyploid. Cell having more than the diploid number of the cell, arising out of endomitosis.

Endoreduplication. Replication of DNA within the nucleus without division.

Euploidy. Cell showing chromosome number as exact multiples of basic number.

Gametogenesis. Nuclear division leading to formation of male and female gametes.

Haploid. Plant with half the number of diploid set of chromosomes.

Heterochromatin. Different from euchromatin in the staining cycle and condensation, made up of mostly of repetitive DNA.

Heteromorphicity. Homologous chromosome in which one partner differs from the other.

Hexaploid. Polyploid representing six sets of chromosomes.

Hybrid. The progeny obtained by crossing or mating two individuals or strains having different genotypes.

Interspecific polyploidy. Polyploid hybrid between two species.

Interphase. The stage of cell cycle between the two successive divisions.

Isogamy. Sexual reproduction through identical gametes.

Karyokinesis. Cell division involving equational separation of chromosomes, also known as mitosis.

Karyotype. Characteristic chromosome types in the somatic complement.

Meiosis. Reductional separation of chromosomes leading to half the complement in the daughter cells.

Metaphase. Mitotic phase with chromosomes/chromatids arranged at the equatorial plane in the spindle.

Multivalent. Pairing of more than one pair of homologous chromosome.

Non-disjunction. Chromosomes/chromatids not separating and going en block to one pole.

Polycentric. Chromosome having more than one centromere.

Polyteny. Replication of DNA without separation of chromosome strands.

Robertsonian fusion. Fusion between two short arm of two telocentric chromosomes.

Somaclonal variation. Chromosomal variation in cells grown in vitro.

Synaptonemal complex. A proteinaceous structure formed between homologous chromosomes during synapsis facilitating crossing over.

Telocentric. Chromosome having centromere at the end.

Tetrad. The group of four daughter cells following second division of meiosis.

Totipotency. Capacity of each cell to regenerate into a whole plant.

Translocation. Transfer of a portion of chromosome to another chromosome other than crossing over.

Translocation heterozygote. Hybrid in which two parents differ from each other in chromosome structure arising out of non homologous translocation.

References

Abbas A, Godward MBE (1963) Cytology of *Tribonema uticulosum* (Kuetz) Itazen. Phykos 2: 49–50

Abbas A, Godward MBE (1964) Cytology in relation to taxonomy of Chaetophorales. J Linn Soc 58: 499

Abbas A, Godward MBE (1965) Chromosome numbers in some members of Chaetophorales. Labdev J Sci Technol 3: 269

Abraham A, Nihan CA (1958) Cytology of *Isoetes*. Curr Sci 27: 60–61

Abraham AP, Mathew PM (1962) Cytological studies in the cycads. Sex chromosomes in *Cycas*. Ann Bot (London) 26: 261–266

Bhatnagar SP, Moitra A (eds) (1996) Gymnosperms, New Age International Limited Publishers, New Delhi, pp. 1–470

Bhojwani SS, Razdan MK (eds) (1992) Plant Tissue Culture—Theory and Practice. Elsevier, Amsterdam pp. 1–598

Bitis GN (1983) Evidence for diffusible mating-type specific trichnogyne attractants in *Neurospora crassa*. Experimental Mycology 7: 292–295

Brandham PE (1965) Polyploidy in desmids. Can J Bot 43: 405–417

Brandham PE, Godward MBE (1965) Meiosis in *Cosmarium botrytis*. Can J Bot 43: 1379–1386

Buffaloe NP (1958) A comparative cytological study of four species of *Chlamydomonas*. Science 12: 25

Calzada JV, Crane CF, Stelly DM (1996) Apomixis: The asexual revolution. Science 274: 1322–1323

Charlesworth S (1991) The evolution of sex chromosome. Science 251: 1030–1033

Chattopadhyay D, Sharma AK (1991) Sex determination in dioecious plants. Feddes Report 102: 29–55

Chinnappa CC, Sreenivasan MS (1968) Cytology of *Hamileia vastatris*. Caryologia 21: 75–82

Cole KM (1990) Chromosomes. *In*: Cole KM, Sheath RG (eds) Biology of the red algae. Cambridge University Press, Cambridge (UK), pp 73–101

Day A (1976) Communication through fimbriae during conjugation in a fungus. Nature 262: 283–284

Egel R, Nielson O, Weilgany D (1990) Sexual differentiation in fission yeast. Trends in Genetics 6: 369–373

Evans AM (1968) The *Polypodium pectinatum plumula*-complex in Florida. Am Fern J 58: 169

Evans DA, Wetter LR, Gamborg OL (1980) Somatic hybrid plants of *Nicotiana glauca* and *Nicotiana tabacum* obtained by protoplast fusion. Physiol Plant 48: 225–230

Evens LV (1962) Cytological studies in the genus *Fucus*. Ann Bot (London) 26: 345–360

Farrar DF, Wagner WH (1968) The gametophyte of *Trichomanes holopterum* Kunze. Bot Gaz 129: 210–219

Friedman WE (1992) Double fertilization in non-flowering seed plants and its relevance to the origin of flowering plants. *In* Rev Cytol 140: 319–355

Friedman WE (1994) The evolution of embryogeny in seed plants and the developmental origin and early history of endosperm. Am J Bot 81: 1468–1486

Gaff J (1981) Chromosome structure and the C value pavadox. J Cell Biol 91: 35–145

Gillissen B, Pergemann J, Sandmann C Schroeer B, Boiker M and Kahmann R (1992) A two-component regulator system for self/non-self-recognition in *Ustilago maydis*. Cell 68: 647–657

Glass NL, Kuldau GA (1992) Mating type and vegetative incompatibility in filamentous Ascomycetes. Annu Rev Phytopathol 30: 201–224

Godward MBE (1953) Geitler's nucleolar substance in *Spirogyra*. Ann Bot (London) NS 17: 403–416

Godward MBE (1961) Meiosis in *Spirogyra crassa*. Heredity 16: 53

Godward MBE (1966) The chromosomes of the algae. Edward Arnold, London, 212 pp

Goff L, Coleman AW (1986) A novel pattern of apical cell polyploidy, sequential polyploidy reduction and intercellular nuclear transfer in the red alga *Polysiphonia*. Am J Bot 73: 1109–1130

Goldstein M (1964) Speciation and mating behavour in *Eudorina*. J Protozool **6**: 249

Gow NAR (1995) Yeast-Hyphal dimorphism. *In*: Gow NAR, Gadd CM (eds) The Growing fungus. Chapman and Hall, London, pp 402–422

Guerlesquin M (1972) Cytologie et nombres chromosomiques. *In*: Corillion R, Guerlesquin M (eds) Recherches sur les Charophycees d'Afrique occidentale: Systematique, phytogeographie, ecology et cytology. Bull Soc Sci Bretag 27: 125

Guiry MD (1987) The evolution of life history types in *Rhodophyla*: An appraisal. Cryptogam Algol 8: 1–12

Haring V, Gray JE, McClure, Anderson MA, Clarke AE (1990) Selfincompatibility: A self-recognition system in plants. Science 250: 937–941

Hastings PJ, Levine EE, Cosbey E, Huddock MO, Gillham NW, Suzyeki SJ, Loppes R, Levine RP (1965) The linkage groups of *Chlamydomonas reinhardii*. Microbiol Gen Bull 23: 17–19

Herskowitz I, Rine J, Strathern J (1992) Mating type determination and mating type interconversion in *Saccharomyces cerevisiae*, In: The Molecular and Cellular Biology of the yeast Saccharomyces: Gene Expression, 1st edn JR Broach, JR Pringle and EWJ Jones (eds) Cold Spring Harbor Laboratory Press Plainview NY, pp 583–656

Johri BM, Ambegaokar KB, Srivastava PS (1992) Comparative embryology of Angiosperms. Vol I: 1–614, Vol 2: 615–1221. Springer, Berlin Heidelberg New York

Jose G, Chowdary YBK (1977) Karyological studies on *Cephaleuros* Kunze, Acta Bot Indica 5: 114

Kanahori T (1971) Cytotaxonomical research on the Characeae: Karyotype on the section *Nitella* of the genus. Bot Mag Tokyo 84: 327

Kasprik W (1973) Beitrage zur karyologie der Desmidiaceen-gattung *Micrasterias* Ag. Beih Nowa Hedwigia 42: 115

Kelly M, Burke J, Smith M et al (1988) Four mating type genes control sexual differentiation in the fission yeast. EMBO J 6: 1537–47

Khanna KR (1960) Cytological studies in some Himalayan mosses. Caryologia 13: 559–618

Khanna KR (1971) Sex chromosomes in bryophytes. Nucleus 14: 14–23

Khoshoo TN (1959) Polyploidy in gymnosperms. Evolution 13: 24–39

Klekowski FJ (1970) Genetical features of ferns as contrasted to seed plants. Ann Mo Bot Gard 59: 138

Klekowski FJ (1973) Sexual and subsexual systems in homosporous pteridophytes: A new hypothesis. Am J Bot 60: 535–544

Klekowski FJ (1976) Homologous pairing in ferns. *In*: Jones K, Brandham PE (eds) Current chromosome research. Elsevier, North Holland and Biomedical Press, Amsterdam, pp 175

Knapp E (1936) Zur Genetik von *Sphaerocarpus* (Tetradenanalytische Untersuchungen). Ber Dtsch Bot Ges 54: 58–69

Koltunow AM (1993) Apomixis: embryo sac and embryos formed with meiosis or fertilization in ovules. Plant Cell 5: 1425–1437

Koop HU (1979) The life cycle of *Acetabularia* Dasycladaceae, Chlorophyceae—A compilation of evidence for meiosis in the primary nucleus. Protoplasma 100: 353–366

Koul AK, Koul Pushpa (1994) Cytology of fungi. *In*: Johri BM (ed) Botany of India, Vol I. Oxford & IBH Publ, New Delhi, pp 327–344

Krichenbauer H (1937) Beitrag zur Kenntnis der Morphologie und Entwicklungsgeschichte der Gattungen *Euglena* and *Phacus*. Arch Protistenkd 90: 88

Krishnamurthy V (1994) Cytology of Algae. *In*: Johri BM (ed) Botany in India: History and progress, vol 1. Oxfored & IBH Publ, New Delhi pp 223–231

Kumar SS (1983) Cytogenetics of Indian Bryophyta. *In*: The Genetical Researches in India. ICAR, New Delhi

Kumar SS, Verma SK (1980) Cytological studies on some west Himalayan species of *Bryum* Hedw. Misc Bryol Lichnol 8: 182–188

Kuriachen PI (1963) Cytology of the genus *Selaginella*. Cytologia 28: 376–380

Lee CL (1954) Sex chromosomes in *Ginkgo biloba*. Am J Bot 41: 545–549

Levine RP, Ebersold WT (1960) The genetics and cytology of *Chlamydomonas*. Annu Rev Microbiol 14: 197–216

Ling HU, Tyler PA (1976) Meiosis, polyploidy and taxonomy of the *Pleurotaenium mamitlatum* Complex (Desmidiaceae). Br Phycol J 11: 315

Lorbeer G (1941) Struktur und Inhalt der Geschlechtschromosomen. Ber Dtsh Bot Ges 59: 369–418

Lovis JD (1977) Evolutionary patterns and processes in ferns. Adv Bot Res 4: 229–415

Lowry RJ (1948) A cytotaxonomic study of the genus *Mnium*. Mem Torrey Bot Club 20: 1–42

Loyal DS, Jairath AK, Bhatia PS (1984) Indian Fern J 1: 1–10

Maguire M (1976) Mitotic and meiotic behaviour of chromosomes of the octel strain of *Chlamydomonas*. Genetika 46: 479–502

Maheshwari P, Vasil V (1961) *Gnetum*. Bot Monoger No. 1. Council Sci Industr Res, New Delhi

Maheshwari Devi H, Johri BM, Rau MA (1994) Embryology of Angiosperms. *In*: Johri BM (ed) Botany in India: History and Progress, vol 2. Oxford & IBH Publ, New Delhi, 59–146

Manton I (1950) Problems of cytology and evolution in the Pteridophyta, Cambridge University Press, London

Mehra PN (1950) Occurrence of hermophrodite flowers and the development of female gametophytes in *Ephedra intermedia* Shrenk et Mey. Ann bot (London) 14: 165–180

Mehra PN, Khitha CS (1981) Karyotype and mechanisms of sex determination in *Ephedra foliata* Boiss. A dioecious gymnosperm. Cytologia 46: 173–181

Mohan Ram HY, Chatterjee J (1970) Gametophyte of *Equisetum ramosissimum*. Phytomorphology 20: 151–172

Ninan CA (1958) Studies on cytology and phylogeny of pteridophytes 5. Observations on the *Isoetaceae*. J Indian Bot Soc 17: 93–102

Ono K (1970) Karyological studies on Mniaceae and Polytrichaceae with special reference to the structural sex chromosomes. J Sci Hiroshima Univ B 13 (Div 2) 91

Ono K, Morita Y, Taguchi M, Inoue S (1977) On the chromosome number in Mniaceae. Kumamoto J Sci Biol 13: 49–65

Pant DD, Srivastava GK (1965) Cytology and reproduction of some Indian species of *Isoetes*. Cytologia 30: 239–251

Power JB, Frearson EM, Hayward C, George D, Evens PK, Berry SF, Cocking EC (1976) Somatic hybridization of *Petunia hybrida* and *P. parodii*. Nature 263: 500–502

Proskauer J (1957) Studies on Anthocerotales 5. Phytomorphology 7: 113–135

Rajendran RB (1967) A new type of nuclear life cycle in *Hemileia vastatrix*. Nature 213: 105–106

Raju NB (1980) Meiosis and ascospore genesis in *Neurospora*. Eur J Cell Biol 23: 208–223

Rao CSP (1956) The life history and reproduction of *Polyides caprinus* (Gunn) Papenf. Ann Bot (London) 20: 211–230

Rao CSP (1971) Sex chromosomes of *Wrangelia argus* Mont. Bot Mar 14: 113–115

Riley HP (1957) Introduction to Genetics and Cytogenetics. Hafner Publishing Company, New York and London, 1–596

Rueness J (1978) Hybridization in red algae. *In*: Irvine DEG, Price JH (eds) Modern approaches to the taxonomy of red and brown algae Academic Press, London pp 247–262.

Sarma YSRK (1983) Algal Karyology and evolutionary trends. *In*: Sharma AK, Sharma A (eds) Chromosomes in evolution of eukaryotic group vol I. CRC Press, Boca Raton Florida, pp 177–223

Schutz D, Bannett F, Dahl M et al. (1990) The *b* alleles of *U. maydis*, whose combinations program pathogenic development, cold for polypetides containing a homeodomain-related motif. Cell 60: 295–306

Sharma AK (1983) Chromosome evolution in the monocotyledons—an overview. *In*: Sharma AK, Sharma A (eds) Chromosome in the evolution of eukaryotic group, vol 2. CRC Press, Boca Raton, Florida, pp 169–184

Sharma BD (1994a) Pteridophyta: morphology, anatomy, reprodutive biology. *In*: Johri BM (ed) Botany in India: history and progress, vol I. Oxford & IBH Publ, New Delhi pp 455–489

Sharma BD (1994b) Gymnosperm: morphology, systematic, reproductive biology. *In*: Johri BM (ed) Botany in India: history and progress, vol 2. Oxford & IBH Publ, New Delhi, pp 1–23

Shashikala J, Sarma YSRK (1980) Karyology of three species of *Euglena*. J Indian Bot Soc 3: 196–201

Shyam R (1980) On the life cycle, cytology and taxonomy of *Cladophora callicoma* from India. Am J Bot 67: 619–624

Sinha JP (1968) Cytotaxohomical studies on *Cladophora glomerata* four freshwater forms. Cytologia 32: 507–518

Singh RA, Pavgi MS (1975) Cytology and teliospore development of *Entyloma thirumalachari*. Caryologia 28: 15–518

Smith AJE (1978) Cytogenetics, biosystematics and evolution in the Bryophyta. Adv Bot Res 6: 196–277

Smith AJE (1983) Chromosomes in the evolution of the Bryophyta. *In*: Sharma AK, Sharma A (eds) Chromosomes in evolution of eukaryotic groups, vol 1. CRC Press Boca Raton, Florida pp 225–254

Srivastava S, Sarma YSRK (1979) Karyological studies on the genus *Oedogonium* Link (Oedogoniales, Chlorophyceae). Phycologia 18: 228–236

Staben C (1995) Sexual reproduction in higher fungi. *In*: Gow NAR, Gadd CM (eds) The growing Fungus. Chapman and Hall, London pp 383–402

Starr RC (1955) Sexuality in *Gonium sociale* (Dujardin) Warming. J Tenn Acad Sci 30: 90–93

Starr RC (1975) Meiosis in *Volvox carferi* f. *nagariensis*. Arch Protistented 117: 187–191

Storms R, Hastings PJ (1977) A fine structure analysis of meiotic pairing in *Chlamydomonas reinhardii*. Exp Cell Res 104: 39–46

Subramanyan R (1946) On somatic division, reduction division, auxospore formation and sex differentiation in *Navicula holophila* (grun) Cl. Indian Bot Soc (Prof MOP Iyengar Comm Vol) 239–266 pp

Tatuno S (1941) Zytologische Untersuchungen über die Lebermoose von Japan. J Sci Hiroshima Univ 48: 1–187

Thirumalachar MJ, Whitehead MD, Boyle JS (1949) Gametogenesis and oospore formation in *Cystopus evolvuli* (*Albugo*). Bot Gaz 110: 487–491

Treimer RE, Brown RM Jr (1974) Cell division in *Chlamydomonas moewusii*. J Physiol 14: 419

Tseng CK, Chang TJ (1955) Studies of *Porphyra*. 3. Sexual reproduction of *Porphyta*. Acta Bot Sin 4: 153–166

Tseng CK, Chang TJ (1989) Studies on the alternation of the nuclear phases and chromosome numbers in the life history of some species of *Porphyra* from China. Bot Mag Jpn 32: 1–8

Tuttle AH (1924) The reproductive cycle of the Characeae. Science 60: 412

Tuttle AH (1926) The location of reduction division in Charophyta. Univ Calif Publ Bot 13: 227

Tyagi SNS, Singh L, Sharma SL (1982) Gametogenesis and oospore formation in *Cystopus* (*Albugo*) *evolvulus*. Acta Bot Indica 10: 233–239

Van der Meer JP, Todd ER (1980) The life history of *Palmaria palmata* in culture: A new type Rhodophyta. Can J Bot 58: 1250–1256

Vasil V (1959) Morphology and embryology of *Gnetum ula* Brongn. Phytomorphology 9: 167–215

Verma SC (1961) Cytology of *Isoetes coramandelina*. Am Fern J 51: 99–104

Von Stosch HA (1973) Observations in vegetative reproduction and sexual life-cycles of two freshwater dinoflagellates, *Gymnodiinium pseudopalustre* Schiller and *Woloszynskia apiculata* sp. nov. Br Phycol J 8: 105

Wagner WH, Wagner FS (1966) Pteridophytes of the Mountain Lake Area. Giles Co. Virginia Biosystematic Studies 1964–65. Castanea 31: 121–140

Walker TC (1966) Apomixis and vegetative reproduction in ferns. *In*: Reproductive Biology and taxonomy of vascular plants. Bot Soc Br Isles Conf Rep 9: 152–161

Walker JG (1984) Chromosomes and evolution in pteridophytes. *In*: Sharma AK, Sharma A (eds) Chromosomes in evolution of eukaryotic groups, vol 2. CRC Press, Boca Raton, Florida, pp 103–141

Wigh K (1972) Cytotaxonomical and modification studies in some Scandinavian mosses. Lindbergia 1: 130

Wik-Sjostedt A (1970) Cytogenetic investigations in *Cladophora*. Hereditas 66: 33–262

Yano K (1957) Cytological studies in Japanese Mosses 3. Mem Takada Branch Nigata Univ 1: 129–159

Zingmark RG (1970) Sexual reproduction in the dinoflagellate *Noctilaca milliarcs* Suriray. J Phycol 6: 122–126

13

Concluding Remarks

B.M. Johri

In the various Chapters, an attempt has been made to discuss the initiation, development and differentiation of sex organs in different groups of plants. In some groups the sex organs lack morphological distinction, but the reproductive cell/s do have physiological/biochemical/genetical distinction. The universal occurrence of two generations—gametophyte and sporophyte—and their functional significance was established through the classic comparative work of Hofmeister (1869). The impact of this discovery stimulated much discussion of life cycle evolution and the origin of morphological differences between the gametophyte and sporophyte.

Alternation of generations are the two multicellular phases—gametophyte and sporophyte in the diplobiontic life cycles of plants. Two hypotheses have been suggested (see Kenrick 1994).

1. The 'Antithetic' Theory proposes a green algal ancestor for land plants with a haploid haplobiontic life cycle resembling the 'charophycean' algae *Coleochaete*. The gametophytic generation is multicellular, but the sporophyte (zygote) is unicellular and the first division is meiotic. The antithetic hypothesis implies that the sporophyte generation of land plants originated de novo through the intercalation of somatic cell division in the zygote before meiosis.

2. The 'Homologous' Theory proposes a green algal ancestor with a diplobiotic life cycle. One consequence is that many of the features of sporophyte and gametophyte generations of land plants could already be present in algal ancestor.

Modern systematic treatments of Chlorobionta offer strong support for the 'Antithetic' Theory.

In a thought-provoking article—Diverted development of reproductive organs: A source of morphological innovation in land plants, Crane and Kenrick (1997) point out that "many of the morphological underlying angiosperm reproductive biology, such as reduction, simplification, and aggregation of sporophylls to form a flower, enclosure of the ovules, truncation of the gametophytic phase of the life cycle, and extreme vegetative flexibility, may have been secondary effects of selection for a "weedy" progenetic life history characterized by precocious reproduction".

In Conclusion, it may be pointed out that all constituents—cells, tissues, organs—of an organism do not evolve at the same rate. While some constituent/s of the organism remain in a primitive state, other constituents (of the same organism) acquire a highly-evolved state. Besides linear evolution, there can be parallel/convergent and/or divergent evolution. These facts are fully supported by the present comparative study.

References

Crane PR, Kenrick P (1997) Diverted development of reproductive organs: A source of morphological innovation in land plants. P1 Syst Evol 206: 161–174

Hofmeister W (1860) On the germination, development and fructification of the higher Cryotogamia and on the fructification of the Coniferae. Published for the Ray Society by Robert Hardwick, London

Kenrick P (1994) Alternation of generations in land plants: New phylogenetic and palaeobotanical evidence. Biol Rev 69: 293–330

Kenrick P, Crane PR (1997) The origin and early evolution of plants on land. Nature 389: 33–39

14

Future Studies

B.M. Johri and P.S. Srivastava

In the 20th century, while the study of structure and function of reproductive organs in plants has brought out many fascinating features, some controversial issues have also arisen. Some of these unanswered questions are mentioned here.

For example, in angiosperms, what is the control mechanism which determines:

(a) the patterns of microspore tetrads—linear, T-shaped, tetrahedral, isobilateral,

(b) the number of nuclei in the mature pollen grains—two and three nuclei,

(c) the development of an embryo sac-like structure from pollen grain in *Ornithogalum, Hyacinthus*, and some other taxa,

(d) distribution of megaspore nuclei to produce tetrasporic embryo sacs—Adoxa, Drusa, Penaea, Plumbago, Fritillaria and some other types,

(e) Nuclear, Helobial and Cellular types of endosperm, and the rare composite type as in Loranthaceae,

(f) division of zygote by a transverse wall in most angiosperms, and by a longitudinal wall in Loranthaceae, and in a few other taxa, and

(g) the division in *Paeonia* zygote is free-nuclear, and this feature is not knwon in any other angiosperm.

In gymnosperms,

(a) in some taxa, e.g. *Pinus*, what factors/mechanisms control/s the long period of rest between pollination and fertilization,

(b) after the rest period, what mechanism reactivates the pollen grain until fertilization occurs, and

(c) in some taxa—*Cycas* and *Ginkgo*—the pollen tube arises from the apical region, grows and branches in the nucellus (to obtain nourishment). In other gymnosperms, and all angiosperms, the tube arises from the basal region of pollen.

Fertilization (syngamy 1884—Strasburger; double fertilization 1898—S.G. Nawashin) in angiosperms was discovered almost a hundred years ago (see Maheshwari 1950). Since then, the pollen, pollen tube, male gametes, fusion of egg with one male gamete, and fusion of the second male gamete with secondary nucleus (fused polars) or triple fusion (two polar nuclei and one male gamete) have been studied in much detail (see Johri 1984).

Spectacular investigations on isolation of male gamete and egg, their fusion, and subsequent growth and development of cob in *Zea mays*—using in vitro culture techniques have—been reported by Kranz and Lörz (1993, 1994, see also Johri et al. 1997). These findings will result in the improvement of crops, especially food crops.

The endosperm in the seed supports the embryo during germination, and early growth of seedling. Endosperm results from the fertilization of the secondary nucleus, it is 2n in plants in which the lower polar nucleus is not formed, e.g. *Butomopsis lanceolata* (Johri 1936). The fertilized upper polar nucleus gives rise to endosperm (Helobial type).

Irrespective of the number of nuclei taking part in fusion and formation of primary endosperm nucleus, the end product is always the endosperm tissue, same type as the mother plant. What is the control mechanism which allows such a behaviour.

In *Butomopsis*, the zygote (2n) and the primary endosperm nucleus (2n) are genetically indentical but the zygote produces the embryo, and the other nucleus the endosperm. What is the control for this differential activity.

The monocot and dicot embryo also pose a problem. The monocot condition results from suptession of one cotyledonary primordium, OR fusion of the two cotyledonary primordia at an early or late stage of development?

Why, in some plants, especially in some monocots, the endosperm remains viable for long periods? In these monocot seed/s, the 'liquid' endosperm is gelatinous.

In some plants, the development of haustoria in (a) megaspore/s, (b) embryo sac, (c) synergid, (d) antipodal/s, (e) endosperm or (f) embryo is rather a persistent feature. What are the mechanisms which control their location and development? It will be interesting to examine if the haustoria can be induced in plants which do not have the haustoria.

The information about (1) Physiology, (2) biochemistry, (3) cell, and (4) molecular biology, and (5) genetics of reproductive organs is negligible. Therefore, these aspects should be studied in detail. An investigator can not have expertise in all the areas, and a team of experts should undertake such studies. The handicap is that a morphologist, and an anatomist do not want to undertake such research.

We recall that Professor P. Maheshwari (as early as 1957/1958) was fully aware about the collaborative/ co-operative research not only in the field of embryology, but also in the field of in vitro tissue culture research. We have no doubt that such approaches will have to be adopted sooner or later.

References

Johri BM (1936) The life history of *Butomopsis lanceolata* Kunth. Proc Indian Acad Sci B 4: 139–162

Johri BM (ed) (1984) Embryology of Angiosperms. Springer, Heidelberg, Berlin, New York, pp 830

Johri BM, Srivastava PS, Purohit M (1997) Plant tissue culture and biotechnology. Palaeobotanist 46: 134–159

Kranz E, Lörz, H (1993) In vitro fertilization of single, isolated gamete result in zygotic embryogenesis and fertile maize plants develop. Plant Cell 5: 739–746

Kranz E, Lörz H (1994) In vitro fertilization of maize by single egg and sperm cell protoplasm fusion mediated by high calcium and high pH. Zygote 2: 125–128

Maheshwari P (1950) An Introduction to the Embryology of Angiosperms. McGraw-Hill, New York, pp 453

Plant Index